全国科学技术名词审定委员会

公　　布

科学技术名词·工程技术卷（全藏版）

9

公路交通科技名词

CHINESE TERMS IN HIGHWAY AND
TRANSPORTATION SCIENCE AND TECHNOLOGY

公路交通科学名词审定委员会

国家自然科学基金资助项目

科学出版社

北　京

内 容 简 介

本书是全国科学技术名词审定委员会审定公布的第一批公路交通科技名词。包括总论,道路工程,交通工程,桥、涵、隧道、渡口工程,汽车运用工程,汽车运输,筑路机械及运用管理,路用材料及性能,材料与结构试验及设备,公路环境保护等十大类,共 4 536 条。是科研、教学、生产、经营以及新闻出版等部门使用的公路交通科技规范名词。

图书在版编目(CIP)数据

科学技术名词. 工程技术卷: 全藏版 / 全国科学技术名词审定委员会审定.
—北京: 科学出版社,2016.01
  ISBN 978-7-03-046873-4

  I. ①科… II. ①全… III. ①科学技术–名词术语 ②工程技术–名词术语
IV. ①N-61 ②TB-61

  中国版本图书馆 CIP 数据核字(2015)第 307218 号

责任编辑: 李玉英 / 责任校对: 陈玉凤
责任印制: 张 伟 / 封面设计: 铭轩堂

科学出版社 出版
北京东黄城根北街 16 号
邮政编码: 100717
http://www.sciencep.com
北京厚诚则铭印刷科技有限公司印刷
科学出版社发行   各地新华书店经销
*
2016 年 1 月第 一 版   开本: 787×1092 1/16
2016 年 1 月第一次印刷   印张: 18
字数: 498 000
定价: 7800.00 元(全 44 册)
(如有印装质量问题,我社负责调换)

全国自然科学名词审定委员会
第三届委员会委员名单

特邀顾问：吴阶平　　钱伟长　　朱光亚

主　　任：卢嘉锡

副 主 任：路甬祥　　章　综　　林　泉　　左铁镛　　马　阳

　　　　　孙　枢　　许嘉璐　　于永湛　　丁其东　　汪继祥

　　　　　潘书祥

委　　员（以下按姓氏笔画为序）：

马大猷	王　夔	王大珩	王之烈	王亚辉
王树岐	王绵之	王窝骧	方鹤春	卢良恕
叶笃正	吉木彦	师昌绪	朱照宣	仲增墉
华茂昆	刘天泉	刘瑞玉	米吉提·扎克尔	
祁国荣	孙家栋	孙儒泳	李正理	李廷杰
李行健	李　竞	李星学	李焯芬	肖培根
杨　凯	吴凤鸣	吴传钧	吴希曾	吴钟灵
吴鸿适	沈国舫	宋大祥	张　伟	张光斗
张钦楠	陆建勋	陆燕荪	陈运泰	陈芳允
范维唐	周　昌	周明煜	周定国	罗钰如
季文美	郑光迪	赵凯华	侯祥麟	姚世全
姚贤良	姚福生	夏　铸	顾红雅	钱临照
徐　傅	徐士珩	徐乾清	翁心植	席泽宗
谈家桢	黄昭厚	康景利	章　申	梁晓天
董　琨	韩济生	程光胜	程裕淇	鲁绍曾
曾呈奎	蓝　天	褚善元	管连荣	薛永兴

公路交通科学名词审定委员会委员名单

顾问委员：张佐周　张　烨

主任委员：郑光迪

副主任委员：曾　戚　刘以成　杨盛福　张叔辉　陈炳麟　唐中一

委　员（按姓氏笔画为序）：

王明仪　王秉刚　王唐生　邓学钧　卢世深　朱玉孚

朱照宏　刘　伟　孙祖望　杨守立　杨佩昆　束明鑫

吴国庆　何光里　张亚明　张祖荫　陈连汉　范立础

金如霆　庞日成　饶鸿雁　洪时言　贾日学　柴仁栋

郭生海　鄂俊泰　熊哲清　戴　竞　戴冠军

秘　书：陈允纬　程丽萍　焦振芳

序

　　科技名词术语是科学概念的语言符号。人类在推动科学技术向前发展的历史长河中,同时产生和发展了各种科技名词术语,作为思想和认识交流的工具,进而推动科学技术的发展。

　　我国是一个历史悠久的文明古国,在科技史上谱写过光辉篇章。中国科技名词术语,以汉语为主导,经过了几千年的演化和发展,在语言形式和结构上体现了我国语言文字的特点和规律,简明扼要,蓄意深切。我国古代的科学著作,如已被译为英、德、法、俄、日等文字的《本草纲目》、《天工开物》等,包含大量科技名词术语。从元、明以后,开始翻译西方科技著作,创译了大批科技名词术语,为传播科学知识,发展我国的科学技术起到了积极作用。

　　统一科技名词术语是一个国家发展科学技术所必须具备的基础条件之一。世界经济发达国家都十分关心和重视科技名词术语的统一。我国早在1909年就成立了科技名词编订馆,后又于1919年中国科学社成立了科学名词审定委员会,1928年大学院成立了译名统一委员会。1932年成立了国立编译馆,在当时教育部主持下先后拟订和审查了各学科的名词草案。

　　新中国成立后,国家决定在政务院文化教育委员会下,设立学术名词统一工作委员会,郭沫若任主任委员。委员会分设自然科学、社会科学、医药卫生、艺术科学和时事名词五大组,聘任了各专业著名科学家、专家,审定和出版了一批科学名词,为新中国成立后的科学技术的交流和发展起到了重要作用。后来,由于历史的原因,这一重要工作陷于停顿。

　　当今,世界科学技术迅速发展,新学科、新概念、新理论、新方法不断涌现,相应地出现了大批新的科技名词术语。统一科技名词术语,对科学知识的传播,新学科的开拓,新理论的建立,国内外科技交流,学科和行业之间的沟通,科技成果的推广、应用和生产技术的发展,科技图书文献的编纂、出版和检索,科技情报的传递等方面,都是不可缺少的。特别是计算机技术的推广使用,对统一科技名词术语提出了更紧迫的要求。

　　为适应这种新形势的需要,经国务院批准,1985年4月正式成立了全国自然科学名词审定委员会。委员会的任务是确定工作方针,拟定科技名词术语审定工作计划、实施方案和步骤，组织审定自然科学各学科名词术语，并予以公布。根据国务院授权，委员会审定公布的名词术语，科研、教学、生产、经营以及新闻出版等各部门，均应遵照使用。

全国自然科学名词审定委员会由中国科学院、国家科学技术委员会、国家教育委员会、中国科学技术协会、国家技术监督局、国家新闻出版署、国家自然科学基金委员会分别委派了正、副主任担任领导工作。在中国科协各专业学会密切配合下，逐步建立各专业审定分委员会，并已建立起一支由各学科著名专家、学者组成的近千人的审定队伍，负责审定本学科的名词术语。我国的名词审定工作进入了一个新的阶段。

这次名词术语审定工作是对科学概念进行汉语订名，同时附以相应的英文名称，既有我国语言特色，又方便国内外科技交流。通过实践，初步摸索了具有我国特色的科技名词术语审定的原则与方法，以及名词术语的学科分类、相关概念等问题，并开始探讨当代术语学的理论和方法，以期逐步建立起符合我国语言规律的自然科学名词术语体系。

统一我国的科技名词术语，是一项繁重的任务，它既是一项专业性很强的学术性工作，又涉及到亿万人使用习惯的问题。审定工作中我们要认真处理好科学性、系统性和通俗性之间的关系；主科与副科间的关系；学科间交叉名词术语的协调一致；专家集中审定与广泛听取意见等问题。

汉语是世界五分之一人口使用的语言，也是联合国的工作语言之一。除我国外，世界上还有一些国家和地区使用汉语，或使用与汉语关系密切的语言。做好我国的科技名词术语统一工作，为今后对外科技交流创造了更好的条件，使我炎黄子孙，在世界科技进步中发挥更大的作用，作出重要的贡献。

统一我国科技名词术语需要较长的时间和过程，随着科学技术的不断发展，科技名词术语的审定工作，需要不断地发展、补充和完善。我们将本着实事求是的原则，严谨的科学态度作好审定工作，成熟一批公布一批，提供各界使用。我们特别希望得到科技界、教育界、经济界、文化界、新闻出版界等各方面同志的关心、支持和帮助，共同为早日实现我国科技名词术语的统一和规范化而努力。

全国自然科学名词审定委员会主任

钱 三 强

1990 年 2 月

前　言

公路交通是跨行业、跨部门的科学技术。其名词术语,使用广泛,百家汇流。近年来,公路交通科学技术又有了很大发展,创立了新的交通工程学等分支学科及一些新理论、新材料、新的测试技术和设备等。它们与国际上的科技名词也未尽一致。为适应公路交通科学技术发展及国际交流日增的需要,1990 年 3 月全国科学技术名词审定委员会(简称全国名词委,原名全国自然科学名词审定委员会)委托中国公路学会负责组建"公路交通科学名词审定委员会",对有关公路交通的科技名词进行审定。经中国公路学会常务理事会研究,并请示交通部领导同意,承担了这一任务。根据全国名词委组织条例,由中国公路学会提名,经交通部领导同意,报全国名词委批准,于 1990 年 11 月组成了"公路交通科学名词审定委员会",并开始工作。

几年来,在全体委员的努力和各有关部门的大力支持下,经过搜集,整理和一审、二审会议的审定和修改。提出了第一批公路交通科技名词的征求意见稿,发至全国公路交通有关各设计、施工、生产、科研、高校等单位以及各省、自治区、直辖市交通厅(局)、公路学会和有关的兄弟学会,广泛征求意见。各单位都很重视,大部分提出了修改意见,又经汇总整理,反复研究修改,提交三审会议审定后,报送全国名词委。全国名词委卢嘉锡主任又委托曾威、王秉刚、孙祖望、范立础、金如霆等专家、教授进行复审,最后由全国名词委批准公布。

第一批审定的公路交通科技名词共分总论,道路工程,交通工程,桥、涵、隧道、渡口工程、汽车运用工程、汽车运输、筑路机械及运用管理、路用材料及性能、材料与结构试验及设备、公路环境保护等 10 大类,共收词 4 536 条。其收词范围及审定原则,系按照全国名词委制定的审定工作条例及有关协调会议的决定进行。

在审定中,我们参照了国家已颁布的有关的名词术语标准、现行有关技术标准与规范、国际路协(PIARC)编制的《公路术语技术词典》、美国 AASHTO 编制的《运输术语汇编》(Transportation Glossary)及 AASHO 编制的《AASHO Highway Definition》、日本高速公路调查会编制的新版《交通工学用语事典》、我国台湾台北名山出版社出版的《土木建筑工程名词辞典》以及汽车运用、筑路机械的国外有关文献书刊等,尽量与之取得一致,互相衔接。对修订的名词,有些加以"又称"、或"曾用名",以便于查对,使之逐步过渡。

名词的审定,以中文词为准,其对应的英文词,由于国外的用词,也不尽规范、统一,且英、美两国用词也不完全一致,因此,对应的英文词有的对应了 2~3 个英文词。有些我国所特有的名词,无对应的英文词,则采取按中文词意译成国外能接受的英文,以便于国际交流。有些组合词中，其基本词的中文定义完全一致，但在各专业中与其他词组合后，其相对应的英文词的习用词不完全一致，则不再强求一致，而按其习用词相对应。

在第一批名词的审定过程中，交通部及全国各公路交通系统有关单位、有关高等院校和有关学科的专家、学者对审定工作给予了大力支持和协助，交通部科技司给予了资助，交通部公路科学研究所承担了大量的具体工作，在此向他们一并表示衷心感谢。

这批名词颁布以后，希望大家在使用中提出宝贵意见。另外，对尚须增订的名词，以及有些技术标准与规范修改后所变更的名词，亦希大家随时反馈，以进一步修订完善。

公路交通科学名词审定委员会
1995 年 5 月

编 排 说 明

一、本批公布的是公路交通的基本名词。

二、本书正文按分支学科,共分为10大类。

三、汉文词按学科的相关概念排列,每一词条后附有与该词相对应的同概念的英文词,非英语的词用( )注明其文种。

四、一个汉文名词,一般只对应一个英文词或最常用的两个英文词,两词中间用","号分开。

五、凡英文词的首字母大、小写均可时,一律小写。

六、对有些必须加以说明的名词,在注释栏中给出简单的定义或说明。

七、[ ]中的字为可省略的字。

八、注释栏中,"又称"为不推荐用名;"曾用名"为被淘汰的旧名。

九、书末所附的英汉索引,按英文名词字母顺序排列。汉英索引按汉语拼音顺序排列。所示号码为该词在正文中的序号。索引中带有"＊"号的词为在注释栏内的条目。

目　　录

序 ……………………………………………………………………………………… i

前言 …………………………………………………………………………………… iii

编排说明 ……………………………………………………………………………… v

正文

　01．总论 …………………………………………………………………………… 1

　02．道路工程 ……………………………………………………………………… 11

　03．交通工程 ……………………………………………………………………… 31

　04．桥、涵、隧道、渡口工程 …………………………………………………… 47

　05．汽车运用工程 ………………………………………………………………… 77

　06．汽车运输 ……………………………………………………………………… 89

　07．筑路机械及运用管理 ………………………………………………………… 104

　08．路用材料及性能 ……………………………………………………………… 121

　09．材料与结构试验及设备 ……………………………………………………… 128

　10．公路环境保护 ………………………………………………………………… 137

附录

　英汉索引 ………………………………………………………………………… 141

　汉英索引 ………………………………………………………………………… 208

01. 总 论

序 码	汉 文 名	英 文 名	注 释
01.001	公路交通	highway and transportation	
01.002	道路工程学	road engineering	
01.003	公路工程学	highway engineering	
01.004	交通工程学	traffic engineering	
01.005	运输工程学	transportation engineering	
01.006	路面工程学	pavement engineering	
01.007	桥梁工程学	bridge engineering	
01.008	水文地质学	hydrogeology	
01.009	工程地质学	engineering geology	
01.010	工程测量学	engineering surveying	
01.011	基础工程学	foundation engineering	
01.012	隧道工程学	tunnel engineering	
01.013	工程材料学	engineering material	
01.014	汽车运用工程	инженерия зксплоатация авто-мобилъного транспорта(俄)	
01.015	运输经济学	transportation economics	
01.016	筑路机械	road machinery	
01.017	环境保护	environmental protection	
01.018	工效学	ergonomics	
01.019	交通心理学	traffic psychology	
01.020	交通生理学	traffic physiology	
01.021	交通统计学	traffic statistics	
01.022	公路美学	highway aesthetics	
01.023	桥梁美学	bridge aesthetics	
01.024	公路建设管理	highway construction administration	
01.025	交通运输网	transportation network	
01.026	公路网	highway network	
01.027	公路网规划	highway network planning	
01.028	公路网密度	density of highway network	
01.029	现有公路	existing highway	
01.030	分期修建	stage construction	
01.031	公路建设基金	highway construction fund	

序码	汉文名	英文名	注释
01.032	过路费	road toll	
01.033	过桥费	bridge toll	
01.034	过隧费	tunnel toll	
01.035	船渡费	ferry toll	
01.036	养路费	highway maintenance fee	
01.037	车辆购置附加费	vehicle purchase additional fee	
01.038	公路基本建设程序	highway capital construction procedure	
01.039	预可行性研究	pre-feasibility study	
01.040	可行性研究	feasibility study	
01.041	计划任务书	planning assignment	
01.042	初步设计	preliminary design	
01.043	技术设计	technical design	
01.044	扩大初步设计	enlarged preliminary design	
01.045	施工图设计	construction drawing design	
01.046	后评价	post-assessment	
01.047	经济效益费用比	economic benefit cost ratio, EBCR	
01.048	经济净现值	economic net present value, ENPV	
01.049	经济内部收益率	economic internal rate of return, EIRR	
01.050	投资回收期	repayment period of investment	
01.051	贷款偿还能力	loan repay ability	
01.052	敏感性分析	susceptivity analysis	
01.053	直接效益	direct benefit	
01.054	间接效益	indirect benefit	
01.055	道路经济分析	road economical analysis	
01.056	费用效益分析	cost/benefit analysis	
01.057	公路使用者费用	highway user's cost	
01.058	公路使用者效益	highway user's benefit	
01.059	缩短里程效益	benefit from distance shortening	
01.060	减少交通事故效益	benefit from accidents reducing	
01.061	公路造价	highway construction cost	
01.062	工程直接费	direct expense of project	
01.063	工程间接费	indirect expense of project	
01.064	工程预备费	reserve fund of project	

序　码	汉　文　名	英　文　名	注　释
01.065	养护费用	maintenance cost	
01.066	大修工程费	major maintenance cost	
01.067	交通管理费	traffic administration cost	
01.068	公路残值	highway residual value	
01.069	公路技术管理	highway technical administration	
01.070	公路工程技术标准	highway technical standard	
01.071	标准设计	standard design	
01.072	公路勘测规程	highway reconnaissance and survey regulations	
01.073	材料试验法	materials test methods	
01.074	设计规范	design specification	
01.075	施工规范	construction specification	
01.076	定额	norm, quota, rating	
01.077	公路工程估算指标	estimate index of highway project	
01.078	公路工程概算定额	approximate estimate norm of highway project	
01.079	公路工程预算定额	budgetary norm of highway project	
01.080	设计交通量	design traffic volume	
01.081	计算行车速度	design speed	又称"设计车速"。
01.082	交通组成	traffic composition	
01.083	交通强度	traffic intensity	
01.084	交通特性	traffic characteristics	
01.085	车辆换算系数	vehicle conversion factor	
01.086	公路功能	highway function	
01.087	公路荷载标准	highway load standard	
01.088	设计荷载	design load	
01.089	验算荷载	check load	
01.090	轴载	axle load	
01.091	轮载	wheel load	
01.092	公路自然区划	highway natural zoning	
01.093	公路施工管理	highway construction management	
01.094	工程质量管理	engineering quality control	
01.095	全面质量管理	total quality control, TQC	
01.096	施工现场管理	construction site management	

序　码	汉　文　名	英　文　名	注　　释
01.097	施工计划	construction plan	
01.098	公路工程监理	highway engineering supervision	
01.099	招标	calling for tenders	
01.100	投标	bidding	
01.101	业主	employer, owner	
01.102	承包人	contractor	
01.103	分包人	subcontractor	
01.104	代理人	agent	
01.105	监理工程师	supervising engineer	
01.106	监理工程师代表	supervising engineer's representative	
01.107	驻地监理工程师	resident supervising engineer	
01.108	招标文件	bidding documents	
01.109	投标书	bid, tender	
01.110	投标人须知	instruction to bidder	
01.111	投标担保	bid bond	
01.112	投标邀请书	invitation to bid	
01.113	投标书有效期	bid validity	
01.114	国际竞争性招标	international competitive bidding, ICB	
01.115	国内竞争性招标	local competitive bidding, LCB	
01.116	联合投标	joint ventures bidder	
01.117	投标截止期	deadline of bid	
01.118	标底	engineer's estimate	
01.119	评标	evaluation of bids	
01.120	开标	bid opening	
01.121	中标	award of contract	
01.122	中标通知书	notification of award	
01.123	国内投标人优惠	preference for domestic bidders	
01.124	履约担保	performance bond	
01.125	协议书	agreement	
01.126	合同	contract	
01.127	分包	subcontracting	
01.128	合同管理	contract management	
01.129	施工合同条款	construction contract terms	
01.130	合同通用条件	general conditions of contract	
01.131	合同专用条件	special conditions of contract	

序　码	汉　文　名	英　文　名	注　释
01.132	工程量清单	bill of quantities, BOQ	
01.133	现场考察	site inspection	
01.134	动员费预付贷款	advance mobilization loan	
01.135	不可预见费	contingencies	
01.136	计量与支付	measurement and payment	
01.137	保留金	retention money	
01.138	价格调整	price adjustment	
01.139	暂定金额	provisional sums	
01.140	开工	starting of construction	
01.141	放样	setting out	
01.142	计日工	daywork labor	
01.143	隐蔽工程验收	hidden work acceptance	
01.144	临时工程	temporary works	
01.145	竣工	completion of construction	
01.146	验收	acceptance	
01.147	验收证书	acceptance certificate	
01.148	工程决算	final account of project	
01.149	工期	construction time limit	
01.150	违约	default	
01.151	延期违约偿金	liquidated damages	
01.152	终止合同	termination of contract	
01.153	延期	extension of time	
01.154	索赔	claims	
01.155	缺陷责任期	defects liability period	
01.156	缺陷责任终止证书	defects liability release certificate	
01.157	仲裁	arbitration	
01.158	经济调查	economy survey	
01.159	交通调查	traffic survey	
01.160	交通量调查	traffic volume survey	
01.161	起迄点调查	origin-destination study, OD study	又称"OD调查"。
01.162	料场调查	materials field investigation	
01.163	土基干湿类型	subsoil moistness classification	
01.164	干燥类型	dry type	
01.165	中湿类型	median dampness type	
01.166	潮湿类型	dampness type	
01.167	过湿类型	excessive dampness type	

序 码	汉 文 名	英 文 名	注 释
01.168	冰冻地区	frost region	
01.169	冰冻深度	frost depth	
01.170	水文地质调查	hydrogeological survey	
01.171	植被	vegetation	
01.172	洪水调查	flood survey	
01.173	历史洪水位	historic flood level	
01.174	测量	survey	
01.175	踏勘	reconnaissance	
01.176	初测	preliminary survey	
01.177	定测	location survey	
01.178	导线	traverse	
01.179	测站	instrument station	
01.180	测点	observation point	
01.181	基线	base line	
01.182	施工测量	construction survey	
01.183	竣工测量	final survey	
01.184	路线测量	route survey	
01.185	路线平面图	route plan	
01.186	平面测量	plan survey	
01.187	中线	center line	
01.188	中线测量	center line survey	
01.189	中线桩	center line stake	
01.190	百米桩	100m stake	
01.191	护桩	reference stake	
01.192	加桩	additional stake	
01.193	断链	broken chain	
01.194	交点	intersection point, IP	
01.195	虚交点	imaginary intersection point	
01.196	转点	turning point, TP	
01.197	偏角	deflection angle	
01.198	方位角	azimuth	
01.199	方向角	direction angle	
01.200	坐标法	coordinate method	
01.201	偏角法	deflection angle method	
01.202	切线支距法	tangent offset method	
01.203	纵断面测量	profile survey	
01.204	横断面测量	cross section survey	

序 码	汉 文 名	英 文 名	注 释
01.205	水准测量	leveling survey	
01.206	水准点	bench mark, BM	
01.207	高程	elevation	又称"标高"。
01.208	绝对基面	absolute datum	
01.209	相对高程	relative elevation	
01.210	设计高程	design elevation	
01.211	地面高程	ground elevation	
01.212	中桩填挖高度	height of cut and fill at center stake	
01.213	地形测量	topographic survey	
01.214	地形图	topographic map	
01.215	带状地形图	belt topographic map	
01.216	等高线	contour line	
01.217	等高距	contour interval	
01.218	地貌	topographic feature	
01.219	地物	culture	
01.220	地形	topography	
01.221	平原区	plain terrain	
01.222	微丘区	rolling terrain	
01.223	重丘区	hilling terrain	
01.224	山岭区	mountain terrain	
01.225	峻岭	steep mountain	
01.226	山脊	ridge	
01.227	盆地	basin	
01.228	峡谷	canyon	
01.229	台地	terrace	
01.230	垭口	pass	
01.231	摄影测量	photogrametry	
01.232	航空摄影测量	aerial photogrametry	
01.233	地面立体摄影测量	ground stereo-photogrametry	
01.234	地面控制点测量	ground control point survey	
01.235	航摄基线	aerophoto base line	
01.236	影像地图	photographic map	
01.237	像片索引图	photo-index	又称"镶辑复照图"。
01.238	航摄像片判读	aerophoto interpretation	
01.239	综合法测图	panimatric photo	

序　码	汉　文　名	英　文　名	注　释
01.240	全能法测图	universal photo	
01.241	微分法测图	differential photo	
01.242	像片镶嵌图	photo-mosaic	
01.243	全球定位系统	global positioning system, GPS	
01.244	数字地面模型	digital terrain model	
01.245	地理信息系统	geographic information system	
01.246	卫星遥感测量	satellite remote sensing	
01.247	工程地质遥感测量	remote sensing of engineering geology	
01.248	土质调查	soil survey	
01.249	勘探	exploration	
01.250	槽探	trench test	
01.251	井探	shaft test	
01.252	坑探	pit test	
01.253	钻探	boring test	
01.254	触探	sounding, cone penetration test	
01.255	电探	electrical resistivity survey	
01.256	物探	geophysical prospecting	
01.257	地下水	ground water	
01.258	地下水位	ground water level	
01.259	含水层	water stratum	
01.260	缝隙水	interstitial water	
01.261	喀斯特	karst	又称"岩溶"。
01.262	岩堆	talus	
01.263	滑坡	landslide	
01.264	崩坍	collapse, fall	
01.265	碎落	dehris, avalanche	
01.266	流砂	quick sand	
01.267	沙丘	sand dune	
01.268	沙漠	desert	
01.269	戈壁	gobi	
01.270	泥石流	mud avalanche, debris flow	
01.271	冻土	frozen soil	
01.272	多年冻土	permafrost	又称"永冻土"。
01.273	冰丘	ice hummock	
01.274	冰锥	ice cone	
01.275	涎流冰	salivary flow ice	

序 码	汉 文 名	英 文 名	注 释
01.276	淤泥	mud, silt	
01.277	沼泽	swamp	
01.278	泥炭	peat	
01.279	软土	soft soil, mollisoil	
01.280	黄土	loess	
01.281	湿陷性黄土	collapsed loess	
01.282	黑棉土	black cotton soil	
01.283	膨胀土	expansive soil, swelling soil	
01.284	红粘土	red clay	
01.285	盐渍土	salty soil	
01.286	有机[质]土	organic soil	
01.287	腐殖土	muck	
01.288	高岭土	kaolin	
01.289	荒漠土	desert soil	又称"沙漠土"。
01.290	裂隙粘土	fissured clay	
01.291	蒙脱土	montmorillonite	又称"蒙脱石"。
01.292	白垩土	chalky clay	
01.293	原状土	parent soil	
01.294	表土	topsoil	
01.295	底土	subsoil	
01.296	风积土	neolian soil	
01.297	冲积土	alluvial soil	
01.298	坡积土	slope wash soil	又称"堆积土"。
01.299	洪积土	diluvial soil	
01.300	地震区	seismic region	
01.301	砂土液化	sand liquefaction	
01.302	桥位勘测	bridge site survey	
01.303	桥位平面图	plan of bridge site	
01.304	桥位地形图	topographic map of bridge site	
01.305	桥轴线纵剖面	profile of bridge axis	
01.306	地质调查	geologic survey	
01.307	地质剖面图	geologic profile	
01.308	工程地质略图	engineering geologic sketch	
01.309	地质测井	geologic logging	
01.310	地质图例	geologic legend	
01.311	钻孔柱状图	boring log	
01.312	硬表层	crust	

序 码	汉 文 名	英 文 名	注 释
01.313	硬盘	hardpan	
01.314	基岩	bedrock	
01.315	硬岩	hardrock	
01.316	软岩	softrock	
01.317	岩层	rock stratum	
01.318	岩缝填充物	joint-filling material	
01.319	岩相	lithofacies	
01.320	构造节理	structure joint	
01.321	沉积节理	sedimentation joint	
01.322	层理	stratification	
01.323	片理	schistosity	
01.324	断层	fault	
01.325	断层走向	fault strike	
01.326	倾斜断层	dip fault	
01.327	褶皱	fold	
01.328	破碎带	fracture zone	
01.329	地层	stratum	
01.330	持力层	bearing stratum	
01.331	下卧层	underlying stratum	
01.332	层状土	stratified soil	
01.333	岩心	core	
01.334	岩心采取率	percentage of coring	
01.335	原状土样	undisturbed sample	
01.336	扰动土样	disturbed sample	
01.337	场地稳定性评价	site stability evaluation	
01.338	公路路政管理	highway administration	
01.339	公路路线编号	code number of highway route	
01.340	国道主干线	national main trunk line	
01.341	国家干线公路	national trunk highway	简称"国道"。
01.342	省干线公路	provincial trunk highway	简称"省道"。
01.343	县公路	county highway	简称"县道"。
01.344	乡公路	township highway	简称"乡道"。
01.345	国防公路	national defence highway	
01.346	公路用地	right-of-way	
01.347	道路建筑限界	boundary line of road construction	
01.348	路产	highway property	
01.349	养路总段	maintenance division	

序 码	汉 文 名	英 文 名	注 释
01.350	养路段	maintenance section	
01.351	养路道班	maintenance gang	
01.352	开放交通	open to traffic	又称"通车"。
01.353	封闭交通	close to traffic	
01.354	交通疏导	traffic dispersion	
01.355	桥梁限载	bridge load limit	
01.356	减速过桥	passing bridge with reduced speed	
01.357	养护对策	maintenance counterproposal	
01.358	收费管理	toll administration	
01.359	公路交通监理	highway traffic supervision	
01.360	公路附属设施	highway appurtenance	
01.361	交通安全设施	traffic safety facilities	
01.362	公路服务设施	highway service facilities	
01.363	公路限界架	highway boundary frame	

02. 道 路 工 程

序 码	汉 文 名	英 文 名	注 释
02.001	道路	road	
02.002	公路	highway	
02.003	城市道路	urban road	
02.004	城间道路	interurban road	
02.005	厂矿道路	factory and mine road	
02.006	林区道路	forest road	
02.007	风景区道路	park-way	
02.008	乡村道路	rural road	
02.009	小道	path	
02.010	自行车道	bike-way, cycle track	
02.011	畜力车道	cart road	又称"大车道"。
02.012	机耕道	tractor road	
02.013	驮道	bridle path	
02.014	高架道路	elevated road	
02.015	栈道	trestle road along cliff	
02.016	直道	straight road [Qin Dynasty]	秦代名。
02.017	驰道	royal road [Qin Dynasty], ancient drive way	秦代名。

序 码	汉 文 名	英 文 名	注 释
02.018	驿道	post road [Tang Dynasty]	唐代名。
02.019	[古]丝绸之路	the silk road	
02.020	高速公路	freeway	全部控制进入设有中央分隔带的汽车专用公路。
02.021	一级公路	first class highway	部分控制进入设有中央分隔带的汽车专用公路。
02.022	汽车专用公路	expressway, motorway	全部或部分控制进入专供汽车行驶的各等级公路的总称,包括高速公路、一级公路和二级汽车专用公路。
02.023	一般公路	ordinary highway, mixed traffic highway	供汽车和其它车辆混合行驶的二、三、四级公路的总称。
02.024	等级公路	standard highway, classified highway	
02.025	等外公路	substandard highway	
02.026	便道	temporary road	
02.027	辐射式公路	radial highway	
02.028	环形道路	ring road	
02.029	绕行公路	bypass	
02.030	过境公路	through highway	
02.031	城市出入口公路	city approach highway	
02.032	辅道	auxiliary road	
02.033	集散道路	collector-distributor road	
02.034	专用公路	accommodation highway	
02.035	干线公路	trunk highway, arterial highway	
02.036	支线公路	branch highway, feeder highway	
02.037	地方公路	local highway	
02.038	收费公路	toll highway	
02.039	有路面公路	paved highway	
02.040	土路	earth road	
02.041	晴雨通车公路	all-weather highway	
02.042	晴通雨阻公路	fine-weather highway	

序 码	汉 文 名	英 文 名	注 释
02.043	断头路	dead end highway	
02.044	联络线	linking-up line	
02.045	路线	route	
02.046	路线设计	route design	
02.047	路线线形	route alignment	
02.048	公路几何设计	highway geometric design	
02.049	选线	route selection	
02.050	线形要素	alignment elements	
02.051	路线控制点	control point of route	
02.052	路线连续性	route continuity	
02.053	线形协调	alignment coordination	
02.054	路线计算机辅助设计	route computer aided design	
02.055	公路透视图	highway perspective view	
02.056	透视图检验	perspective examination	
02.057	路线三维空间设计	route three-dimensional space design	
02.058	路线优化设计	route optimum design	
02.059	平面示意图	plan sketch	
02.060	公路功能设计	highway functional design	
02.061	公路美学设计	highway esthetic design	
02.062	公路景观设计	highway landscape design	
02.063	沿溪线	valley line	
02.064	山脊线	ridge line	
02.065	山坡线	hill side line	
02.066	越岭线	ridge crossing line	
02.067	比较线	alternative line	
02.068	展线	route development	
02.069	定线	line location	
02.070	野外定线	field location	
02.071	纸上定线	paper location	
02.072	土方累积图	mass diagram	
02.073	土方调配图	cut-fill transition diagram	
02.074	平面线形	horizontal alignment	
02.075	直线	straight line	
02.076	平曲线	horizontal curve	
02.077	平曲线最小半径	minimum radius of horizontal	

序　码	汉　文　名	英　文　名	注　释
		curve	
02.078	平曲线极限最小半径	ultimate minimum radius of horizontal curve	
02.079	[汽车]最小转弯半径	minimum turning radius	
02.080	平曲线最小长度	minimum length of horizontal curve	
02.081	平曲线横净距	lateral clear distance of horizontal curve	
02.082	圆曲线	circular curve	
02.083	复曲线	compound curve	
02.084	同向曲线	same direction adjacent curve	
02.085	反向曲线	reverse curve	
02.086	断背曲线	broken back curve	
02.087	回头曲线	switch back curve, reverse loop	
02.088	发针形曲线	hairpin curve	
02.089	切线长	tangent length	
02.090	曲线长	curve length	
02.091	外距	external distance	
02.092	缓和曲线	transition curve	
02.093	回旋线	clothoid	
02.094	螺旋线	spiral curve	
02.095	麦康奈尔螺旋线	McConnell's curve	
02.096	直圆点	point of curve, PC	
02.097	圆直点	point of tangent, PT	
02.098	直缓点	point of tangent to spiral, TS	
02.099	缓圆点	point of spiral to curve, SC	
02.100	圆缓点	point of curve to spiral, CS	
02.101	缓直点	point of spiral to tangent, ST	
02.102	复曲线点	point of compound curve, PCC	
02.103	曲线中点	mid point of curve, MC	
02.104	视线	sight line	
02.105	视线不良弯道	blind curve	
02.106	视距	sight distance	
02.107	停车视距	stopping sight distance	
02.108	超车视距	overtaking sight distance	
02.109	视线障碍物	sight obstruction	

序 码	汉 文 名	英 文 名	注 释
02.110	纵面线形	vertical alignment	
02.111	纵断面	profile	
02.112	纵坡	longitudinal gradient	
02.113	纵坡折减	grade compensation	
02.114	缓和坡段	transitional grade	
02.115	平均纵坡	average gradient	
02.116	坡度损失	loss in grade	
02.117	坡长限制	grade length limit	
02.118	合成坡度	resultant gradient	
02.119	变坡点	grade change point	
02.120	竖曲线	vertical curve	
02.121	竖曲线最小长度	minimum length of vertical curve	
02.122	凸形竖曲线	convex vertical curve	
02.123	凹形竖曲线	concave vertical curve	
02.124	横断面	cross section	
02.125	净空	clearance	
02.126	净高	vertical clearance	
02.127	净宽	horizontal clearance	
02.128	半填半挖断面	cut and fill section	
02.129	台口式断面	benched section	
02.130	全挖式断面	full cut section	
02.131	零断面	zero section	
02.132	平曲线加宽	curve widening	
02.133	加宽缓和段	transition zone of curve widening	
02.134	断面渐变段	transition zone of cross section	
02.135	路拱	crown	
02.136	路拱曲线	crown curve	
02.137	路拱横坡	crown slope, cross fall	
02.138	超高	superelevation	
02.139	超高横坡度	superelevation slope	
02.140	最大超高率	maximum superelevation rate	
02.141	超高缓和段	superelevation run-off	
02.142	路肩	shoulder, verge	
02.143	土路肩	earth shoulder	
02.144	硬路肩	hard shoulder	
02.145	分隔带	separator	
02.146	外侧分隔带	outer separator	

序　码	汉　文　名	英　文　名	注　释
02.147	中间带	median	
02.148	中央分隔带	median separator	
02.149	中央分隔带开口	median opening	
02.150	路缘带	marginal strip	
02.151	[路]缘石	curb	
02.152	平缘石	flush curb	
02.153	立缘石	vertical curb	
02.154	栏式缘石	barrier curb	
02.155	缓坡缘石	lip curb	
02.156	绿化带	green belt	
02.157	侧向余宽	lateral clearance	
02.158	路幅	roadway	
02.159	车道	lane	
02.160	车行道	carriageway	又称"行车道"。
02.161	分隔式车行道	divided carriageway	
02.162	双幅车行道	dual carriageway	
02.163	单车道	single lane	
02.164	双车道	dual lane	
02.165	多车道	multi-lane	
02.166	单行路	one-way road	
02.167	双向路	two-way road	
02.168	转弯车道	turning lane	
02.169	变速车道	speed change lane	
02.170	加速车道	acceleration lane	
02.171	减速车道	deceleration lane	
02.172	爬坡车道	climbing lane	
02.173	错车道	passing bay	
02.174	停车车道	parking lane	
02.175	应急停车带	emergency parking strip	
02.176	内侧车道	fast lane	又称"快车道"。
02.177	中间车道	center lane	
02.178	外侧车道	near-side lane	
02.179	附加车道	auxiliary lane	又称"辅助车道"。
02.180	可变向车道	reversible lane	
02.181	超车车道	overtaking lane, passing lane	
02.182	专用车道	accommodation lane	
02.183	回车道	turn around loop	

序　码	汉　文　名	英　文　名	注　释
02.184	港湾式停车处	parking bay	
02.185	停车场	parking area, parking	
02.186	公共汽车站	bus station	
02.187	公共汽车终点站	bus terminal	
02.188	乘车安全岛	loading island	
02.189	休息区	rest area	
02.190	服务区	service area	
02.191	休息娱乐区	recreation area	
02.192	观景台	sightseeing stand	
02.193	里程标	mileage post	
02.194	里程碑	mile stone, kilometer stone	
02.195	自行车车道	bike lane	
02.196	公路交叉	highway intersection	
02.197	[平面]交叉口	intersection, crossing, junction	
02.198	交叉角	intersection angle	
02.199	平面交叉	at-grade intersection	
02.200	正交叉	right-angled intersection	
02.201	斜交叉	skew intersection	
02.202	十字形交叉	cross road	
02.203	丁字形交叉	T-intersection	
02.204	Y 形交叉	Y-intersection	
02.205	错位交叉	staggered intersection, offset intersection	
02.206	环形交叉	roundabout, rotary intersection	
02.207	渠化交叉口	channelized intersection	
02.208	拓宽路口式交叉口	flared intersection	
02.209	加宽转角式交叉口	intersection with widened corners	
02.210	渐变段	tapered section	
02.211	分道角区	gore area	
02.212	交叉口视距	sight distance of intersection	又称"路口视距"。
02.213	视距三角形	sight triangle	
02.214	信号控制交叉口	signalized crossing	
02.215	无信号交叉口	non-signalized crossing	
02.216	交通岛	traffic island	
02.217	导流岛	channelization island	

序 码	汉 文 名	英 文 名	注 释
02.218	中心岛	center island	
02.219	安全岛	refuge island	
02.220	路岛端	approach nose	又称"接近端"。
02.221	汇合端	merging end	又称"合流端"。
02.222	立体交叉	grade separation	简称"立交"。
02.223	分离式立交	grade separation without ramps	又称"非互通式立交"。
02.224	上跨式立交	overpass	
02.225	下穿式立交	underpass	
02.226	互通式立交	interchange	
02.227	多层立交	multi-level interchange	
02.228	苜蓿叶形立交	clover-leaf interchange	
02.229	半苜蓿叶形立交	partial clover-leaf interchange	
02.230	菱形立交	diamond interchange	
02.231	定向式立交	directional interchange	
02.232	半定向式立交	semi-directional interchange	
02.233	喇叭形立交	trumpet interchange, 3-leg interchange	
02.234	环形立交	rotary interchange	
02.235	全部控制进入	full control of access	曾用名"全封闭"。
02.236	部分控制进入	partial control of access	曾用名"半封闭"。
02.237	匝道	ramp	
02.238	驶入匝道	entrance ramp	
02.239	驶出匝道	exit ramp	
02.240	单向匝道	one-way ramp	
02.241	双向匝道	two-way ramp	
02.242	左转弯匝道	left turn ramp	
02.243	右转弯匝道	right turn ramp	
02.244	环形匝道	loop ramp	
02.245	铁路平交道口	railway crossing	
02.246	公路铁路立交	highway-railway grade separation	
02.247	路基	subgrade	
02.248	路基工程	subgrade engineering	
02.249	路堤	embankment	
02.250	透水路堤	permeable embankment	
02.251	路堑	cutting	
02.252	台口式路基	benched subgrade	

序　码	汉　文　名	英　文　名	注　释
02.253	半山洞	half tunnel	
02.254	分离式路基	separated subgrade	
02.255	引道	approach	
02.256	边坡	side slope	
02.257	坡顶	top of slope	
02.258	坡脚	toe of slope	
02.259	护坡	slope protection	
02.260	护坡道	berm	
02.261	反压护道	loading berm	
02.262	护墙	guard wall	
02.263	边坡平台	plain stage of slope	
02.264	碎落台	stage for heaping debris	
02.265	挡土墙	retaining wall	
02.266	墙趾	wall toe	
02.267	墙踵	wall heel	
02.268	重力式挡土墙	gravity retaining wall	
02.269	衡重式挡土墙	balance weight retaining wall	
02.270	悬臂式挡土墙	cantilever retaining wall	
02.271	扶臂式挡土墙	counterfort retaining wall	
02.272	柱板式挡土墙	pile-plank retaining wall	
02.273	锚杆式挡土墙	tie rod anchored retaining wall	
02.274	锚锭板式挡土墙	anchored bulkhead retaining wall	
02.275	加筋土挡土墙	reinforced earth retaining wall	
02.276	上挡墙	top retaining wall	
02.277	下挡墙	lower retaining wall	
02.278	路基稳定性	subgrade stability	
02.279	路基强度	subgrade strength	
02.280	土基承载能力	subsoil bearing capacity	
02.281	路基设计高程	design elevation of subgrade	
02.282	最小填土高度	minimum height of fill	
02.283	路基含水量	subgrade moisture content	
02.284	路基临界高度	critical height of subgrade	
02.285	路基防护	subgrade protection	
02.286	路基加固	subgrade strengthening	
02.287	土方[工程]	earth work	
02.288	填方	fill	
02.289	挖方	cut, excavation	

序 码	汉 文 名	英 文 名	注 释
02.290	借土	borrow earth	
02.291	弃土	waste	
02.292	取土坑	borrow pit	
02.293	弃土堆	waste bank	
02.294	回填土	back fill	
02.295	换土	replacement of earth	
02.296	压实	compaction	
02.297	压实度	degree of compaction	
02.298	压实系数	compacting factor	
02.299	夯实	tamping	
02.300	重锤夯实法	heavy tamping method	
02.301	碾压	rolling	
02.302	振动压实	vibrating compaction	
02.303	有效碾压深度	effective rolling depth	
02.304	线压力	linear pressure	
02.305	碾压速度	rolling speed	
02.306	压实能量	compaction energy	
02.307	松铺厚度	loose laying depth	
02.308	压实厚度	compaction depth	
02.309	松铺系数	coefficient of loose laying	
02.310	预留沉落量	reserve settlement	
02.311	天然地基	natural subsoil	
02.312	加固地基	consolidated subsoil	
02.313	软弱地基	soft ground	
02.314	固结	consolidation	
02.315	动力固结法	dynamic consolidation method	
02.316	预压法	preloading method	
02.317	袋装砂井	sand bag well	
02.318	排水板法	sheet drainage	
02.319	排水砂垫层	drainage sand mat	
02.320	超载预压	surcharge preloading	
02.321	砂桩	sand pile	
02.322	石灰桩	lime pile	
02.323	砂井	sand well	
02.324	石方爆破	rock blasting	
02.325	爆破作业	blasting operation	
02.326	爆破漏斗	blasting crater	

序　码	汉　文　名	英　文　名	注　　释
02.327	抛掷爆破	throwing blasting	
02.328	松动爆破	loosening blasting	
02.329	毫秒爆破	millisecond delay blasting	
02.330	抛坍爆破	collapse blasting	
02.331	定向爆破	directional blasting	
02.332	多面临空爆破	open face blasting	
02.333	路面	pavement	
02.334	高级路面	high class pavement	
02.335	次高级路面	sub-high class pavement	
02.336	中级路面	intermediate class pavement	曾用名"过渡式路面"。
02.337	低级路面	low class pavement	
02.338	路面结构	pavement structure	
02.339	面层	surface course	
02.340	联结层	binder course	
02.341	应力吸收薄膜	stress absorbing membrane	
02.342	粘层	tack coat	
02.343	透层	prime coat	
02.344	磨耗层	wearing course	
02.345	透水磨耗层	pervious wearing course	
02.346	保护层	protection course	
02.347	松散保护层	loose protection course	
02.348	稳定保护层	stabilized protection course	
02.349	防滑层	skid resistant course	
02.350	封层	seal coat	
02.351	罩面	overlay	
02.352	补强层	strengthening course	
02.353	承重层	supporting course	
02.354	基层	base course	
02.355	底基层	subbase	
02.356	垫层	bed course	
02.357	隔水层	water insulation course	
02.358	隔温层	thermal insulation course	
02.359	路床	road bed	
02.360	路槽	road trough	
02.361	整平层	leveling course	
02.362	路面设计	pavement design	

序　码	汉　文　名	英　文　名	注　释
02.363	路面特性	pavement characteristics	
02.364	表面平整度	surface evenness	
02.365	表面粗糙度	surface roughness	
02.366	表面摩擦系数	surface friction coefficient	
02.367	表面构造	surface texture	曾用名"表面纹理"。
02.368	表面构造深度	surface texture depth	曾用名"纹理深度"。
02.369	表面宏观构造	surface macro texture	
02.370	表面微观构造	surface micro texture	
02.371	表面巨观构造	surface mega texture	
02.372	水滑现象	hydroplaning phenomena	
02.373	路面疲劳寿命	fatigue life of pavement	
02.374	现有服务指数	present service index, PSI	
02.375	弹性层状体系理论	elastic layer system theory	
02.376	粘弹性理论	viscoelasticity theory	
02.377	弹性半无限地基	elastic semi-infinite foundation	
02.378	双轮荷载	dual wheel loading	
02.379	当量圆直径	diameter of equivalent circle	
02.380	标准轴载	standard axle load	
02.381	当量轴次	equivalent axles	
02.382	累计当量轴次	accumulative equivalent axles	
02.383	特重交通	very heavy traffic	
02.384	重交通	heavy traffic	
02.385	中等交通	medium traffic	
02.386	轻交通	light traffic	
02.387	路面强度	pavement strength	
02.388	回弹弯沉	rebound deflection	
02.389	弯沉值	deflection value	
02.390	动弯沉	dynamic deflection	
02.391	动态回弹弯沉	dynamic rebound deflection	
02.392	理论弯沉系数	theoretical deflection coefficient	
02.393	容许回弹弯沉值	allowable rebound deflection value	
02.394	弯沉系数	deflection coefficient	
02.395	当量厚度	equivalent thickness	
02.396	弯沉盆	deflection basin	
02.397	形变模量	modulus of deformation	
02.398	弹性模量	modulus of elasticity	

序　码	汉　文　名	英　文　名	注　释
02.399	回弹模量	modulus of resilience	
02.400	模量比	modulus ratio	
02.401	界面连续接触	continuous contact of interface	
02.402	界面滑动接触	sliding contact of interface	
02.403	轮迹横向分布系数	coefficient of wheel tracking transverse distribution	
02.404	车道系数	coefficient of lanes	
02.405	抗折强度	rupture strength	
02.406	弯拉应力	flexural-tensile stress	
02.407	弯拉应变	flexural-tensile strain	
02.408	弯拉强度	flexural-tensile strength	
02.409	临界荷载	critical load	
02.410	板体温度翘曲应力	slab stress due to thermal warping	
02.411	路面补强设计	pavement strengthening design	
02.412	路面结构组合设计	pavement structural composition design	
02.413	加州承载比	California bearing ratio, CBR	
02.414	柔性路面	flexible pavement	
02.415	沥青路面	bituminous pavement	
02.416	沥青混凝土路面	bituminous concrete pavement	
02.417	沥青碎石路面	bituminous macadam pavement	
02.418	沥青贯入式路面	bituminous penetration pavement	
02.419	上拌下贯式路面	penetration macadam with coated chips	
02.420	沥青表面处治	bituminous surface treatment	
02.421	全厚式沥青路面	full depth asphalt pavement	
02.422	渣油路面	residue oil pavement	
02.423	稀浆封层	slurry seal	
02.424	块料路面	block pavement	
02.425	砖块路面	brick pavement	
02.426	细琢石路面	dressed stone pavement	
02.427	嵌花式路面	mosaic pavement	
02.428	圆石路面	cobble stone pavement	
02.429	弹街路面	pitching pavement	
02.430	水结碎石路面	water bound macadam	曾用名"马克当路面"。

序 码	汉 文 名	英 文 名	注 释
02.431	泥结碎石路面	clay bound macadam	
02.432	泥灰结碎石路面	clay-lime bound macadam	
02.433	级配路面	graded aggregate pavement	
02.434	粒料改善土路	aggregate treated earth road	
02.435	层铺法	layer spread method	
02.436	路拌法	road-mixing method	
02.437	厂拌法	plant-mixing method	
02.438	热拌法	hot mixing method	
02.439	冷拌法	cold mixing method	
02.440	热铺法	hot laid method	
02.441	冷铺法	cold laid method	
02.442	贯入法	penetration method	
02.443	拌和	mixing	
02.444	摊铺	paving	
02.445	撒布	spreading	
02.446	喷洒沥青	asphalt distribution	
02.447	灌浆	slurry penetration	
02.448	行车碾压	traffic bound, free bound	
02.449	铺砌	pitching	
02.450	刚性路面	rigid pavement	
02.451	水泥混凝土路面	cement concrete pavement	
02.452	钢筋混凝土路面	reinforced concrete pavement	
02.453	连续配筋混凝土路面	continuous reinforced concrete pavement	
02.454	纤维混凝土路面	fiber concrete pavement	
02.455	碾压混凝土路面	rolling compacted concrete pavement, RCCP	
02.456	复合式水泥混凝土路面	composite type concrete pavement	
02.457	接缝	joint	
02.458	纵缝	longitudinal joint	
02.459	横缝	transverse joint	
02.460	胀缝	expansion joint	
02.461	缩缝	contraction joint	
02.462	施工缝	construction joint	
02.463	企口缝	tongue and groove joint, rabbet	
02.464	平缝	flat joint	

序　码	汉　文　名	英　文　名	注　释
02.465	真缝	true joint	
02.466	假缝	dummy joint	
02.467	拉杆	tie bar	
02.468	传力杆	dowel bar	
02.469	锯缝	sawn joint	
02.470	填缝料	joint filler	
02.471	填缝板	joint plate	
02.472	边缘钢筋	edge reinforcement	
02.473	角隅钢筋	corner steel	
02.474	渐变板	transition slab	
02.475	整形	shaping	
02.476	抹光	finishing	
02.477	混凝土表面泌水	surface weeping of concrete	
02.478	真空脱水工艺	vacuum dewatering technique	
02.479	拉毛	broom-finish	
02.480	压槽	rolling groove	
02.481	刻槽	carving groove	
02.482	铣槽	milling groove	
02.483	湿法养生	moist curing	
02.484	薄膜养生	membrane curing	
02.485	漏浆	leakage	
02.486	蒸汽养生	steam curing	
02.487	贫混凝土基层	lean concrete base	
02.488	块石基层	Telford base	
02.489	碎石基层	macadam base	
02.490	级配集料基层	graded aggregate base	
02.491	半刚性基层	semi-rigid base	
02.492	稳定材料基层	stabilized material base	
02.493	石灰土基层	lime-soil base	
02.494	水泥土基层	cement-soil base	
02.495	石灰粉煤灰土基层	lime-flyash-soil base	
02.496	综合稳定基层	comprehensive stabilized base	
02.497	矿渣基层	slag base	
02.498	粒料稳定土基层	aggregate stabilized soil base	
02.499	公路排水	highway drainage	
02.500	地表水	surface water	

序　码	汉　文　名	英　文　名	注　释
02.501	洪水	flood	
02.502	毛细水	capillary water	
02.503	渗透水	seepage water	
02.504	透水性	water permeability	
02.505	排水能力	drainage ability	
02.506	排水系统	drainage system	
02.507	排水设施	drainage facility	
02.508	排水设计	drainage design	
02.509	排水工程	drainage works	
02.510	地下排水	subsoil drainage	
02.511	路基排水	subgrade drainage	
02.512	路面排水	pavement drainage	
02.513	中间带排水	median drainage	
02.514	纵向排水	longitudinal drainage	
02.515	横向排水	transverse drainage	
02.516	竖向排水	vertical drainage	
02.517	边沟	side ditch	
02.518	天沟	gutter	
02.519	截水沟	intercepting ditch	
02.520	排水沟	drainage ditch	
02.521	明渠排水	gutter drainage	
02.522	管道排水	pipe drainage	
02.523	渗井	seepage well	
02.524	蒸发池	evaporation pond	
02.525	沉淀池	sedimentation basin	
02.526	泵站排水	pumping drainage	
02.527	[路面]单向排水	single-slope drainage	
02.528	[路面]双向排水	double-slope drainage	
02.529	界面排水	interface drainage	
02.530	反滤层	inverted filter	
02.531	人字形排水系统	herringbone drainage system	
02.532	盲沟	blind drain, blind ditch	
02.533	集中排水	concentrated drainage	
02.534	集水槽	gully	
02.535	急流槽	chute	
02.536	跌水	hydraulic drop, water drop	
02.537	进水口	inlet	又称"雨水口"。

序　码	汉　文　名	英　文　名	注　释
02.538	栅式进水口	grated inlet	
02.539	出水口	outlet	
02.540	检查井	manhole	
02.541	泄水孔	weep-hole	
02.542	泄水口	drain opening	
02.543	过水路面	ford	
02.544	公路养护	highway maintenance	
02.545	养护管理	maintenance management	
02.546	养护工程	maintenance project	
02.547	公路里程	highway mileage	
02.548	通车里程	mileage open to traffic	
02.549	养护里程	maintenance mileage	
02.550	绿化里程	planting mileage, greening mileage	
02.551	路面铺装率	paved ratio	
02.552	路况	highway condition	
02.553	路况调查	highway condition investigation	
02.554	路况登记	highway condition registration	
02.555	路况检查	highway condition inspection	
02.556	路况巡视	highway condition patrol	
02.557	日常检查	routine inspection	
02.558	定期检查	periodical inspection	
02.559	险情检查	examination of dangerous situation	
02.560	路容	highway appearance	
02.561	路面调查	pavement investigation	
02.562	路面随机调查	pavement random investigation	
02.563	路面定期调查	pavement periodical investigation	
02.564	公路技术档案	highway technical file	
02.565	公路数据库	highway data bank	
02.566	预防性养护	preventive maintenance	
02.567	初期养护	initial maintenance	
02.568	巡回养护	patrol maintenance	
02.569	定期养护	periodical maintenance	
02.570	公路小修[保养]	highway routine maintenance	
02.571	公路中修	highway intermediate maintenance	
02.572	公路大修	highway major maintenance	
02.573	大中修周期	maintenance period	
02.574	公路改善	highway improvement	

序 码	汉 文 名	英 文 名	注 释
02.575	拓宽	widening	
02.576	调坡	adjusting gradient	
02.577	调整路拱	adjusting crown	
02.578	旧路技术改造	technical reformation of existing road	
02.579	公路改线	highway relocation	
02.580	路面翻修	pavement recapping	
02.581	路面补强	pavement strengthening	
02.582	抢修工程	rush-repair work	
02.583	修复工程	rehabilitation work	
02.584	公路灾害	highway disaster	
02.585	公路灾害防治	prevention and cure against highway disaster	
02.586	不利季节	unfavourable season	
02.587	道班养护	gang maintenance	
02.588	道班房	maintenance gang house	
02.589	行道树	roadside trees	
02.590	苗圃	nursery	
02.591	标志服	safety marked coat	
02.592	路障	roadblock	
02.593	堆料台	roadside material terrace	
02.594	路面积水	surface gathered water	
02.595	路面滑溜	surface skidding	
02.596	公路病害	highway distress	
02.597	裂缝	crack	
02.598	[路面]裂缝率	cracking rate	
02.599	[路面]裂缝度	cracking ratio	
02.600	收缩裂缝	contraction crack	
02.601	反射裂缝	reflection crack	
02.602	网裂	net-shaped crack	
02.603	龟裂	alligator crack	
02.604	发裂	hair-like crack	
02.605	纵向裂缝	longitudinal crack	
02.606	横向裂缝	transverse crack	
02.607	麻面	surface pockmark	
02.608	松散	surface loosening	
02.609	集料剥落	stripping of aggregate	

序 码	汉 文 名	英 文 名	注 释
02.610	露骨	bare surface	
02.611	泛油	bleeding	
02.612	油包	oil pox·	
02.613	拥包	upheaval	
02.614	翻浆	frost boiling	
02.615	弹簧现象	springing	
02.616	车辙	rut	
02.617	坑槽	pot holes	
02.618	搓板	corrugation	
02.619	啃边	edge failure	
02.620	脱皮	scaling	
02.621	磨光	polishing	
02.622	磨损	wearing	
02.623	接缝破损	joint failure	
02.624	填缝料破损	joint filler failure	
02.625	角裂	corner break	
02.626	错台	slab staggering	
02.627	板端错台	faulting of slab ends	
02.628	板体翘曲	slab warping	
02.629	板体断裂	slab rupture	
02.630	拱胀	blow up	
02.631	[桥基]沉降	settlement	
02.632	路面沉陷	pavement depression	
02.633	[路基]沉降	subsidence	
02.634	路面板唧泥	slab pumping	
02.635	碱性集料反应	alkali-aggregate reaction	
02.636	硫侵蚀	sulphur corrosion	
02.637	盐侵蚀	salt corrosion	
02.638	盐胀	salt heaving	
02.639	边坡冲蚀	slope erosion	
02.640	路肩缺口	shoulder gap	
02.641	水侵蚀	water erosion	
02.642	坍方	landslide	
02.643	冲刷	scouring, erosion	
02.644	管涌	piping	
02.645	水毁	flood damage, washout	
02.646	雪害	snow hazard	

序　码	汉　文　名	英　文　名	注　释
02.647	冻害	frost damage	
02.648	冻胀	frost heaving	
02.649	冻融	frost thawing	
02.650	溶陷	melt sinking	
02.651	冰障	ice jam	
02.652	风蚀	wind erosion	
02.653	沙害	sand hazard	
02.654	震害	seismic hazard	
02.655	植草	grass planting	
02.656	铺草皮	sodding	
02.657	清理边沟	ditch cleaning out	
02.658	边坡修整	slope trimming	
02.659	铺砂	sanding	
02.660	回砂	sand sweeping	
02.661	匀砂	sand brooming	
02.662	裂缝灌注	crack grouting	
02.663	填缝	joint filling	
02.664	补坑	patching	
02.665	铲除拥包	upheaval leveling	
02.666	毛接法	coarse joining method	
02.667	滚浆法	rolling slurry process	
02.668	[保护层]水养法	water curing	
02.669	防滑处理	anti-skid treatment	
02.670	沥青罩面	asphalt overlay	
02.671	全厚翻修	full depth resurfacing	
02.672	剥层翻修	peeled resurfacing	
02.673	路面再生	pavement recycling	
02.674	清除坍方	removing landslide	
02.675	防沙设施	sand protection facilities	
02.676	沙障	sand barrier	
02.677	防沙林	sand protection green	
02.678	风力加速堤	wind accelerating dike	
02.679	防沙坝	sand protection dike	
02.680	消冰	deicing	
02.681	除雪	snow removing	
02.682	防雪设施	snow protection facilities	
02.683	雪崩防治	snow slide protection	

序　码	汉　文　名	英　文　名	注　释
02.684	防雪栅	snow-fence	
02.685	人造雪崩	artificial snow slide	
02.686	养护质量	maintenance quality	
02.687	路面使用质量指标	pavement operating quality index	
02.688	路面破损率	pavement damage ratio	
02.689	公路养护质量等级	quality level of highway maintenance	
02.690	好路率	rate of good level highway	
02.691	公路标准化美化工程	highway standardized and beautified project	又称"GBM 工程"。
02.692	公路景观	highway landscape, vista	
02.693	坡面景观	slope landscape	
02.694	景观栽植	landscape planting	
02.695	公路绿化	highway planting, highway greening	
02.696	路旁绿化带	roadside green belt	
02.697	防护林带	shelter belt	
02.698	路面管理系统	pavement management system, PMS	
02.699	路面检测评价系统	pavement monitor and evaluation system	
02.700	路面状况指数	pavement condition index	
02.701	路面评价模型	pavement evaluation model	
02.702	国际平整度指数	international roughness index, IRI	
02.703	路面抗滑性能	pavement skid resistance condition, PSRC	

03. 交 通 工 程

序　码	汉　文　名	英　文　名	注　释
03.001	车队	platoon	
03.002	混合交通	mixed traffic	
03.003	交通流	traffic flow	
03.004	交通流理论	traffic flow theory	
03.005	流体动力学理论	hydrodynamics theory	

序 码	汉 文 名	英 文 名	注 释
03.006	跟车理论	car-following theory	
03.007	拓扑学理论	topologic theory	
03.008	微观交通模型	micro-traffic model	
03.009	宏观交通模型	macro-traffic model	
03.010	连续性模型	continuity model	
03.011	确定性模型	determinacy model	
03.012	随机性模型	stochastic model	
03.013	集合模型	aggregation model	
03.014	非集合模型	disaggregation model	
03.015	灰色系统理论	grey system theory	
03.016	灰色控制系统	grey control system	
03.017	交通仿真	traffic simulation	
03.018	车流返回波	backward wave in traffic flow	
03.019	车流停驶波	stopping wave in traffic flow	
03.020	交通行为	traffic behavior	
03.021	车流起动波 、	starting wave in traffic flow	
03.022	交通稳定性	traffic stability	
03.023	局部稳定性	local stability	
03.024	渐近稳定性	asymptotic stability	
03.025	加速噪声	acceleration noise	
03.026	可插车空档	acceptance gap	
03.027	临界空挡	critical gap	
03.028	排队理论	queue theory	
03.029	排队规则	queue rule	
03.030	单路通道排队	single-channel queue	
03.031	多路通道排队	multiple-channel queue	
03.032	交通需求	traffic demand	
03.033	交通网络	traffic network	
03.034	交通拥挤	traffic congestion	
03.035	交通汇合	traffic converging	
03.036	交通分流	traffic diverging	
03.037	交通分析	traffic analysis	
03.038	交通瓶颈	traffic bottleneck	
03.039	交通系统	traffic system	
03.040	交通线网优化	optimization of traffic line and net-work	
03.041	自由交通流	free traffic flow	

序　码	汉　文　名	英　文　名	注　释
03.042	稳定交通流	stable traffic flow	
03.043	不稳定交通流	unstable traffic flow	
03.044	强制性交通流	forced traffic flow	
03.045	交织交通流	weaving traffic flow	
03.046	非交织交通流	nonweaving traffic flow	
03.047	连续性交通流	continuous traffic flow	
03.048	中断性交通流	interrupted traffic flow	
03.049	交通分类	traffic classification	
03.050	交通量	traffic volume	
03.051	交通量估计	traffic volume estimating	
03.052	交通因素	traffic factor	
03.053	交通流率	traffic flow rate	
03.054	方向分布	direction distribution	
03.055	车道分布	lane distribution	
03.056	高峰小时系数	peak hour factor, PHF	
03.057	年最大小时交通量	annual maximum hourly traffic volume	
03.058	年平均日交通量	annual average daily traffic volume, AADT volume	
03.059	月平均日交通量	mouthly average daily traffic volume, MADT volume	
03.060	年第三十位最大小时交通量	annual thirtieth highest hourly traffic volume, 30HV	
03.061	换算交通量	equivalent traffic volume	又称"当量交通量"。
03.062	交通量计量	traffic volume measurement	
03.063	交通干扰	traffic interference	
03.064	平均日交通量	average daily traffic volume, ADT	
03.065	设计小时交通量	design hourly traffic volume, DHV	
03.066	定向设计小时交通量	directional design hourly traffic volume, DDHV	
03.067	车速	speed	
03.068	临界车速	critical speed	
03.069	点速度	spot speed	
03.070	时间平均速度	time mean speed	
03.071	空间平均速度	space mean speed	
03.072	推荐速度	advisory speed	

序 码	汉 文 名	英 文 名	注 释
03.073	自由流车速	free flow speed	
03.074	平均车速	average speed	
03.075	行驶速度	running speed	
03.076	行程车速	travel speed, journey speed	
03.077	运行车速	operating speed	
03.078	最高车速	maximum speed	
03.079	经济车速	economic speed	
03.080	期望车速	desired speed	
03.081	行驶时间	running time	
03.082	行程时间	travel time, journey time	
03.083	最佳速度	optimum speed	
03.084	交通密度	traffic density	
03.085	最佳密度	optimum density	
03.086	阻塞密度	jam density	
03.087	临界密度	critical density	
03.088	车头间距	space headway	
03.089	车头时距	[time] headway	
03.090	车间净距	vehicular gap	
03.091	临界 V/C 比	critical V/C ratio	
03.092	延误	delay	
03.093	行驶延误	operational delay	
03.094	起动延误	starting delay	
03.095	交叉口延误	intersection delay	
03.096	停车线延误	stop line delay	
03.097	停车延误	stop delay	
03.098	加减速延误	acceleration-deceleration delay	
03.099	行程时间比	travel-time ratio	
03.100	车道占有率	lane occupancy ratio	
03.101	时间占有率	time occupancy ratio	
03.102	随车观测法	moving observer method	
03.103	暂停	standing	
03.104	停止	stopping	
03.105	停车	parking	又称"存车","泊车"。
03.106	存车换乘	park and ride	
03.107	存车搭乘	park and drive	
03.108	公路交通规划	highway and transportation plan-	

序　码	汉　文　名	英　文　名	注　释
		ning	
03.109	公路交通方式	highway traffic mode	
03.110	交通方式划分	[traffic] modal split	
03.111	土地利用预测模型	land-use forecast model	
03.112	境内交通	local traffic	
03.113	区内交通	intra-zone traffic	
03.114	过境交通	through traffic	
03.115	交通生成	trip generation	又称"交通发生"。
03.116	交通量分配	traffic [volume] assignment	
03.117	交通量分布	traffic [volume] distribution	
03.118	交通量预测	traffic volume forecast	
03.119	交通需要预测	traffic demand forecast	
03.120	增长率法	growth rate method	
03.121	连续式交通量观测站	continuous traffic count station	
03.122	间歇式交通量观测站	intermittent traffic count station	
03.123	增长交通量	increment of traffic [volume]	
03.124	诱增交通量	induced traffic [volume]	
03.125	转移交通量	diverted traffic [volume]	
03.126	吸引交通量	absorbed traffic [volume]	
03.127	出行	trip	
03.128	出行率	trip rate	
03.129	出行长度	trip length	
03.130	出行时间	trip time	
03.131	出行分布	trip-distribution	
03.132	徒步出行	pedestrian trip	
03.133	境内出行	local trip	
03.134	出行交换模型	trip interchange model	
03.135	出行端点模型	trip end model	
03.136	交通可达性	traffic accessibility	
03.137	区界交通调查	cordon traffic survey, cordon counts	
03.138	交通调查表	traffic survey chart	
03.139	渠化交通	channelized traffic	
03.140	综合运输系统	comprehensive transportation sys-	

序　码	汉　文　名	英　文　名	注　释
		tem	
03.141	快速公共交通系统	rapid transit system	
03.142	交通方式选择模型	traffic modal choice model	
03.143	停车车位	parking set	
03.144	停车率	parking rate	
03.145	停车持续时间	parking duration	
03.146	停车周转次数	parking turnover	
03.147	停车车位需要	demand for parking spaces	
03.148	需要停车次数	parking demand	
03.149	停车车辆总数	parking accumulation	
03.150	停车车位短缺	parking deficiency	
03.151	停车车位供应量	parking supply	
03.152	停车车位过剩	parking surplus	
03.153	路边停车	side parking, curb parking	
03.154	路外停车	off-road parking	
03.155	停车计划	parking plan	
03.156	停车管理计划	parking management program	
03.157	停车饱和度	degree of parking saturation	
03.158	停车费	parking fee	
03.159	公路通行能力	highway capacity	
03.160	饱和流量	saturation volume	
03.161	饱和流率	saturation volume rate	
03.162	服务水平	level of service, LOS	
03.163	服务水平量度参数	effectiveness measures of LOS	
03.164	基本通行能力	basic capacity	
03.165	可能通行能力	possible capacity	
03.166	设计通行能力	design capacity	
03.167	路段通行能力	road section capacity	
03.168	交叉口通行能力	intersection capacity	
03.169	车道通行能力	lane capacity	
03.170	匝道通行能力	ramp capacity	
03.171	交织区通行能力	weaving section capacity	
03.172	经济通行能力	economical capacity	
03.173	储备通行能力	reserve capacity	

序 码	汉 文 名	英 文 名	注 释
03.174	施工区通行能力	construction section capacity	
03.175	通行能力指数	capacity index	
03.176	最大服务交通量	maximum service volume	
03.177	最大服务流率	maximum service flow rate	
03.178	拥挤度	degree of congestion	
03.179	慢速车混入率	slow vehicle mixed rate	
03.180	匝道连接处	ramp junction	
03.181	交织区	weaving area	
03.182	非约束运行	unconstrained operation	
03.183	约束运行	constrained operation	
03.184	分流交通量	diverging traffic volume	
03.185	合流交通量	converging traffic volume	
03.186	车道平衡	lane balance	
03.187	车速限制	speed limit	
03.188	车道偏移	lane shift	
03.189	优先通行车道	priority lane	
03.190	四向停车	four-way stop	
03.191	匝道交通调节	ramp metering	
03.192	交通系统管理	transportation system management, TSM	
03.193	交通指挥	traffic guidance	
03.194	交通巡逻	traffic patrolling	
03.195	禁止左转弯	no left turn	
03.196	禁止通行	traffic prohibited	
03.197	禁止停车	parking prohibited	
03.198	纵列式停车	parallel parking	
03.199	斜列式停车	diagonal parking	
03.200	路上停车点	road parking lot	
03.201	车辆停放指数	parking index	
03.202	停储长度	storage length	
03.203	走廊交通管理计划	corridor traffic management program	
03.204	交通分隔	segregation of traffic	
03.205	综合交通管理	comprehensive traffic management	
03.206	施工路段交通管理	construction section traffic management	
03.207	行人管制	pedestrian control	

序　码	汉　文　名	英　文　名	注　释
03.208	交通规则	traffic rules, traffic regulations	
03.209	交通违章	traffic violation	
03.210	交通违章者	traffic violator	
03.211	停车规则	parking regulation	
03.212	交通安全	traffic safety	
03.213	安全措施	safety measure	
03.214	交通事故	traffic accident	
03.215	交通事故率	traffic accident rate	
03.216	交通肇事罪	traffic accident crime	
03.217	事故死亡率	death rate of accident	
03.218	事故地点图	accident spot map	
03.219	撞人事故	pedestrian accidents	
03.220	事故易发地点	[accident] black spot	
03.221	侧向碰撞	side collision	
03.222	一维碰撞	one dimension collision	
03.223	二维碰撞	two dimension collision	
03.224	一次碰撞	primary collision	
03.225	二次碰撞	secondary collision	
03.226	多次碰撞	multiple collision	
03.227	迎面碰撞	head-on collision	
03.228	头尾相撞	rear-end collision	
03.229	交通冲突	traffic conflict	
03.230	驶出路外事故	run-off-road accident	
03.231	交通法庭	traffic court	
03.232	雨天事故	wet traffic accident	
03.233	非雨天事故	dry traffic accident	
03.234	交通事故隐患	hidden peril of accident	
03.235	交通事故预测	traffic accident prediction	
03.236	多角度交通事故预测法	multiplex traffic accident prediction method	
03.237	同步交通事故预测法	simultaneous traffic accident prediction method	
03.238	多乘员车辆车道	high occupancy vehicle lane	
03.239	合乘车辆车道	car pool lane	
03.240	信号控制	signal control	
03.241	灯光信号	light signal	
03.242	信号周期	signal cycle	

序 码	汉 文 名	英 文 名	注 释
03.243	周期长	cycle length	
03.244	相位	phase	
03.245	绿灯时间	green time	
03.246	红灯时间	red time	
03.247	有效绿灯时间	effective green time	
03.248	绿信比	split, green ratio	
03.249	绿波带	green wave band	
03.250	全红信号	all red signal	
03.251	相位差	offset, phase offset	又称"[信号]时差"。
03.252	优先时差	priority phase offset	
03.253	交叉口负荷系数	load factor of intersection	
03.254	闪光信号	flashing signal	
03.255	绿灯时差型式	offset pattern	
03.256	清尾[车]时间	clearance time	
03.257	行人信号相位	pedestrian signal phase	
03.258	联动交通信号	coordinated traffic signals	
03.259	同步式系统	synchronous system	
03.260	交互式系统	alternate system	
03.261	车辆感应信号	vehicle actuated signal	
03.262	交通通道	traffic aisle	
03.263	实时交通控制	real-time traffic control	
03.264	交通感应式控制	traffic responsive control	
03.265	全感应式信号控制	fully-actuated signal control	
03.266	半感应式信号控制	semi-actuated signal control	
03.267	交通量修正系数	traffic volume adjustment factor	
03.268	自适应信号控制系统	adaptive traffic signal control system	
03.269	多路停车控制	multiway stop control	
03.270	交通信号脱机控制	traffic signal off-line control	
03.271	联机控制	on-line control	
03.272	定时信号控制	fixed-time control	
03.273	出口匝道控制	exit ramp control	
03.274	入口匝道控制	entrance ramp control	
03.275	匝道集成系统控	ramp integrated system control	

序　码	汉　文　名	英　文　名	注　释
	制		
03.276	高速公路监控	freeway surveillance and control	
03.277	高速公路主线控制	freeway mainline control	
03.278	交通监控中心	traffic surveillance and control center	
03.279	交通监控系统	traffic surveillance and control system	
03.280	交通信息系统	traffic information system	
03.281	可变限速控制	variable speed-limit control	
03.282	警察巡逻	police patrol	
03.283	服务巡逻	service patrol	
03.284	道路引导系统	route guidance system	
03.285	行人检测器	pedestrian detector	
03.286	最大安全速度	maximum safe speed	
03.287	车辆检测器	vehicle detector	
03.288	环形线圈式车辆检测器	loop vehicle detector	
03.289	磁感应式车辆检测器	magnetic vehicle detector	
03.290	超声波车辆检测器	ultra-sonic vehicle detector	
03.291	红外线车辆检测器	infra-red vehicle detector	
03.292	存在型车辆检测器	presence vehicle detector	
03.293	通过型车辆检测器	passage vehicle detector	
03.294	雷达测速器	radar speedometer	
03.295	应急电话	emergency telephone	
03.296	路侧广播	roadside radio broadcasting	
03.297	点控制	spot control, isolated signal control	
03.298	线控制	line control, linked control	
03.299	联动控制	coordinated control	
03.300	地图板显示	map display	
03.301	信号控制器	signal controller	
03.302	感应信号控制机	traffic-actuated controller	

序 码	汉 文 名	英 文 名	注 释
03.303	定时信号控制机	pretimed controller, timing controller	
03.304	主控制机	master controller	
03.305	行人按钮信号控制机	pedestrian push-button signal controller	
03.306	行人过街信号灯	pedestrian crossing signal lamp	
03.307	交通信号灯	traffic signal lamp	
03.308	车型检测器	vehicle type detector	
03.309	轴重检测器	axle weight detector	
03.310	气象检测器	weather detector	
03.311	雾检测器	fog detector	
03.312	动态轴重检测器	dynamic axle weight detector	
03.313	静态轴重检测器	static axle weight detector	
03.314	车辆高度检测器	vehicle height detector	
03.315	收费车道检测器	toll lane detector	
03.316	智能车路系统	intelligent vehicle highway system, IVHS	
03.317	先进的交通管理系统	advanced traffic management system, ATMS	智能车路系统的子系统之一。
03.318	先进的汽车控制系统	advanced vehicle control system, AVCS, IVCS	又称"智能汽车控制系统"。
03.319	先进的驾驶员信息系统	advanced driver information system, ADIS	
03.320	先进的城市[公共]运输系统	advanced public transportation system, APTS	
03.321	先进的长途运输系统	advanced rural transportation system, ARTS	
03.322	商用车辆运行管理系统	commercial vehicle operation management system, CVOM	
03.323	应急管理系统	emergency management system	
03.324	收费显示器	toll display	
03.325	车道收费机	lane toll machine	
03.326	收费通行券	toll pass ticket	
03.327	磁卡收费机	magnetic card toll machine	
03.328	磁卡通行券	magnetic card toll pass ticket	
03.329	收费车道信号灯	toll lane signal lamp	
03.330	发卡机	card sender	

序　码	汉　文　名	英　文　名	注　释
03.331	读卡机	card reader	
03.332	道路收费系统	road toll system	
03.333	收费卡门	toll gate	
03.334	收费亭	toll booth, toll house	
03.335	收费站	toll station	
03.336	收费岛	toll island	
03.337	收费棚	toll canopy	
03.338	收费广场	toll plaza	
03.339	主线收费站	mainline toll station	
03.340	互通式立交收费站	interchange toll station	
03.341	停车计时器	parking meter	
03.342	收费弹性	toll elasticity	
03.343	收费中心	toll center	
03.344	收费制式	toll mode	
03.345	统一收费系统	unified toll system	
03.346	开放式收费系统	open toll system	
03.347	封闭式收费系统	closed toll system	
03.348	混合式收费系统	mixed toll system	
03.349	电子收费系统	electronic toll system	
03.350	交通管理设施	traffic management devices	
03.351	交通设施	traffic facilities	
03.352	标志	sign	
03.353	交通标志	traffic sign	
03.354	警告标志	warning sign, caution sign	
03.355	禁令标志	prohibitory sign	
03.356	指示标志	mandatory sign	
03.357	指路标志	directional sign, guide sign	
03.358	辅助标志	auxiliary sign	
03.359	标志视认性	sign legibility	
03.360	地名标志	place name sign	
03.361	分界标志	boundary sign	
03.362	诱导标志	induction sign	
03.363	可变标志	variable sign	
03.364	可变限速标志	variable speed-limit sign	
03.365	可变信息标志	changeable message sign	
03.366	固定信息标志	fixed-message sign	

序 码	汉 文 名	英 文 名	注 释
03.367	停车场标志	parking lot sign	
03.368	让路标志	yielding sign	
03.369	前置标志	advance sign	
03.370	门式交通标志	overhead traffic sign	
03.371	悬臂式交通标志	cantilever traffic sign	
03.372	立柱式交通标志	post traffic sign	
03.373	管制标志	regulatory traffic sign	
03.374	停车标志	stop sign	
03.375	行人过街标志	pedestrian crossing sign	
03.376	限重标志	weight limit sign	
03.377	弯道标志	curve sign	
03.378	岔道标志	cross-buck sign	
03.379	前方停车标志	stop ahead sign	
03.380	前方让路标志	yield ahead sign	
03.381	前方信号标志	signal ahead sign	
03.382	汇流标志	converging sign	
03.383	窄桥标志	narrow bridge sign	
03.384	分隔行驶公路标志	divided highway sign	
03.385	坡道标志	slope sign	
03.386	限速标志	stated-speed sign	
03.387	路滑标志	skid sign	
03.388	最低限速标志	minimum stated-speed sign	
03.389	禁止调头标志	no U-turn sign	
03.390	禁止右转弯标志	no right turn sign	
03.391	禁止左转弯标志	no leftturn sign	
03.392	禁止超车标志	overtaking prohibited sign	
03.393	爬坡车道标志	climbing lane sign	
03.394	禁止驶入标志	no entry sign	
03.395	让车道标志	passing bay sign	
03.396	禁止通行标志	traffic prohibited sign	
03.397	单行线标志	one-way sign	
03.398	禁止停放标志	parking ban sign	
03.399	低净空标志	low clearance sign	
03.400	绕行标志	detour sign	
03.401	休息区标志	rest area sign	
03.402	服务区标志	service area sign	

序　码	汉　文　名	英　文　名	注　释
03.403	应急电话标志	emergency telephone sign	
03.404	出口标志	exit sign	
03.405	出口预告标志	advance exit sign	
03.406	发光标志	luminous sign	
03.407	反光标志	reflecting sign	
03.408	光纤标志	optic fiber sign	
03.409	灯光照明标志	illuminated sign	
03.410	学童过街标志	children crossing sign	
03.411	学校标志	school sign	
03.412	铁路道口标志	railway crossing sign	
03.413	道路标识	road markings	
03.414	双黄线	double amber lines	
03.415	车道分界线	lane line	
03.416	边缘线	edge line	
03.417	斑马线	zebra crossing	
03.418	左转弯导向线	left turn guide line	
03.419	右转弯导向线	right turn guide line	
03.420	轮廓标线	delineation line	
03.421	可变向中心车道线	reversible center lane line	
03.422	渠化标线	channelizing marking	
03.423	中心岛标线	center island marking	
03.424	停放车位标线	parking space limit marking	
03.425	停车线	stop line	
03.426	匝道标线	ramp marking	
03.427	立面标记	object marking	
03.428	路面文字标记	pavement lettering marking	
03.429	缘石标记	curb marking	
03.430	定向行车道标记	directional carriageway marking	
03.431	交通护栏	traffic guardrail	
03.432	钢索护栏	cable guardrail	
03.433	管式护栏	pipe guardrail	
03.434	箱梁型护栏	box-girder fence	
03.435	W 型护栏	W-type guardrail	
03.436	缓冲护栏	cushion guardail	
03.437	钢板护栏	steel guardrail	
03.438	波形梁护栏	corrugated beam barrier	

序　码	汉　文　名	英　文　名	注　释
03.439	刚性护栏	stiff safety fence	
03.440	柔性护栏	flexible safety fence	
03.441	钢筋混凝土护栏	reinforced concrete fence	
03.442	玻璃钢护栏	glass fiber reinforced plastic fence	
03.443	阻挡式护栏	block out type safety fence	
03.444	液压缓冲护栏	hydraulic cushion guardrail	
03.445	中央分隔带护栏	median barrier	
03.446	防撞墩	crash bearer	
03.447	实物分隔带	physical separator	
03.448	突起分隔带	raised separator	
03.449	新泽西护栏	New Jersey safety barrier	
03.450	护柱	guard post	
03.451	交通隔离墩	traffic divided blocks	
03.452	柱式轮廓标	post delineator	
03.453	可调头交叉口	turn crossing	
03.454	示警柱	warning post	
03.455	照明	lighting	
03.456	缓和照明	adaptation lighting	又称"适应照明"。
03.457	高杆照明	high mast lighting	
03.458	照明灯柱	lighting standard	
03.459	隧道照明	tunnel lighting	
03.460	照明过渡	lighting transition	
03.461	照明过渡段	lighting transition section	
03.462	照明适应段	lighting adaptation section	
03.463	人工照明过渡	artificial lighting transition	
03.464	隧道入口区亮度	tunnel entrance brilliance	
03.465	隧道适宜亮度	suitable brilliance in tunnel	
03.466	纵向连续带光源	longitudinal continuous band illu-minant	
03.467	横向间断光源	laterally intermittent illuminant	
03.468	周期性闪光	cyclic flashing	
03.469	黑洞效应	black-hole effect	
03.470	光度控制系统	light control system	
03.471	照明灯具	lamps and lanterns	
03.472	照明设备	lighting facilities	
03.473	道路反光镜	road reflecting mirror	
03.474	反光标识	reflecting marking	

序　码	汉文名	英文名	注　释
03.475	路钮	button	
03.476	反光路钮	reflecting button	
03.477	猫眼	cat eyes	反光路钮之一种。
03.478	防眩板	anti-glare panel	
03.479	防眩屏	anti-glare screen	
03.480	积雪标杆	snow[deposit] markerpost	
03.481	交通指向钮	traffic direction stud	
03.482	交通指路牌	traffic direction block	
03.483	锥形交通路标	traffic cone	
03.484	行人安全设施	pedestrian safety devices	
03.485	行人横穿设施	pedestrian crossing devices	
03.486	人行横道	cross walk	
03.487	人行地道	pedestrian underpass	
03.488	人行天桥	pedestrian overpass	
03.489	行人护栏	pedestrian barrier	
03.490	防护栅	guard fence, safety fence	
03.491	隔离栅	separation fence	
03.492	隔离网	separation net	
03.493	防牲畜护栏	cattle fence	
03.494	动物通道	animal corridor	
03.495	人机调节系统	man-machine processing system	
03.496	神经机能测定	nerve function measurement	
03.497	生理机能测定	physiological function measurement	
03.498	血中酒精浓度	blood-alcohol concentration	
03.499	生物节律	biorhythms	
03.500	驾驶信息	driving information	
03.501	视野	field of vision	
03.502	前方视野	field of front vision	
03.503	侧向视野	field of lateral vision	
03.504	直接视野	direct field of vision	
03.505	间接视野	indirect field of vision	
03.506	能见度	visibility	
03.507	人动视力	vision with driver moving	
03.508	物动视力	vision with object moving	
03.509	全动视力	vision with both driver and object moving	

序　码	汉　文　名	英　文　名	注　　释
03.510	[司机]反应距离	[driver] reaction distance	
03.511	门口效应	gate effect	视觉上收缩路幅的效应。
03.512	[司机]判断时间	[driver] judgement time	
03.513	[司机]识别距离	[driver] decipherment distance	
03.514	[司机]感觉反应时间	[driver] perception-reaction time	
03.515	[司机]感觉反应距离	[driver] perception-reaction distance	
03.516	制动反应时间	brake reaction time	
03.517	制动距离	brake distance	
03.518	道路催眠状态	road hypnosis	
03.519	生理疲劳	physiological fatigue	
03.520	驾驶疲劳	driving fatigue	
03.521	驾驶操纵能力	driving operation ability	
03.522	驾驶习惯	driving habits	
03.523	驾驶行为	driving behavior	
03.524	驾驶员行为特性	driver behavior pattern	
03.525	驾驶员心理和生理反应	driver psychological and physiological reaction	
03.526	驾驶适应性	driving adaptability	
03.527	视力适应性	adaptation of vision	
03.528	颜色适应性	color adaptability	
03.529	视觉敏锐度	visual acuity	
03.530	动视觉敏锐度	dynamic visual acuity	
03.531	静视觉敏锐度	static visual acuity	

04．桥、涵、隧道、渡口工程

序　码	汉　文　名	英　文　名	注　　释
04.001	桥梁总体规划	bridge overall planning	
04.002	桥型	bridge type	
04.003	桥名	bridge name	
04.004	桥位	bridge site	
04.005	桥位选择	bridge site selection	

序 码	汉 文 名	英 文 名	注 释
04.006	桥梁标准设计	standard design of bridge	
04.007	主桥	main bridge	
04.008	引桥	approach bridge, approach span	
04.009	跨径	span	
04.010	主跨	main span	
04.011	通航桥孔	navigable span	
04.012	非通航桥孔	non-navigable span	
04.013	桥梁全长	total length of bridge	
04.014	跨径长	span length	
04.015	计算跨径	calculated span	
04.016	净跨径	clear span	
04.017	经济跨径	economic span	
04.018	桥梁建筑高度	construction height of bridge	
04.019	桥面宽度	bridge deck width	
04.020	桥梁总宽度	total width of bridge	
04.021	宽跨比	width-span ratio	
04.022	桥面净空	clearance above bridge deck	
04.023	桥下净空	clearance under span	
04.024	通航净空	navigable clearance	
04.025	净空高度	clearance height	
04.026	梁高	beam depth, girder depth	
04.027	矢高	rise of arch	
04.028	计算矢高	calculated rise of arch	
04.029	高跨比	depth-span ratio	
04.030	矢跨比	rise-span ratio	
04.031	桥面标高	deck elevation	
04.032	桥面纵坡	deck profile grade	
04.033	桥面横坡	deck transverse slope	
04.034	桩入土长度	embeded length of pile	
04.035	基础埋置深度	embedment depth of foundation	
04.036	斜交角	skew angle	
04.037	水道测量	hydrographic survey	
04.038	河道调查	river survey	
04.039	水文测量	hydrologic survey	
04.040	水文资料	hydrologic data	
04.041	气象资料	meteorologic data	
04.042	水文分析	hydrologic analysis	

序　码	汉　文　名	英　文　名	注　释
04.043	水力计算	hydraulic computation	
04.044	桥址水文观测	hydrologic observation at bridge site	
04.045	水文站	hydrologic station	
04.046	气象站	meteorologic station	
04.047	水工构造物	hydraulic structure	
04.048	淹水地区	inundated district	
04.049	泛滥区	flooded area	
04.050	水网地区	dense waterway net region	
04.051	过水断面	discharge flow cross section	
04.052	过水孔径	discharge opening of bridge	
04.053	阻水面积	current-obstruction area	
04.054	流速	current velocity	
04.055	设计流速	design current velocity	
04.056	表观流速	apparent velocity	
04.057	行近流速	approach velocity	
04.058	泥沙起动流速	sediment-moving incipient velocity	
04.059	止冲流速	non-scouring velocity	
04.060	水面比降	water surface gradient	
04.061	河床比降	river bed gradient	
04.062	湿周	wetted perimeter	
04.063	水力半径	hydraulic radius	
04.064	糙率	coefficient of roughness	
04.065	流向	current direction	
04.066	流量	flow discharge	
04.067	洪水流量	flood discharge	
04.068	洪峰流量	peak flow discharge	
04.069	设计流量	design discharge	
04.070	流量过程线	flow duration curve	
04.071	洪水频率	flood frequency	
04.072	设计洪水频率	design flood frequency	
04.073	洪水重现期	return period of flood	
04.074	洪水位过程线	flood level duration curve	
04.075	流域	river basin	
04.076	分水岭	dividing ridge	
04.077	汇水面积	catchment area	
04.078	径流	run-off	

序 码	汉 文 名	英 文 名	注 释
04.079	暴雨径流	rainstorm run-off	
04.080	暴雨强度	rainstorm intensity	
04.081	洪水径流	flood run-off	
04.082	径流面积	run-off area	
04.083	径流系数	run-off coefficient	
04.084	河床	river bed	
04.085	河槽	river channel	
04.086	主槽	main channel	
04.087	沙洲	shoal	
04.088	河滩	flood land	
04.089	河岸	river bank	
04.090	水位	water level	
04.091	水头	water head	
04.092	最高水位	highest water level, HWL	
04.093	最低水位	lowest water level, LWL	
04.094	平均最高水位	mean highest water level	
04.095	平均最低水位	mean lowest water level	
04.096	常水位	mean water level	
04.097	枯水位	low water level	
04.098	设计水位	design water level	
04.099	通航水位	navigable water level, NWL	
04.100	施工水位	construction water level	
04.101	潮水位	tidal level	
04.102	平滩水位	flood land line water level	
04.103	河槽压缩	channel contraction	
04.104	桥孔压缩	bridge opening contraction	
04.105	壅水	back-water	
04.106	桥前壅水	back-water in front of bridge	
04.107	墩前壅水	back-water at pier	
04.108	壅水高度	back-water height	
04.109	浪高	wave height	
04.110	天然演变冲刷	natural scour	
04.111	河槽天然冲刷	natural scour of channel	
04.112	堤岸冲刷	bank erosion	
04.113	桥基冲刷	scouring at bridge foundation	
04.114	桥下一般冲刷	general scour at bridge opening	
04.115	桥墩局部冲刷	local scour around pier	

序 码	汉 文 名	英 文 名	注 释
04.116	清水冲刷	scour without sediment motion	
04.117	浑水冲刷	scour with sediment motion	
04.118	冲刷深度	scouring depth	
04.119	冲刷系数	coefficient of scouring	
04.120	淤积	silting	
04.121	漂浮物	drifter	
04.122	悬移质	suspended load	
04.123	推移质	bed material load	
04.124	流冰	drift ice	
04.125	水流导向	current deflecting	
04.126	调治构造物	regulating structure	
04.127	导流堤	diversion dike	
04.128	顺坝	longitudinal dike	
04.129	丁坝	spur dike	
04.130	淹没丁坝	submerged spur dike	
04.131	梨形丁坝	pear-ended spur dike	
04.132	勾头丁坝	bend-ended spur dike	
04.133	防波堤	break-water	
04.134	河堤	levee	
04.135	[桥头]锥坡	conical slope	
04.136	护岸	bank protection, bank revetment	
04.137	河床铺砌	river bed paving	
04.138	桥墩分水尖	pier break-water	
04.139	消能槛	baffle sill	
04.140	护坦	apron	
04.141	柴排	mattress, willow fascine	
04.142	柴捆	fagot	
04.143	柴捆坝	fagot dam	
04.144	石笼	gabion	
04.145	抛石防护	riprap protection	
04.146	桥[梁]	bridge	
04.147	公路桥	highway bridge	
04.148	公铁两用桥	combined highway and railway bridge, bi-purposed bridge	
04.149	人行桥	pedestrian bridge	
04.150	跨河桥	river-crossing bridge	
04.151	跨线桥	flyover	

序 码	汉 文 名	英 文 名	注 释
04.152	军用桥	military bridge	
04.153	港口立交桥	estuarial crossing	
04.154	高架桥	viaduct, trestle	又称"栈桥"、"旱桥"。
04.155	上承式桥	deck bridge	
04.156	中承式桥	half-through bridge	
04.157	下承式桥	through bridge	
04.158	正交桥	right bridge	
04.159	斜交桥	skew bridge	
04.160	弯桥	curved bridge	又称"曲线桥"。
04.161	坡桥	bridge on slope	
04.162	匝道桥	ramp bridge	又称"坡道桥"。
04.163	旧桥	existing bridge	
04.164	危桥	bridge in danger	
04.165	改建桥	reconstructed bridge	
04.166	收费桥	toll bridge	
04.167	特大桥	grand bridge	
04.168	大桥	great bridge	
04.169	大跨径桥	long span bridge	
04.170	中桥	medium bridge	
04.171	小桥	small bridge	
04.172	高水位桥	high water level bridge	
04.173	中水位桥	medium water level bridge	
04.174	低水位桥	low water level bridge	
04.175	漫水桥	submersible bridge	
04.176	溢洪桥	spillway bridge	
04.177	排涝桥	flood relief bridge	
04.178	永久性桥	permanent bridge	
04.179	半永久性桥	semi-permanent bridge	
04.180	临时性桥	temporary bridge	
04.181	便桥	detour bridge	
04.182	施工便桥	temporary bridge for construction	
04.183	钢筋混凝土桥	reinforced concrete bridge	
04.184	预应力混凝土桥	prestressed concrete bridge	
04.185	钢骨混凝土桥	rolled shape steel reinforced concrete bridge	
04.186	劲性骨架混凝土	skeleton reinforced concrete bridge	

序　码	汉　文　名	英　文　名	注　释
	桥		
04.187	钢管混凝土桥	steel pipe-encased concrete bridge	
04.188	钢桥	steel bridge	
04.189	铆接钢桥	riveted steel bridge	
04.190	焊接钢桥	welded steel bridge	
04.191	栓接钢桥	bolted steel bridge	
04.192	栓焊钢桥	bolted and welded steel bridge	
04.193	圬工桥	masonry bridge	
04.194	圬工拱桥	masonry arch bridge	
04.195	石桥	stone bridge	
04.196	石拱桥	stone arch bridge	
04.197	料石拱桥	ashlar arch bridge	
04.198	块石拱桥	block stone arch bridge	
04.199	木桥	timber bridge, wooden bridge	
04.200	胶合木桥	glued timber bridge	
04.201	钉板梁桥	pin-connected plywood girder bridge	
04.202	玻璃钢桥	glass fiber reinforced plastic bridge	
04.203	梁桥	beam bridge, girder bridge	
04.204	简支梁桥	simple-supported beam bridge	
04.205	连续梁桥	continuous beam bridge	
04.206	悬臂梁桥	cantilever beam bridge	
04.207	板桥	slab bridge	
04.208	空心板桥	hollow slab bridge	
04.209	肋板桥	ribbed slab bridge	
04.210	钢板梁桥	steel plate girder bridge	
04.211	T[形]梁桥	T-beam bridge	
04.212	箱[形]梁桥	box-girder bridge	
04.213	组合梁桥	composite beam bridge	
04.214	桁架桥	truss bridge	
04.215	拱桥	arch bridge	
04.216	空腹拱桥	open spandrel arch bridge	
04.217	实腹拱桥	filled spandrel arch bridge	
04.218	无铰拱桥	hingless arch bridge	
04.219	三铰拱桥	three-hinged arch bridge	
04.220	二铰拱桥	two-hinged arch bridge	
04.221	单铰拱桥	single-hinged arch bridge	

序 码	汉 文 名	英 文 名	注 释
04.222	连续拱桥	continuous arch bridge	
04.223	刚架拱桥	rigid-framed arch bridge	
04.224	双曲拱桥	two-way curved arch bridge	
04.225	肋拱桥	ribbed arch bridge	
04.226	板肋拱桥	slab-rib arch bridge	
04.227	桁架拱桥	trussed arch bridge	
04.228	系杆拱桥	bowstring arch bridge, tied arch bridge	
04.229	扁壳桥	shell bridge	
04.230	刚构桥	rigid frame bridge	
04.231	T形刚构桥	T-shaped rigid frame bridge	
04.232	桁式T形刚构桥	T-shaped truss rigid frame bridge	
04.233	斜腿刚构桥	slant legged rigid frame bridge	
04.234	连续刚构桥	continuous rigid frame bridge	
04.235	斜拉桥	cable stayed bridge	
04.236	混凝土斜拉桥	concrete deck cable stayed bridge	
04.237	钢斜拉桥	steel deck cable stayed bridge	
04.238	组合梁式斜拉桥	composite deck cable stayed bridge	
04.239	双索面斜拉桥	double plane cable stayed bridge	
04.240	单索面斜拉桥	single plane cable stayed bridge	
04.241	独塔斜拉桥	single pylon cable stayed bridge	
04.242	双塔斜拉桥	double pylon cable stayed bridge	
04.243	多塔斜拉桥	multi pylon cable stayed bridge	
04.244	悬索桥	suspension bridge	又称"吊桥"。
04.245	悬带桥	suspended ribbon bridge	
04.246	微弯板组合梁桥	composite shell-slab and I-beam bridge	
04.247	闸门桥	water gate bridge	
04.248	舟桥	pontoon bridge, bateau bridge	
04.249	浮桥	floating bridge	
04.250	开启桥	movable bridge	又称"活动桥"。
04.251	平转桥	swing bridge	又称"平旋桥"。
04.252	单叶平转桥	single-leaf swing bridge	
04.253	双叶平转桥	double-leaf swing bridge	
04.254	竖旋桥	bascule bridge	
04.255	单叶竖旋桥	single-leaf bascule bridge	

序 码	汉 文 名	英 文 名	注 释
04.256	双叶竖旋桥	double-leaf bascule bridge	
04.257	升降桥	lift bridge	
04.258	装配式桥	fabricated bridge	
04.259	装拆式钢桥	fabricated and detachable steel bridge	
04.260	贝雷桥	Bailey bridge	
04.261	上部结构	superstructure	
04.262	梁	beam, girder	
04.263	主梁	main beam, main girder	
04.264	箱梁	box girder	
04.265	单室箱梁	single cell box girder	
04.266	多室箱梁	multi cell box girder	
04.267	组合梁	composite beam	
04.268	工形梁	I[-shaped] beam	
04.269	T形梁	T[-shaped] beam	
04.270	U形梁	U[-shaped] beam	
04.271	矩形梁	rectangular beam	
04.272	挂梁	suspended beam	
04.273	内纵梁	interior beam, interior stringer	
04.274	外纵梁	exterior beam, exterior stringer	
04.275	横梁	transverse beam, cross beam	
04.276	横隔板	transverse diaphragm	
04.277	柱	column	
04.278	拱	arch	
04.279	板	slab	
04.280	桥面板	deck slab	
04.281	正交异性板	orthotropic slab	
04.282	少筋微弯板	under-reinforced slab with slightly curved bottom	
04.283	肋腋板	slab with haunched ribs	
04.284	单向板	one-way slab	
04.285	双向板	two-way slab	
04.286	悬臂板	cantilever slab	
04.287	桥面系	bridge deck system	
04.288	桥面	deck	
04.289	连续桥面	continuous deck	
04.290	桥面排水系统	deck drainage system	

序　码	汉　文　名	英　文　名	注　释
04.291	桥面铺装	deck pavement	
04.292	防水层	water proof layer	
04.293	防水涂层	water proof coating	
04.294	沥青薄膜防水层	water proof asphalt membrane	
04.295	聚合织物防水层	water proof polymer fabrics	
04.296	油毡防水层	water proof asphalt-felt	
04.297	沥青胶砂防水层	water proof asphalt mastic	曾用名"沥青玛琋脂防水层"。
04.298	桥面伸缩缝	deck expansion joint	
04.299	桥面伸缩装置	deck expansion installation	
04.300	对接式伸缩装置	butt type expansion installation	
04.301	支承式伸缩装置	support type expansion installation	
04.302	填缝材料	joint sealing materials	
04.303	填缝板接缝	joint with sealing plate	
04.304	埋式接缝	buried joint	
04.305	楔形缝	keyed joint	
04.306	角钢镶边接缝	angle steel edged joint	
04.307	橡胶接缝	rubber joint	
04.308	搭接钢板接缝	lapped steel plate joint	
04.309	梳式接缝	steel comb joint	
04.310	双梳式接缝	double steel comb joint	
04.311	加强氯丁橡胶板缝	strengthened neoprene plate joint	
04.312	金属滚轴带氯丁橡胶板接缝	metal rollers with neoprene plate joint	
04.313	戴马格接缝	Demag joint	
04.314	多姆克蜂房接缝	Domke honeycomb joint	
04.315	毛尔伸缩装置	Mauer expansion installation	
04.316	无纺布伸缩装置	non-woven fabrics expansion installation	
04.317	密封条三维伸缩装置	three-dimension expansion installation with sealing strip	
04.318	桥侧人行道	bridge sidewalk	
04.319	桥梁栏杆	bridge railing	
04.320	桥头堡	bridge head	
04.321	桥塔	bridge pylon, bridge tower	
04.322	索塔	cable support tower	

序 码	汉 文 名	英 文 名	注 释
04.323	门形索塔	portal framed tower	
04.324	A 形索塔	A-framed tower	
04.325	倒 Y 形索塔	inverted Y tower	
04.326	锚座	anchor socket	
04.327	索鞍	cable saddle	
04.328	斜索	stay cable	
04.329	锚索	anchor cable	
04.330	竖琴形拉索	harp-type stay cable	
04.331	扇形拉索	fan-type stay cable	
04.332	吊杆	suspender	
04.333	锚[碇]跨	anchor span	
04.334	锚碇	anchorage	
04.335	锚碇体	anchor body, anchor block	
04.336	锚碇板	anchor plate	
04.337	重力式锚碇	gravity anchor	
04.338	山洞式锚碇	anchor in rock gallery	
04.339	缆索散开室	[steel] cable splay chamber	
04.340	拱圈	arch ring	又称"拱环"。
04.341	拱肋	arch rib	
04.342	拱顶	arch crown	
04.343	拱座	arch support	
04.344	拱腹[线]	intrados, soffit	
04.345	拱背[线]	extrados	
04.346	起拱点	springing	
04.347	护拱	back haunch fillet of arch	
04.348	拱上结构	spandrel structure	
04.349	拱上侧墙	spandrel wall	
04.350	拱上横墙	spandrel cross wall	
04.351	拱上立柱	spandrel column	
04.352	腹拱	spandrel arch	
04.353	拱波	two way curved arch tile	
04.354	抗风支撑	wind bracing	
04.355	扰流板	spoiler	
04.356	桥头引道	bridge approach	
04.357	桥头搭板	bridge end transition slab	
04.358	下部结构	substructure	
04.359	桥墩	pier	

序　码	汉　文　名	英　文　名	注　释
04.360	墩身	pier body	
04.361	墩帽	pier cap, pier coping	
04.362	台帽	abutment cap, abutment coping	
04.363	盖梁	bent cap	又称"帽梁"。
04.364	重力式[桥]墩	gravity pier	
04.365	实体[桥]墩	solid pier	
04.366	空心[桥]墩	hollow pier	
04.367	柱式[桥]墩	column pier, shaft pier	
04.368	单柱式[桥]墩	single-columned pier, single shaft pier	
04.369	双柱式[桥]墩	two-columned pier, two shaft pier	
04.370	排架桩墩	pile-bent pier	
04.371	Y 形[桥]墩	Y-shaped pier	
04.372	柔性墩	flexible pier	
04.373	制动墩	braking pier, abutment pier	
04.374	单向推力墩	single direction thrusted pier	
04.375	抗撞墩	anti-collision pier	
04.376	锚墩	anchor pier	
04.377	辅助墩	auxiliary pier	
04.378	破冰体	ice apron	
04.379	防震挡块	anti-knock block, restrain block	
04.380	桥台	abutment	
04.381	台身	abutment body	
04.382	前墙	front wall	又称"胸墙"。
04.383	翼墙	wing wall	又称"耳墙"。
04.384	U 形桥台	U-abutment	
04.385	八字形桥台	flare wing-walled abutment	
04.386	一字形桥台	head wall abutment	
04.387	T 形桥台	T-abutment	
04.388	箱形桥台	box type abutment	
04.389	拱形桥台	arched abutment	
04.390	重力式桥台	gravity abutment	
04.391	埋置式桥台	buried abutment	
04.392	扶壁式桥台	counterfort abutment, buttressed abutment	
04.393	衡重式桥台	weight-balanced abutment	
04.394	锚碇板式桥台	anchored bulkhead abutment	

序　码	汉　文　名	英　文　名	注　释
04.395	支撑式桥台	supported type abutment	又称"轻型桥台"。
04.396	组合式桥台	composite abutment	
04.397	[桥]台后回填	back filling behind abutment	
04.398	桥梁基础	bridge foundation	
04.399	浅基础	shallow foundation	
04.400	深基础	deep foundation	
04.401	明挖基础	open cut foundation	
04.402	扩大基础	spread foundation	
04.403	沉井基础	open caisson foundation	
04.404	沉井刃脚	caisson cutting edge	
04.405	气压沉箱基础	pneumatic caisson foundation	
04.406	管柱基础	colonnade foundation	
04.407	双壁钢围堰钻孔桩基础	double-walled steel cofferdam and bored pile foundation	
04.408	墩式基础	pier-foundation	
04.409	桩基础	pile foundation	
04.410	排架桩基础	pile-bent foundation	
04.411	高承台桩基础	high-rise platform pile foundation	
04.412	桩	pile	
04.413	单桩	individual pile, single pile	
04.414	群桩	pile group	又称"桩群"。
04.415	斜桩	battered pile, raked pile	
04.416	锚桩	anchor pile	
04.417	主桩	key pile	
04.418	预制桩	precast pile	
04.419	[钢筋]混凝土桩	concrete pile	
04.420	预应力混凝土桩	prestressed concrete pile	
04.421	木桩	timber pile	
04.422	钢桩	steel pile	
04.423	钢管桩	steel pipe pile	
04.424	混凝土管桩	concrete pipe pile	
04.425	钻孔桩	bored pile	
04.426	就地灌注桩	cast-*in-situ* pile	
04.427	沉管灌注桩	tube-sinking cast-in-situ pile	
04.428	板桩	sheet pile	
04.429	微型桩	mini-pile	又称"小型桩"。
04.430	扩底桩	under-reamed pile	

序 码	汉 文 名	英 文 名	注 释
04.431	支挡桩	soldier pile	
04.432	抗滑桩	anti-slide pile	
04.433	排土桩	displacement pile	
04.434	非排土桩	non-displacement pile	
04.435	摩擦桩	friction pile	
04.436	支承桩	bearing pile	
04.437	嵌岩桩	socketed pile	
04.438	挖孔桩	cast-*in-situ* pile by excavation	
04.439	打入桩	driven pile	
04.440	承台	bearing platform, pile cap	
04.441	灌浆截水墙	grouted cut-off wall	
04.442	灌浆帷幕	grouted curtain	
04.443	地下连续墙	continuous slurry wall, diaphragm wall	
04.444	支座	bearing	
04.445	固定支座	fixed bearing	
04.446	活动支座	expansion bearing, movable bearing	
04.447	合成橡胶支座	elastomeric pad bearing	
04.448	板式橡胶支座	laminated rubber bearing	
04.449	盆式橡胶支座	pot rubber bearing	
04.450	聚四氟乙烯支座	polytetrafluoroethylene bearing, PTFE bearing	
04.451	铅橡胶支座	lead-rubber bearing	
04.452	滚轴支座	roller bearing	
04.453	摆式支座	rocker bearing	
04.454	平板支座	plate bearing	
04.455	球形支座	spherical bearing	
04.456	垫块支座	pad bearing	
04.457	活动伸缩缝支座	sliding expansion bearing	
04.458	伸缩滚轴	expansion roller	
04.459	拉压支座	tension-compression bearing	
04.460	铰式支座	knuckle bearing	
04.461	桥梁设计	bridge design	
04.462	桥梁方案设计	bridge conceptual design	
04.463	桥梁结构设计	bridge structure design	
04.464	桥梁细部设计	bridge detail design	

序 码	汉 文 名	英 文 名	注 释
04.465	荷载	load	
04.466	实用荷载	service load	
04.467	施工荷载	construction load	
04.468	开裂荷载	cracking load	
04.469	疲劳荷载	fatigue load	
04.470	容许荷载	allowable load	
04.471	破坏荷载	failure load	
04.472	桥梁试运行荷载	test run loading for bridge	
04.473	静载	static load	
04.474	动载	dynamic load	
04.475	活载	live load	
04.476	恒载	dead load	
04.477	永久荷载	permanent load	
04.478	可变荷载	variable load	
04.479	偶然荷载	accidental load	
04.480	非常荷载	abnormal load	
04.481	环境荷载	environmental load	
04.482	均布荷载	uniform load, distributed load	
04.483	等代均布荷载	equivalent uniform distributed load	
04.484	集中荷载	concentrated load	
04.485	车道荷载	lane load	
04.486	人群荷载	pedestrian load	
04.487	风荷载	wind load	
04.488	地震荷载	earthquake load, seismic force	又称"地震力"。
04.489	离心力	centrifugal force	
04.490	冲击力	impact force	
04.491	制动力	braking force	
04.492	船舶撞击力	ship collision force	
04.493	土压力	earth pressure	
04.494	冰压力	ice pressure	
04.495	流水压力	flowing water pressure	
04.496	浮力	buoyancy	
04.497	下拉荷载	downdrag, negative friction of pile	又称"桩负摩擦力"。
04.498	拱推力	arch thrust	
04.499	基础隆胀力	heave force of foundation	
04.500	临界压曲荷载	critical buckling load	
04.501	屈服破坏荷载	failure load by buckling	

序 码	汉 文 名	英 文 名	注 释
04.502	荷载组合	loading combinations	
04.503	荷载效应	load effect	
04.504	标准车辆荷载	standard truck loading	
04.505	桥梁容许应力设计	allowable stress design of bridge	
04.506	桥梁极限状态设计	limit state design of bridge	
04.507	桥梁承载能力极限状态	ultimate limit state of bridge carrying capacity	
04.508	桥梁使用能力极限状态	serviceability limit state of bridge	
04.509	桥梁荷载系数设计法	load factor design method of bridge	
04.510	桥梁概率极限状态设计法	probabilistic limit state design method of bridge	
04.511	桥梁优化设计	optimum design of bridge	
04.512	桥梁结构可靠度	reliability of bridge structure	
04.513	可靠度基准期	reliability datum period	
04.514	地震危险性分析	seismic risk analysis	
04.515	桥梁抗震设计	aseismatic design of bridge	
04.516	桥梁抗震稳定性	aseismatic stability of bridge	
04.517	地震基本烈度	basic earthquake intensity	
04.518	设计烈度	design intensity of earthquake	
04.519	地震震级	earthquake magnitude	
04.520	地震加速度	seismic acceleration	
04.521	疲劳设计	fatigue design	
04.522	桥头回填设计	bridge end backfilling design	
04.523	桥梁平面结构	plane structure for bridge	
04.524	桥梁空间结构	space structure for bridge	
04.525	静定桥梁结构	statically determinate bridge structure	
04.526	超静定桥梁结构	statically indeterminate bridge structure	
04.527	结构体系转换	structure system transform	
04.528	结构完整性	structural integrity	
04.529	安全系数	safety factor	
04.530	冲击系数	impact factor	又称"动力系数"。

序 码	汉 文 名	英 文 名	注 释
04.531	荷载系数	load factor	
04.532	膨胀系数	coefficient of expansion	
04.533	构件	member, element	又称"杆件"。
04.534	节间	panel	
04.535	梁腹	web	又称"腹板"。
04.536	翼缘	flange	又称"法兰"。
04.537	加劲杆	stiffener	又称"加劲肋"。
04.538	加劲梁	stiffening girder	
04.539	加劲腋	haunch	又称"梁腋"。
04.540	托架	bracket	俗称"牛腿"。
04.541	结点板	gusset plate	又称"节点板"。
04.542	剪力键	shear key	
04.543	横系杆	lateral bracing	
04.544	风嘴	wind fairing	
04.545	伸缩板	expansion plate	
04.546	应力集中	stress concentration	
04.547	应力腐蚀	stress corrosion	
04.548	热应力	thermal stress	又称"温差应力"。
04.549	支撑应力	bearing stress	
04.550	组合应力	combined stress	
04.551	容许应力	allowable stress	又称"许用应力"。
04.552	极限应力	ultimate stress	
04.553	稳定性	stability	
04.554	刚度	rigidity, stiffness	
04.555	位移	displacement	
04.556	挠度	deflection	
04.557	力矩	moment, bending moment	又称"弯矩"。
04.558	剪力	shear	
04.559	双力矩	bi-moment	
04.560	力矩分配法	moment distribution method	又称"弯矩分配法"。
04.561	力矩包络线	moment envelope	
04.562	剪力分布	shear distribution	
04.563	剪力滞后	shear lag	又称"剪滞"。
04.564	影响线	influence line	
04.565	影响面积	influence area	
04.566	压曲	buckling	又称"压屈"。
04.567	扭曲	distortion	又称"畸变"。

序 码	汉 文 名	英 文 名	注 释
04.568	翘曲	warp	
04.569	颤振	flutter	
04.570	驰振	galloping	
04.571	抖振	buffeting	又称"击振"。
04.572	涡流效应	vortex shedding effect	
04.573	风致摆动	wind induced oscillation	
04.574	挠度横向分布	transverse distribution of deflection	
04.575	荷载横向分布	transverse distribution of load	
04.576	杠杆原理法	lever principle method	
04.577	偏心受压法	eccentric compression method	
04.578	刚性横梁法	rigid cross beam method	
04.579	铰接梁法	hinge connected beam method	
04.580	刚接梁法	rigid connected beam method	
04.581	弹性梁支承法	elastic supported beam method	
04.582	准正交各向异性板法	quasi-orthotropic slab method	
04.583	G M 法	Guyon-Massonet method	
04.584	抗风稳定性	wind resisting stability	
04.585	空气动力稳定性	aerodynamic stability	
04.586	风压分布图	wind pressure distribution graph	
04.587	临界风速	critical wind speed	
04.588	温度梯度	thermal gradient	
04.589	温差	temperature difference	
04.590	年均温差	annual mean temperature difference	
04.591	连拱作用	continuous arch effect	
04.592	土－结构共同作用	soil-structure interaction	
04.593	桥梁承载能力	load carrying capacity of bridge	
04.594	桥梁耐久性	bridge durability	
04.595	地基承载力	bearing capacity of subsoil	
04.596	单桩承载力	bearing capacity of pile	
04.597	抗裂性	crack resistance	
04.598	裂缝控制	crack control	
04.599	裂缝宽度限值	limit of crack opening	
04.600	裂缝分布图	crack pattern	

序　码	汉　文　名	英　文　名	注　释
04.601	疲劳限度	fatigue limit	
04.602	磨损限值	wear allowance	又称"容许磨耗"。
04.603	使用状态挠度限值	deflection limit for serviceability	
04.604	结构延展性	structure ductility	
04.605	桥梁破损	bridge failure	
04.606	桥梁破坏	bridge collapse	
04.607	毁灭性破坏	catastrophic collapse	
04.608	钢筋锈蚀	steel-bar corrosion	
04.609	桥梁计算机辅助设计	computer aided design for bridge, CAD for bridge	
04.610	格构比拟	grillage analogy	
04.611	预应力度	degree of prestressing	
04.612	全预应力混凝土	fully prestressed concrete	
04.613	部分预应力混凝土	partially prestressed concrete	
04.614	有限预应力混凝土	limited prestressed concrete	
04.615	A类部分预应力混凝土	type A partially prestressed concrete	
04.616	B类部分预应力混凝土	type B partially prestressed concrete	
04.617	体内预应力	prestressing with bond	
04.618	体外预应力	external prestressing	
04.619	预弯法预应力	prestressing without tendon by pre-bending	
04.620	复传力法预应力	prestressing by subsequent bond	
04.621	压拉双作用预应力	prestressing with subsequent compression and tension	
04.622	后张自锚法预应力	self-anchored post tensioning prestressing	
04.623	双向预应力	two-dimension prestressing	
04.624	三向预应力	three-dimension prestressing	
04.625	非预应力钢筋	non-prestressed reinforcement	
04.626	粘结的预应力筋	bonded tendon	
04.627	消除粘结的预应力筋	debonded tendon	

序　码	汉　文　名	英　文　名	注　释
04.628	无粘结的预应力筋	non-bonded tendon	
04.629	预应力损失	loss of prestress	
04.630	管道摩擦损失	loss due to duct friction	
04.631	锚具变形损失	loss due to anchorage deformation	
04.632	锚具温差损失	loss due to anchorage temperature difference	
04.633	混凝土收缩损失	loss due to concrete shrinkage	
04.634	混凝土弹性压缩损失	loss due to elastic compression of concrete	
04.635	混凝土徐变损失	creep loss of concrete	
04.636	钢筋预应力松弛损失	loss due to tendon relaxation	
04.637	杆件缩短损失	loss due to concrete member shortening	
04.638	杆件弯曲损失	loss due to bending of concrete member	
04.639	锚具	anchorage	又称"工作锚具"。
04.640	夹具	clamper	又称"工具锚具"。
04.641	锥形锚具	cone anchorage, Freyssinet cone anchorage	又称"弗氏锚"。
04.642	夹片式锚具	strand tapered anchorage	
04.643	VSL 锚具	VSL anchorage	瑞士 VSL 厂生产的国际通用夹片锚具。
04.644	YM 锚具	Y-typed anchorage, post-tensioning strand group anchorage	公路规划设计院与新津筑路机械厂研制的 Y 型锥形三夹片式群锚。
04.645	XM 锚具	X-typed anchorage	X 型三夹片式群锚。
04.646	OVM 锚具	oriental cone anchorage	东方(orient)公司与郴州厂研制的锥形二夹片式锚具。
04.647	JM12 型锚具	JM12 anchorage	
04.648	镦头锚	bulb-end anchorage, BBRV anchorage	又称"BBRV 锚"。
04.649	精轧螺纹锚	thread bar anchorage, Dywidag anchorage system	又称"迪维达格锚固系统"。

序　码	汉　文　名	英　文　名	注　释
04.650	连接器	coupler	
04.651	锚固区	anchorage zone	
04.652	横向承载桩地基系数法	subsoil reaction modulus method for laterally loaded pile	
04.653	桩基 m 值法	subsoil reaction modulus m method for laterally loaded pile	
04.654	桩基 c 值法	subsoil reaction modulus c method for laterally loaded pile	
04.655	p-y 曲线法	p-y curve method	
04.656	应力波方程法	stress wave equation method	
04.657	基底应力扩散	stress dispersal beneath footing	
04.658	基础沉降	foundation settlement	
04.659	沉降差	differential settlement	
04.660	桥梁施工	bridge construction	
04.661	缆索吊装法	erection with cableway	又称"无支架吊装法"。
04.662	悬臂拼装法	cantilever erection method	
04.663	平衡悬拼法	precast balancing cantilever method	
04.664	悬臂浇筑法	cast-in-place cantilever method	
04.665	平衡悬浇法	cast-in-place balancing cantilever method	
04.666	逐跨施工法	span by span construction method	
04.667	膺架式架设法	erection with scaffolding method	
04.668	膺架浇筑法	cast on scaffolding method, cast-in-place method	又称"就地浇筑法"。
04.669	顶推法	incremental launching method	
04.670	多点同步顶推	incremental launching construction by multipoint jacking	
04.671	节段预制拼装法	segmental precast erection method	
04.672	纵向拖拉法	erection by longitudinal pulling method	
04.673	转体架桥法	bridge erection by swinging method	
04.674	浮运架桥法	bridge erection by floating method	
04.675	桥涵顶入法	jacking-in method of culvert or subsurface bridge	

序　码	汉 文 名	英 文 名	注　释
04.676	连续砌拱	laying arch continuously	
04.677	分段砌拱	laying arch by sections	
04.678	分环砌拱	laying arch by rings	
04.679	分段分环结合砌拱	laying arch by sections and rings	
04.680	横向悬砌拱法	laying arch by transverse overhanging method	
04.681	拱圈应力调整	arch ring stress adjustment	
04.682	拱圈施工缝	construction joint for arch ring	
04.683	拱圈封顶	closure of arch ring	
04.684	尖拱法封顶	closure by wedging-in crown	又称"刹尖"。
04.685	千斤顶法封顶	closure by jacking and sealing-off crown	
04.686	拱架	arch centering, arch center	
04.687	梁式钢拱架	steel beam centering	
04.688	拱式钢拱架	steel arch centering	
04.689	排架式满布木拱架	full span wooden bent centering	
04.690	斜撑式满布木拱架	full span wooden inclined strut centering	
04.691	桁式木拱架	wood-truss centering	
04.692	叠桁式拱架	double-layer-truss centering	
04.693	夹板拱架	centering by clamped planks in arch shape	
04.694	土牛拱架	earthen centering	
04.695	拱架卸落	centering unloading	
04.696	裸拱卸架	unloading bare rib, unloading bare ring	
04.697	拱架预压	preloading centering	
04.698	支拱板条	lagging	
04.699	卸架木楔	wood wedge for centering unloading	
04.700	卸架砂筒	sand cylinder for centering unloading	
04.701	拱砌块	voussoir	
04.702	拱顶石	arch crown block	
04.703	拱脚斜块	arch skew block	

序 码	汉 文 名	英 文 名	注 释
04.704	脚手架	scaffold, falsework	又称"膺架"、"工作架"。
04.705	施工步道	catwalk	俗称"猫道"。
04.706	风缆	wind-cable	
04.707	导梁	launching nose	
04.708	挂篮	cradle	
04.709	预制场	fabrication yard, casting yard	
04.710	混凝土浇筑	concreting	
04.711	混凝土泵送法	concreting by pumping	
04.712	混凝土养生	concrete curing	
04.713	混凝土蒸汽养生	steam curing of concrete	
04.714	混凝土电热养生	electric curing of concrete	
04.715	镶合浇筑	match casting	
04.716	模板	formwork	
04.717	自升式模板	self-climbing formwork	
04.718	滑模浇筑混凝土	slip-form concreting	
04.719	预拱度	camber	
04.720	电热[法]张拉	electric-heat prestressing	
04.721	油压千斤顶张拉	hydraulic jack prestressing	
04.722	先张法	pretensioning method	
04.723	后张法	post-tensioning method	
04.724	钢丝束制作	wire grouping	
04.725	张拉程序	stressing sequence, tensioning procedure	
04.726	应力控制张拉	prestressing under stress control	
04.727	冷拉率控制张拉	prestressing by cold-drawn rate	
04.728	时效处理	time effect treatment	
04.729	张拉控制应力	control stress for prestressing	
04.730	预应力台座	prestressing bed	
04.731	双作用千斤顶	double acting jack	
04.732	三作用千斤顶	triple acting jack	
04.733	管道灌浆	duct grouting	
04.734	封锚	sealing-off and covering anchorage	
04.735	铆接	riveting	
04.736	栓接	bolting	
04.737	钢梁结构试装配	trial steel work fixing	
04.738	钢桥拼装	assembling of steel bridge	

序 码	汉 文 名	英 文 名	注 释
04.739	钢桥就位	seating steel bridge on supports	
04.740	螺栓扭矩终拧法	bolt final twisting by torque method	
04.741	螺栓扭角终拧法	bolt final twisting by twist angle method	
04.742	浮运沉井	floating caisson	
04.743	筑岛沉井	sinking open caisson on built island	
04.744	不排水下沉沉井	sinking open caisson by undrained dredging	
04.745	泥浆套法下沉沉井	sinking open caisson by slurry coating	
04.746	空气幕法下沉沉井	sinking open caisson by injected air curtain	
04.747	套管护壁钻孔法	casing hole-boring method	
04.748	泥浆护壁钻孔法	slurry hole-boring method	
04.749	静水护壁钻孔法	static water head hole-boring method	
04.750	预钻插桩法	socketing pile in prebored hole	
04.751	先钻后扩法	reaming-after-boring method	
04.752	钻孔扩端法	boring-and-underreaming method	
04.753	挖孔灌桩法	installing pile shaft by excavation	
04.754	冲孔成桩法	pile by percussion drill method	
04.755	冲抓法	percussion and grabbing method	
04.756	正循环钻孔法	circulation boring method	
04.757	反循环钻孔法	reverse circulation boring method	
04.758	静压沉桩法	pile jacking-in method	
04.759	打桩	pile driving	
04.760	锤击沉桩法	pile driving method by hammer	
04.761	震动沉桩法	pile vibrosinking method	
04.762	射水沉桩法	pile jetting method	
04.763	拨桩	pile extracting	
04.764	围堰	cofferdam	
04.765	板桩围堰	sheet pile cofferdam	
04.766	土围堰	earth cofferdam	
04.767	套箱围堰	precast-boxed cofferdam	
04.768	钢围图	steel waling	
04.769	双层钢围图	double-walled steel waling	

序　码	汉　文　名	英　文　名	注　释
04.770	水下浇筑混凝土法	under water concreting	
04.771	[直升]导管浇筑混凝土法	tremie concreting	
04.772	混凝土封底	bottom sealing by concreting	
04.773	基坑井点排水法	foundation pit well point drainage method	
04.774	桥梁合拢	closure of bridge structure	
04.775	桥梁安装监测	bridge erection monitoring	
04.776	桥梁结构安装控制	bridge structure erection control	
04.777	索力控制	cable force control	
04.778	顺序拼装分析	progressive analysis for bridge erection	又称"正向分析"。
04.779	逆向拆除分析	retrogressive analysis for bridge erection	又称"反向分析"。
04.780	桥梁安装容许误差	bridge erection tolerance	
04.781	桥梁养护	bridge maintenance	
04.782	桥梁数据库	bridge data bank	
04.783	桥梁管理系统	bridge management system	
04.784	桥梁评价系统	bridge evaluation system	
04.785	桥梁监测系统	bridge monitoring system	
04.786	桥梁技术档案	bridge technical file	
04.787	桥梁检查	bridge inspection	
04.788	桥梁检查规则	bridge inspection regulation	
04.789	桥梁检查类别	bridge inspection category	
04.790	桥梁检查周期	bridge inspection cycle	
04.791	桥梁编号	bridge numbering	
04.792	桥梁病害整治	bridge fault repairing	
04.793	桥梁浅基防护	bridge shallow foundation protection	
04.794	桥梁墩台防撞	collision prevention of pier and abutment	
04.795	桥梁水毁	bridge disaster by flood	
04.796	预防性修理	preventive repair	
04.797	桥梁加固	bridge strengthening	

序　码	汉　文　名	英　文　名	注　释
04.798	桥梁抗震加固	bridge aseismatic strengthening	
04.799	桥梁翻新	bridge retrofitting	
04.800	构件更换	component replacement	
04.801	涵[洞]	culvert	
04.802	明涵	open culvert	
04.803	暗涵	buried culvert	
04.804	箱涵	box culvert	
04.805	管涵	pipe culvert	
04.806	拱涵	arch culvert	
04.807	盖板涵	slab culvert	
04.808	压力式涵洞	outlet submerged culvert, pressure culvert	
04.809	半压力式涵洞	inlet submerged culvert, partial pressure culvert	
04.810	无压力涵洞	inlet unsubmerged culvert, non-pressure culvert	
04.811	波纹管涵	corrugated-metal pipe culvert	
04.812	四铰管涵	quadri-hinge-pipe culvert	
04.813	倒虹吸涵	inverted siphon culvert	
04.814	陡坡涵洞	culvert on steep grade	
04.815	斜交涵洞	skew culvert	
04.816	阶梯式涵洞	stepped culvert	
04.817	涵洞一字墙	culvert straight end wall	
04.818	涵洞八字墙	culvert wing wall	
04.819	涵洞进水口	culvert inlet	
04.820	涵洞出水口	culvert outlet	
04.821	涵洞跌水	culvert water drop	
04.822	涵洞沟床	gully bed neighboring to culvert	
04.823	涵洞口铺砌	culvert inlet/outlet apron	
04.824	渡槽	aqueduct	
04.825	涵口沉降缝	settlement joint in culvert	
04.826	涵洞口隆起	culvert end lift	
04.827	涵洞洞身长度	culvert barrel length	
04.828	涵洞孔径	culvert aperture	
04.829	隧道工程	tunneling	
04.830	隧道	tunnel	
04.831	隧道级别	tunnel class	

序　码	汉　文　名	英　文　名	注　　释
04.832	公路隧道	highway tunnel	
04.833	市区隧道	urban tunnel	
04.834	土质隧道	tunnel in earth	
04.835	岩石隧道	tunnel in rock	
04.836	水底隧道	subaqueous tunnel	
04.837	服务隧道	service tunnel, acess adit	又称"辅助隧道"。
04.838	深埋隧道	deep-buried tunnel	
04.839	浅埋隧道	shallow-buried tunnel	
04.840	双[孔]隧道	twin tunnel	
04.841	横洞	transverse gallery	
04.842	错车洞	passing bay in tunnel	
04.843	竖井	shaft	
04.844	斜井	inclined shaft	
04.845	明洞	open cut tunnel	
04.846	导洞	heading, pilot	
04.847	隧道洞门	tunnel portal	又称"隧道口"。
04.848	隧道洞身	tunnel trunk	
04.849	隧道埋深	buried depth of tunnel	
04.850	隧道长度	length of tunnel	
04.851	隧道净空	clearance of tunnel	
04.852	隧道口净空	portal clearance	
04.853	特长隧道	super long tunnel	
04.854	长隧道	long tunnel	
04.855	中长隧道	medium tunnel	
04.856	短隧道	short tunnel	
04.857	衬砌隧道	trimmed tunnel	
04.858	未衬砌隧道	rough tunnel	
04.859	围岩	surrounding rock	
04.860	围岩压力	surrounding rock pressure	
04.861	围岩稳定	stability of surrounding rock	
04.862	围岩自承能力	surrounding rock self-supporting capacity	
04.863	围岩分类	surrounding rock classification	
04.864	掘进方法	tunneling method	
04.865	明挖法	open cut method	
04.866	暗挖法	undermining method	
04.867	随挖随盖法	cut-and-cover method	

序　码	汉　文　名	英　文　名	注　释
04.868	矿山法	mine tunneling method	又称"钻爆法"。
04.869	沉埋法	immersed tunneling method	
04.870	盾构法	shield method	
04.871	新奥法	New Austrian Tunneling Method, NATM	
04.872	全断面开挖法	full-face tunneling method	
04.873	导洞法	pilot-tunneling method	
04.874	中央导洞法	central pilot tunneling method	
04.875	上下导洞法	top and bottom pilot tunneling method	
04.876	先墙后拱法	wall in advance of arched roof tunneling method	
04.877	先拱后墙法	arched roof in advance of wall tunneling method	
04.878	蘑菇形开挖法	mushroom-type tunneling method	
04.879	导洞与层阶法	heading and bench tunneling method	
04.880	侧壁导洞法	side heading method	
04.881	双侧壁导洞法	twin-side heading method	
04.882	出碴	tunnel muck hauling	
04.883	隧道测量	tunnel survey	
04.884	隧道激光导向	tunnel alignment by laser	
04.885	隧道衬砌	tunnel lining	
04.886	整体式衬砌	integral lining	
04.887	拼装式衬砌	precast lining	
04.888	组合衬砌	composite lining	
04.889	喷射混凝土衬砌	shotcrete lining	
04.890	隧道支撑	tunnel support	
04.891	构件支撑	member support	
04.892	喷锚支护	shotcrete and rock bolt support	
04.893	锚杆支护	anchor bolt support	
04.894	木支撑	timbering support	
04.895	钢拱支撑	steel-arched support	
04.896	隧道顶板	tunnel roof	
04.897	隧道底板	tunnel floor	
04.898	隧道边墙	tunnel side wall	
04.899	隧道坍顶	tunnel roof fall	

序　码	汉　文　名	英　文　名	注　释
04.900	隧道贯通	tunnel through	
04.901	隧道贯通误差	tunnel through error	
04.902	隧道施工防尘	tunnel construction dust controlling	
04.903	隧道施工通风	tunnel construction ventilation	
04.904	隧道通风	tunnel ventilation	
04.905	通风方式	ventilation type	
04.906	通风设计	ventilation design	
04.907	自然通风	natural ventilation	
04.908	机械通风	mechanical ventilation	
04.909	纵向通风	longitudinal ventilation	
04.910	横向通风	transverse ventilation	
04.911	半横向通风	semi-transverse ventilation	
04.912	组合式纵向通风	composite longitudinal ventilation	
04.913	排风机通风	ventilation by exhaust fan	
04.914	射流纵向通风	ventilation by force draft	
04.915	轴流式通风	ventilation by axial flow fan	
04.916	送风竖井	blowing-in shaft	
04.917	排风竖井	blowing-out shaft	
04.918	通风塔	ventilation tower	
04.919	通风效率	ventilation efficiency	
04.920	通风量	ventilation volume	
04.921	通风监测	ventilation monitoring	
04.922	通风装置	ventilation installation	
04.923	车辆活塞作用	piston action of moving vehicles	
04.924	隧道消防系统	tunnel fire protection system	
04.925	隧道报警装置	tunnel warning installation	
04.926	隧道防灾设施	tunnel anti-disaster facilities	
04.927	隧道火灾监测	tunnel fire monitoring	
04.928	隧道火灾	tunnel fire hazard	
04.929	隧道瓦斯爆炸	gas explosion in tunnel	
04.930	隧道瓦斯泄出	gas emission in tunnel	
04.931	隧道注浆止水	tunnel water sealing by injection	
04.932	隧道防水	tunnel water proofing	
04.933	隧道信号	tunnel signal	
04.934	隧道监控中心	tunnel operation control center	
04.935	过隧经历时间	duration of tunnel passage	

序 码	汉 文 名	英 文 名	注 释
04.936	隧道交通监测系统	tunnel traffic monitoring system	
04.937	隧道电话系统	tunnel telephone system	
04.938	隧道供电系统	tunnel power supply system	
04.939	隧道广播系统	tunnel broadcasting system	
04.940	盾构	tunnel shield	
04.941	全气压盾构	all-round pressurized shield	
04.942	半盾构	semi-shield	
04.943	盾构切口环	cutting ring of shield	
04.944	盾构支承环	supporting ring of shield	
04.945	紧急灭火机	emergency fire extinguisher	
04.946	自动喷水灭火机	automatic water spraying fire extinguisher	
04.947	隧道养护	tunnel maintenance	
04.948	渡口	ferry, ferry crossing	
04.949	公路渡口	highway ferry	
04.950	汽车轮渡	car ferrying	
04.951	轮渡站	highway ferry station	
04.952	码头	wharf	
04.953	渡运码头	ferry wharf	
04.954	锯齿形码头	saw-tooth type wharf	
04.955	直线形码头	straight line type wharf	
04.956	码头引道	approach to ferry	
04.957	码头引桥	bridge approach to ferry	
04.958	渡轮	ferry boat	
04.959	机动渡车船	powered ferry boat	
04.960	人力渡车船	manually operated ferry boat	
04.961	双层渡车船	double-decked ferry boat	
04.962	双体渡车船	twin ferry boat	
04.963	端靠式渡船	front-mooring ferry boat	
04.964	侧靠式渡船	side-mooring ferry boat	
04.965	门桥	portal frame	
04.966	多用浮箱	multi-function pontoon	
04.967	防撞垫	bumper	又称"缓冲垫"。
04.968	跳板	gangboard	
04.969	撑杆	boat pole	
04.970	渡口管理	ferry management	

序 码	汉 文 名	英 文 名	注 释
04.971	渡口安全规则	ferry safety regulation	
04.972	渡口管理所	ferry house	

05. 汽车运用工程

序 码	汉 文 名	英 文 名	注 释
05.001	公路运输车辆	highway transport vehicle	
05.002	机动车辆	power-driven vehicle	
05.003	非机动车辆	non-power-driven vehicle	
05.004	汽车	motor vehicle, automobile	
05.005	载货汽车	[motor] truck	简称"货车"。
05.006	特种载货汽车	special truck	
05.007	厢式货车	van truck	
05.008	保温货车	thermal insulated vehicle	
05.009	冷藏货车	refrigerated vehicle	
05.010	液罐货车	tank truck	
05.011	自卸货车	dump truck	
05.012	集装箱货车	container carrier	
05.013	专用汽车	special purpose vehicle	
05.014	农用运输汽车	agricultural truck	
05.015	客车	bus	
05.016	城市客车	city bus	
05.017	长途客车	intercity bus	
05.018	卧铺客车	motor coach sleeper	
05.019	出租汽车	taxi	
05.020	牵引车	tractor	
05.021	半挂牵引车	tractor-truck	
05.022	挂车	trailer	
05.023	全挂车	full trailer	
05.024	半挂车	semi-trailer	
05.025	栏板挂车	dropside trailer	
05.026	平板挂车	platform trailer, flat trailer	
05.027	低平板挂车	low-bed trailer	
05.028	凹式平板挂车	cranked platform trailer	
05.029	重型平板挂车	heavy haul trailer	
05.030	特种挂车	special trailer	

序　码	汉　文　名	英　文　名	注　释
05.031	厢式挂车	van trailer	
05.032	集装箱挂车	container trailer	
05.033	集装箱专用挂车	container flatframe trailer	
05.034	集装箱平板挂车	container platform trailer	
05.035	集装箱栏板挂车	container dropside trailer	
05.036	自卸挂车	dump trailer	
05.037	液罐挂车	tank trailer	
05.038	保温挂车	thermal insulated trailer	
05.039	冷藏挂车	refrigerated trailer	
05.040	长货挂车	pole trailer	
05.041	预制件挂车	prefab trailer	
05.042	底卸式挂车	bottom-dump trailer	
05.043	粉状货挂车	bulk tanker trailer	
05.044	牲畜家禽挂车	livestock trailer	
05.045	纵伸挂车	adjustable-wheelbase trailer	
05.046	横伸挂车	adjustable track trailer	
05.047	集装箱自装卸挂车	self-loading container trailer	
05.048	轿车运载[半挂]车	car carrier semitrailer	
05.049	汽车列车	tractor-trailer combination	
05.050	全挂汽车列车	full trailer train	
05.051	半挂汽车列车	semi-trailer train	
05.052	双挂汽车列车	double trailer train	
05.053	民用车辆	civilian vehicle	
05.054	在用汽车	vehicle in use	
05.055	商用汽车	commercial vehicle	
05.056	营运汽车	for-hire vehicle	
05.057	自用车辆	private vehicle	
05.058	公路运输车辆管理	highway transport vehicle management	
05.059	汽车管理制度	vehicle management system	
05.060	车辆技术档案	vehicle technical file	
05.061	汽车运行材料	vehicle operational consumption materials	
05.062	汽车技术经济定额	techno-economic rating for vehicle operation	

序 码	汉 文 名	英 文 名	注 释
05.063	汽车燃料消耗定额	vehicle fuel consumption rating	
05.064	轮胎行驶里程定额	tyre mileage rating	
05.065	汽车维修费用定额	cost quota of vehicle maintenance and repair	
05.066	汽车大修费用定额	cost quota of vehicle major repair	
05.067	汽车大修间隔里程	average mileage between vehicle major repairs	
05.068	汽车小修频率	frequency of vehicle current repair	
05.069	轮胎翻新率	tyre recapping rate	
05.070	车辆调拨	vehicle allotment	
05.071	车辆停驶	vehicle lay-off	
05.072	车辆封存	vehicle storing up	
05.073	车辆折旧里程	vehicle depreciation mileage	
05.074	车辆完好率	vehicle avail ability rate	
05.075	车辆平均技术等级	mean grade of vehicle technical condition	
05.076	车辆租赁	vehicle leasing	
05.077	车辆报废	vehicle scrapping	
05.078	车辆折旧率	vehicle depreciation rate	
05.079	车辆更新	vehicle replacement	
05.080	车辆装备	vehicle outfitting	
05.081	车辆改造	vehicle remoulding	
05.082	车辆使用寿命	vehicle service life	
05.083	车辆合理使用寿命	vehicle rational service life	
05.084	车辆经济使用寿命	vehicle economic service life	
05.085	车辆技术使用寿命	vehicle technical service life	
05.086	汽车驾驶	automobile driving	
05.087	预热	preheating	
05.088	起步	starting	
05.089	换档	gear shifting	
05.090	加速	speeding up	

序 码	汉 文 名	英 文 名	注 释
05.091	减速	slowing down	
05.092	转向	turning	
05.093	掉头	turning around	
05.094	倒车	backing	
05.095	制动	braking	
05.096	紧急制动	emergency braking	
05.097	[轮胎]压印	tyre scuffs	
05.098	[轮胎]拖印	tyre skid	
05.099	超车	overtaking	
05.100	会车	meeting	
05.101	侧向最小安全间距	minimum safe lateral clearance	
05.102	车轮滑转	wheel spinning	
05.103	抛锚	breakdown on the way	
05.104	[列车]折叠	jack knifing	
05.105	滑行	coasting	
05.106	脱档滑行	coasting in neutral	
05.107	熄火滑行	coasting with engine off	
05.108	分离离合器滑行	coasting with clutch disengaged	
05.109	加速滑行	accelerating-coasting	
05.110	驾驶能力	driving ability	
05.111	驾驶技能	driving skill	
05.112	驾驶技术	driving technique	
05.113	涉水驾驶	fording drive	
05.114	冰雪路面驾驶	driving on snowy and icy road	
05.115	走合期驾驶	driving in running-in period	
05.116	驾驶模拟器	driving simulator	
05.117	汽车使用	vehicle operation	
05.118	汽车使用条件	vehicle operation condition	
05.119	汽车运行工况	vehicle operational mode	
05.120	怠速工况	idling mode	
05.121	强制怠速工况	forced idling mode	
05.122	加速工况	accelerating mode	
05.123	等速工况	cruising mode	
05.124	减速工况	decelerating mode	
05.125	滑行工况	coasting mode	
05.126	全负荷工况	full load mode	

序　码	汉　文　名	英　文　名	注　释
05.127	部分负荷工况	partial load mode	
05.128	汽车使用强度	vehicle operation intensity	
05.129	汽车使用效率	vehicle operation efficiency	
05.130	汽车合理使用	vehicle rational utilization	
05.131	汽车平均技术速度	vehicle average technical speed	
05.132	汽车经济车速	vehicle economical speed	
05.133	汽车合理拖载	vehicle rational towing load	
05.134	汽车使用性能	vehicle operation performance	
05.135	汽车容载量	vehicle load carrying capacity	
05.136	汽车重量利用系数	vehicle weight efficiency	
05.137	汽车使用可靠性	vehicle operational reliability	
05.138	汽车耐久性	vehicle durability	
05.139	汽车动力性	vehicle dynamic quality	
05.140	汽车经济性	vehicle economy	
05.141	汽车安全性	vehicle safety	
05.142	主动安全性	active safety	
05.143	被动安全性	passive safety	
05.144	汽车使用方便性	vehicle utilization conveniency	
05.145	汽车行驶平顺性	vehicle running smoothness	
05.146	汽车乘坐舒适性	vehicle ride comfort	
05.147	汽车操纵轻便性	easiness of vehicle control	
05.148	汽车机动性	vehicle mobility	
05.149	汽车操纵稳定性	vehicle handling stability	
05.150	汽车通过性	vehicle passability	
05.151	汽车维修性	vehicle maintainability	
05.152	汽车节能	vehicle energy saving	
05.153	汽车节油技术	vehicle fuel saving technique	
05.154	公司平均燃料经济性	corporation average fuel economy, CAFE	
05.155	汽车燃料利用率	vehicle fuel utilization factor	
05.156	汽车能量利用率	vehicle energy utilization factor	
05.157	汽车能量平衡	vehicle energy balance	
05.158	汽车百车公里油耗	vehicle fuel consumption per hundred kilometers	
05.159	汽车百吨公里油	vehicle fuel consumption per hun-	

序　码	汉　文　名	英　文　名	注　释
	耗	dred ton-kilometers	
05.160	汽车代用燃料	alternative motor fuels	
05.161	汽车节能装置	fuel saving device	
05.162	汽车技术状况	technical condition of vehicle	
05.163	汽车完好技术状况	good technical condition of vehicle	
05.164	汽车不良技术状况	bad technical condition of vehicle	
05.165	汽车工作能力	working ability of vehicle	
05.166	汽车技术状况参数	parameters of vehicle technical condition	
05.167	工作过程参数	parameters of operation process	
05.168	伴随过程参数	accompanying parameters of operation process	
05.169	汽车极限技术状况	limiting technical condition of vehicle	
05.170	汽车技术状况变化规律	trend of change of vehicle technical conditions	
05.171	汽车损耗	vehicle wear-out	
05.172	汽车零件磨损	wear of vehicle parts	
05.173	磨损过程	wear process	
05.174	正常磨损	normal wear	
05.175	极限磨损	wear limit	
05.176	允许磨损	permissible wear	
05.177	磨损率	wear rate	
05.178	擦伤	scratching	
05.179	刮伤	scoring	
05.180	点蚀	pitting	
05.181	粘附	adhesion	
05.182	咬粘	seizure	
05.183	烧伤	burning	
05.184	穴蚀	cavitation	
05.185	缺陷	defect	
05.186	损伤	damage	
05.187	汽车故障	vehicle failure	
05.188	完全故障	complete failure	
05.189	局部故障	partial failure	

序　码	汉　文　名	英　文　名	注　释
05.190	致命故障	critical failure	
05.191	严重故障	major failure	
05.192	一般故障	minor failure	
05.193	渐变故障	gradual failure	
05.194	突然故障	sudden failure	
05.195	随机故障	random failure	
05.196	故障机理	failure mechanism	
05.197	故障模式	failure mode	
05.198	异响	abnormal noise	
05.199	泄漏	leakage	
05.200	过热	overheating	
05.201	气阻	vapor lock	
05.202	松动	loosening	
05.203	失调	maladjustment	
05.204	失控	out of control	
05.205	乏力	lack of power	
05.206	污染超限	over-limit pollution	
05.207	费油	excessive fuel consumption	
05.208	振抖	fluttering	
05.209	窜油	oil pumping	
05.210	爆燃	detonation	又称"爆震"。
05.211	敲缸	piston slap	
05.212	回火	back fire	
05.213	缺火	misfiring	
05.214	断火	interruption of firing	
05.215	拉缸	cylinder scoring	
05.216	离合器滑转	clutch slippage	
05.217	脱档	spontaneous out-of-gear	
05.218	车轮抱死	wheel lock up	
05.219	跑偏	pulling to one side	
05.220	侧滑	side skidding	
05.221	拖滞	dragging	
05.222	故障率	failure rate	
05.223	故障分布	failure distribution	
05.224	故障树	failure tree, failure branch chart	
05.225	汽车诊断	vehicle diagnosis	
05.226	汽车检测	vehicle inspection and test	

序　码	汉　文　名	英　文　名	注　　释
05.227	汽车状况监控	vehicle condition monitoring	
05.228	诊断工艺	diagnostic technology	
05.229	诊断方法	diagnostic method	
05.230	参数诊断	parameter diagnosis	
05.231	统计诊断	statistical diagnosis	
05.232	逻辑诊断	logistic diagnosis	
05.233	诊断专家系统	diagnostic expert system	
05.234	综合诊断	general inspection and diagnosis	
05.235	单元诊断	unit diagnosis	
05.236	动态诊断	dynamic diagnosis	
05.237	诊断参数	diagnostic parameters	
05.238	诊断规范	diagnostic norms	
05.239	诊断周期	interval between diagnosis	
05.240	汽车诊断站	vehicle diagnostic station	
05.241	汽车检测站	vehicle inspection and test station	
05.242	汽车维修	vehicle maintenance and repair	
05.243	汽车维修制度	vehicle maintenance and repair system	
05.244	计划预防维修制度	scheduled preventive maintenance and repair system	
05.245	汽车维修指标	vehicle maintenance and repair indices	
05.246	汽车维修企业	vehicle maintenance and repair enterprise	
05.247	汽车维修网点	vehicle maintenance and repair network	
05.248	汽车维护	vehicle maintenance	又称"汽车保养"。
05.249	汽车维护工程	vehicle maintenance engineering	
05.250	汽车维护类别	vehicle maintenance classification	
05.251	汽车维护级别	vehicle maintenance grade	
05.252	技术服务	technical service	
05.253	状况监测维护	status monitoring maintenance	
05.254	汽车定期维护	vehicle periodic maintenance	
05.255	季节性维护	seasonal maintenance	
05.256	走合维护	running-in maintenance	
05.257	预防维护	preventive maintenance	
05.258	视情维护	condition-based maintenance	

序 码	汉 文 名	英 文 名	注 释
05.259	事后维护	maintenance after failure	
05.260	计划外维护	unscheduled maintenance	
05.261	汽车维护工艺	vehicle maintenance technology	
05.262	汽车维护作业	operation of vehicle maintenance	
05.263	汽车维护工艺过程	vehicle maintenance technological process	
05.264	汽车维护规范	vehicle maintenance norms	
05.265	汽车维护方法	vehicle maintenance method	
05.266	汽车维护流水作业法	flow method for vehicle maintenance	
05.267	汽车维护定位作业法	vehicle maintenance on universal post	
05.268	汽车维护生产纲领	production program of vehicle maintenance	
05.269	汽车维护周期	interval between vehicle maintenances	
05.270	汽车维护场	vehicle maintenance depot	
05.271	汽车维护站	vehicle maintenance station	
05.272	汽车修理	vehicle repair	
05.273	汽车修理类别	vehicle repair classification	
05.274	汽车大修	vehicle major repair	
05.275	汽车中修	vehicle medium repair	
05.276	汽车小修	vehicle current repair	
05.277	事前修理	repair before failure	
05.278	事后修理	repair after failure	
05.279	总成修理	unit repair	
05.280	总成大修	unit major repair	
05.281	总成小修	unit current repair	
05.282	发动机大修	engine rebuilding, engine remanufacture	
05.283	[发动机]检修	tune-up	
05.284	整形修理	form correction repair	
05.285	零件修理	parts repair	
05.286	基础件修理	basic part repair	
05.287	修复	reconditioning	
05.288	计划修理	scheduled repair	
05.289	视情修理	repair on technical conditions	

序 码	汉 文 名	英 文 名	注 释
05.290	视情检修	inspection and repair only as necessary, IROAN	
05.291	非计划修理	unscheduled repair	
05.292	汽车修理工艺	vehicle repair technology	
05.293	汽车修理工艺过程	vehicle repair technological process	
05.294	技术检验	technical checking	
05.295	检视	visual inspection	
05.296	零件检验分类	inspection and classification of parts	
05.297	走合	running-in	
05.298	磨合	breaking-in	
05.299	冷磨合	cold breaking-in	
05.300	热磨合	hot breaking-in	
05.301	修理尺寸	repair size	
05.302	极限间隙	clearance limit	
05.303	允许间隙	permissible clearance	
05.304	汽车修理技术标准	vehicle repair technical standard	
05.305	汽车修理方法	vehicle repair method	
05.306	汽车修理流水作业法	flow method for vehicle repair	
05.307	汽车修理定位作业法	vehicle repair on universal post	
05.308	总成互换修理法	unit exchange repair method	
05.309	周转总成	replacement unit	
05.310	混装修理法	depersonalized repair method	
05.311	就车修理法	personalized repair method	
05.312	汽车修理生产纲领	production program of vehicle repair	
05.313	汽车大修平均在厂车日	average days in plant during major repair of vehicle	
05.314	汽车大修平均在修车日	average days during major repair of vehicle	
05.315	汽车维修平均工时	average man-hours of vehicle maintenance and repair	
05.316	汽车维修平均费	average costs of vehicle mainte-	

序　码	汉　文　名	英　文　名	注　　释
	用	nance and repair	
05.317	汽车大修返修率	return rate of vehicle major repair	
05.318	汽车修理厂	vehicle repair plant	
05.319	汽车总成修理厂	unit repair plant	
05.320	汽车专项修理厂	specialty vehicle repair shop	
05.321	汽车维修工艺设备	technological equipment of vehicle maintenance and repair	
05.322	汽车维修工具	instrument of vehicle maintenance and repair	
05.323	汽车安全检测线	vehicle safety inspection and test line	
05.324	汽车综合诊断线	vehicle general inspection and diagnosis line	
05.325	全自动汽车检测系统	computerized vehicle inspection system	
05.326	底盘测功机	chassis dynamometer	
05.327	制动试验台	brake tester	
05.328	反力式制动试验台	reaction type brake tester	
05.329	惯性式制动试验台	inertial type brake tester	
05.330	轴重仪	axle load meter	
05.331	倾斜角度试验机	overturn angle tester	
05.332	[汽车]侧滑试验台	side slip tester	
05.333	[汽车]速度[表]试验台	speedometer tester	
05.334	[汽车]前大灯试验仪	head light tester	
05.335	[汽车]排气分析仪	exhaust gas analyzer, emission analyzer	
05.336	柴油机烟度计	diesel smoke meter	
05.337	[汽车]车轮定位仪	wheel alignment meter	
05.338	轮胎拆装机	tire changer	
05.339	车轮动平衡仪	wheel dynamic balancer	
05.340	发动机诊断仪	engine analyzer	

序　码	汉　文　名	英　文　名	注　释
05.341	正时灯	timing light	
05.342	柴油机喷油定时测定器	diesel injection timing tester	
05.343	喷嘴试验器	nozzle tester	
05.344	压缩压力计	compression gauge	又称"气缸压力表"。
05.345	柴油机喷油泵试验台	diesel fuel injection pump tester	
05.346	燃油流量计	fuel consumption gauge	
05.347	声级计	sound level meter	
05.348	[制动]踏板力计	foot brake pedal pressure gauge	
05.349	镗缸机	cylinder boring machine	
05.350	主轴承镗床	main bearing boring machine	
05.351	连杆校正器	connecting rod aligner	
05.352	连杆轴承镗床	connecting rod bearing boring machine	
05.353	汽缸珩磨机	cylinder honing machine	
05.354	磨气门座机	valve seat grinder	
05.355	磨气门机	valve refacer	
05.356	制动鼓车床	brake drum lathe	
05.357	制动蹄磨床	brake shoe grinder	
05.358	汽车举升机	car lift	
05.359	柱式举升机	post lift	
05.360	架式举升机	crossing lift	
05.361	举升泊车机	parking lift	
05.362	汽车清洗机	car washer	
05.363	汽车门式自动清洗机	gate type automatic car washer	
05.364	汽车底部自动清洗机	automatic underbody washer	
05.365	零件清洗机	parts washer	
05.366	发动机清洗机	engine cleaner	
05.367	[轿车]车身测量整形机	body and frame measure and correct system	
05.368	车身矫正机	body and frame straightener	
05.369	喷漆房	spray booth	
05.370	红外线烘干装置	infrared oven stand	
05.371	调[配]漆机	color matching system	

06. 汽车运输

序　码	汉　文　名	英　文　名	注　　释
06.001	道路运输	road transport	
06.002	公路运输	highway transport	
06.003	机动车运输	power-driven vehicle transport	
06.004	非机动车运输	non-power-driven vehicle transport	
06.005	汽车运输	motor vehicle transport	
06.006	运输结构	transport structure	
06.007	运输体系	transport system	
06.008	运输能力	transport capacity	
06.009	营运方式	operation manner	
06.010	营运范围	operation area	
06.011	联合运输	intermodal transport	又称"联运"。
06.012	接力运输	relay transport	
06.013	门到门运输	door-to-door transport	
06.014	直达运输	through transport	
06.015	国际运输	international transport	又称"出入境运输"。
06.016	过境运输	transit transport	
06.017	货物运输	freight transport	
06.018	旅客运输	passenger transport	
06.019	营业性运输	for-hire transport	
06.020	非营业性运输	own account transport	又称"自用运输"。
06.021	生产过程运输	transport in production process	
06.022	流通过程运输	transport in circulation	
06.023	计划运输	planned transport	
06.024	合同运输	contract transport	
06.025	双程运输	two-way loading transport	
06.026	单程运输	one-way loading transport	
06.027	迂回运输	roundabout transport	
06.028	对流运输	counter flow transport	
06.029	重复运输	reflux of freight	
06.030	分流运输	diversion transport	
06.031	拖挂运输	tractor-trailer transport	
06.032	甩挂运输	trailer pick-up transport	
06.033	滚装运输	roll on/roll off transport	

序 码	汉 文 名	英 文 名	注 释
06.034	公路旅客运输	highway passenger transport	
06.035	旅客	passenger, traveler	又称"乘客"。
06.036	客流	passenger traffic, passenger flow	
06.037	跨区客流	inter-regional passenger traffic	又称"区间客流"。
06.038	区内客流	intra-regional passenger traffic	
06.039	持续性客流	continuous passenger traffic	
06.040	阵发性客流	intermittent passenger traffic	
06.041	季节性客流	seasonal passenger traffic	
06.042	往返性客流	round trip passenger traffic	
06.043	单向性客流	unidirectional passenger traffic	
06.044	集结性客流	aggregate passenger traffic	
06.045	分散性客流	dispersive passenger traffic	
06.046	旅客流量	passenger traffic volume	
06.047	旅客流时	passenger traffic time	
06.048	旅客流向	passenger traffic flow direction	
06.049	客流图	passenger traffic diagram	
06.050	客源	passenger source	
06.051	客运站站务工作量	passenger terminal work load	
06.052	旅客日发送量	daily dispatched passenger number	
06.053	旅客最高聚集人数	maximum gathered passenger number	
06.054	旅客波动系数	passenger number fluctuation coefficient	
06.055	营运线路	operation route	
06.056	班车线路	regular service route	
06.057	班车客运	scheduled bus transport	
06.058	直达客运	through bus transport	
06.059	包车客运	chartered bus transport	
06.060	城市公共客运	urban bus transport	
06.061	出租汽车客运	taxi transport	
06.062	公路客运网	bus transport network	
06.063	客车定员	rated passenger number	
06.064	班次	number of runs	
06.065	车次	serial number of bus run	
06.066	营运里程	operation mileage	
06.067	班次密度	density of runs	

序　码	汉　文　名	英　文　名	注　释
06.068	客运班期	bus schedule	
06.069	定班运行	scheduled run	
06.070	非定班运行	non-scheduled run	
06.071	正点	on schedule	
06.072	晚点	behind schedule	
06.073	客车调度	bus dispatching	
06.074	客车运行周期	bus run cycle	
06.075	大循环运行	full circulation operation	
06.076	小循环运行	partial circulation operation	
06.077	定线运行	fixed line operation	
06.078	套班运行	package run	
06.079	双班运行	two shifts run	
06.080	行包	luggage	
06.081	一般行包	normal luggage	
06.082	计件行包	piece luggage	
06.083	轻浮行包	light luggage	
06.084	贵重行包	precious belongings	
06.085	托运行包	consigned luggage	
06.086	自理行包	hand luggage	
06.087	行包托运	luggage consignment	
06.088	行包受理	luggage acceptance	
06.089	行包承运	acceptance of luggage consignment	
06.090	行包保管	luggage storage service	
06.091	行包交付	luggage delivery	
06.092	开启查验	luggage opening for inspection	
06.093	原件收付	original piece receipt and delivery	
06.094	运达期限	time limit of shipment arrival	
06.095	客运单证	document of passenger transport	
06.096	定站客票	fixed ticket	
06.097	定额客票	valued ticket	
06.098	补充客票	additional ticket	
06.099	电子自动售票系统	automatic electronic ticket system	
06.100	行包标签	luggage label	
06.101	客运包车票	charter bill	
06.102	客运动态记录单	bus running record	
06.103	旅客运输质量	bus service quality	

序　码	汉　文　名	英　文　名	注　释
06.104	旅客一条龙服务	bus streamlined service	
06.105	客运质量管理	quality management of passenger transport	
06.106	标志服装	uniform	
06.107	漏乘	missing a bus	
06.108	错乘	taking wrong bus	
06.109	甩客	denial of passenger	
06.110	旅客意外伤害	passenger unexpected injury	
06.111	旅客意外伤害保险	insurance of passenger unexpected injury	
06.112	行包丢失	loss of luggage	
06.113	行包损坏	damage of luggage	
06.114	客运索赔	passenger claims	
06.115	客运理赔	actions for passenger claims	
06.116	客运拒赔	reject of passenger claims	
06.117	客运索赔时限	period of limitation for passenger claims	
06.118	按质赔偿	compensation for loss	
06.119	限额赔偿	limited compensation	
06.120	客车正班率	rate of on-schedule bus runs	
06.121	定站停靠率	rate of stops at scheduled bus stops	
06.122	发车正点率	rate of on-time departure	
06.123	旅客正运率	passengers transport regularity rate	
06.124	行包正运率	luggages transport regularity rate	
06.125	行包赔偿率	luggage compensation rate	
06.126	停靠站	bus stop	
06.127	代办站	agency	
06.128	自办站	self-owned terminal	
06.129	行包货位	luggage lot	
06.130	班次时刻表	time table of runs	
06.131	营运线路图	operational route map	
06.132	调查表调查	inquiry form survey	
06.133	随车调查	on vehicle survey	
06.134	普通货物	general goods	
06.135	特种货物	special goods	
06.136	散装货物	bulk goods	

序　码	汉　文　名	英　文　名	注　释
06.137	裸装货物	naked goods	
06.138	包装货物	packed goods	
06.139	件货	piece goods	
06.140	危险货物	hazardous goods	
06.141	放射性货物	radioactive goods	
06.142	轻浮货物	light goods	
06.143	长大笨重货物	heavy and bulky goods	
06.144	超限货物	oversize or overweight goods	
06.145	鲜活货物	fresh and living goods	
06.146	易腐货物	perishable goods	
06.147	贵重货物	valuable goods	
06.148	扬尘性货物	dusty goods	
06.149	呼吸性货物	respiratory goods	
06.150	吸湿性货物	humidity absorbing goods	
06.151	易碎货物	fragile goods	
06.152	重点货物	priority goods	
06.153	大宗货物	mass goods	
06.154	禁运货物	contraband goods	
06.155	限运货物	restricted goods	
06.156	混装	mixed loading	
06.157	忌装	avoidance of mixed loading	
06.158	货物包装	goods packing	
06.159	运输包装	packing for transport	
06.160	货物标志	goods mark	
06.161	收发货标志	shipping mark	
06.162	运输标志	transport mark	
06.163	包装储运标志	package indicative mark	
06.164	危险货物包装标志	package mark for hazardous goods	
06.165	整车货物运输	truck-load transport	
06.166	零担[货物]运输	less-than-truck-load transport	
06.167	特种车辆货物运输	special truck transport	
06.168	成组运输	unitized transport	
06.169	集装箱运输	container transport	
06.170	保价运输	insured transport	
06.171	货运业务	freight transport business	

序　码	汉　文　名	英　文　名	注　　释
06.172	托运人	consignor	
06.173	承运人	carrier	
06.174	发货人	shipper	
06.175	收货人	consignee	
06.176	托运计划	consignment plan	
06.177	托运	consignment	
06.178	承运	acceptance of consignment	
06.179	发运	pre-departure operation	
06.180	分运	partite transport	
06.181	货物运单	bill of lading	又称"提单"。
06.182	货物清单	freight list	
06.183	货票	freight bill	
06.184	验货	inspection of goods	
06.185	理货	tallying of cargoes	
06.186	验道	road inspection	
06.187	押运	escorting	
06.188	到达交付	delivery on arrival	
06.189	运输责任期	transport liability period	
06.190	逾期交付	delayed delivery	
06.191	逾期提货	delayed pick-up	
06.192	货源	freight source	
06.193	货源调查	freight source survey	
06.194	货运生产平衡	freight production balance	
06.195	回程系数	return factor	
06.196	运量波动系数	freight volume fluctuation coef-ficient	
06.197	季节性货源	seasonal freight source	
06.198	货源信息	information of freight source	
06.199	货流	freight flow	
06.200	货流图	freight traffic diagram	
06.201	货物流量	freight traffic volume	
06.202	货物流向	freight traffic direction	
06.203	货物流时	freight traffic time	
06.204	货运调度	dispatching of freight transport	
06.205	计划调度	planned dispatching	
06.206	执行调度	operation dispatching	
06.207	现场调度	on-site dispatching	

序 码	汉 文 名	英 文 名	注 释
06.208	值班调度	routine dispatching	
06.209	调度日志	dispatching log	
06.210	调度方法	dispatching method	
06.211	调度命令	dispatching order	
06.212	图上作业法	graphic dispatching method	
06.213	表上作业法	table dispatching method	
06.214	循环调度法	cyclic dispatching method	
06.215	货车动态	operation status of truck	
06.216	货运质量	freight transport quality	
06.217	质量事故	quality accident	
06.218	货差	freight shortage	
06.219	货损	freight damage	
06.220	错装	misloading	
06.221	漏装	neglected loading	
06.222	错运	misshipment	
06.223	定额赔偿	rated compensation	
06.224	收回赔偿	reclaim of compensation	
06.225	货运索赔	freight claims	
06.226	货运理赔	actions for freight claims	
06.227	货运拒赔	reject of freight claims	
06.228	货运索赔时限	period of limitation for freight claims	
06.229	货运质量指标	freight quality indicator	
06.230	货运质量事故频率	frequency of freight quality accidents	
06.231	货损率	damage rate of goods	
06.232	货差率	shortage rate of goods	
06.233	赔偿率	rate of freight compensation	
06.234	货运及时率	rate of timely freight	
06.235	装卸质量合格率	loading and unloading quality conformity rate	
06.236	拖挂重量	trailing load	
06.237	车厢容积	carriage box volume capacity	
06.238	集装箱内容积	container volume capacity	
06.239	集装箱自重系数	coefficient of container dead weight	
06.240	[集装]箱容系数	container specific volume capacity	又称"集装箱比容"。

序　码	汉　文　名	英　文　名	注　释
		coefficient	
06.241	[集装]箱容利用率	container volume capacity utilizing ratio	
06.242	[集装]箱载重利用率	container load capacity utilizing ratio	
06.243	[集装]箱位	container lot	
06.244	[货物]托盘	pallet	
06.245	装货网络	loading network	
06.246	集装袋	contain bag	
06.247	汽车运输站场设施	terminal and yard facilities of motor transport	
06.248	货运站	freight terminal	
06.249	零担货运站	less-than-truck-load terminal	
06.250	货运枢纽站	freight hub terminal	
06.251	货运信息中心	freight information center	
06.252	货运联运站	intermodal freight terminal	
06.253	公路集装箱中转站	highway container transfer terminal	
06.254	[货物]仓库	warehouse	
06.255	高站台	high platform	
06.256	低货位	low freight lot	
06.257	堆货场	stock yard	
06.258	车位	loading/unloading lot	
06.259	货位	freight lot	
06.260	仓库面积使用率	storage area utilization ratio	
06.261	掏箱	unstuffing	
06.262	装箱	stuffing	
06.263	拼箱	mixed stuffing	
06.264	提取空箱	pick-up empty container	
06.265	回送空箱	return empty container	
06.266	内陆还箱站	inland container depot	
06.267	适箱货	container load freight	
06.268	整箱货	full container load, FCL	
06.269	拼箱货	less than container load, LCL	
06.270	汽车载箱行程	vehicle payload-container mileage	
06.271	汽车空箱行程	vehicle empty container mileage	
06.272	汽车重箱行程	vehicle loaded container mileage	

序　码	汉　文　名	英　文　名	注　释
06.273	货物装卸	goods handling	
06.274	装卸方式	handling mode	
06.275	人力装卸	manual handling	
06.276	机械装卸	mechanical handling	
06.277	装卸机械	handling machinery	
06.278	固定式装卸机械	fixed handling machinery	
06.279	移动式装卸机械	mobile handling machinery	
06.280	随车装卸机械	escort handling machinery	
06.281	集装箱装卸机械	container handling machinery	
06.282	水平搬运机械	horizontal handling machinery	
06.283	堆垛机械	stowing machinery	
06.284	吊货工夹具	hoisting fixture	
06.285	随车液压起重臂	escort hydraulic arm	
06.286	装卸工艺	handling technology	
06.287	装卸机械化系统	handling mechanization system	
06.288	装卸机械化方案	handling mechanization scheme	
06.289	装卸工艺设计	handling technology design	
06.290	装卸工艺流程	handling technological process	
06.291	起重量	hoisting capacity	
06.292	汽车货运装卸站	motor transport handling station	
06.293	装卸工序	handling procedure	
06.294	装卸质量标准	handling quality standard	
06.295	辅助装卸作业	auxiliary handling operation	
06.296	货车装卸安全间距	truck handling safety interval distance	
06.297	挂摘钩	coupling and uncoupling of hooks	
06.298	稳钩	stabilize hooks	
06.299	超远作业	over-distance stacking operation	
06.300	超高作业	over-height stacking operation	
06.301	装卸作业线	handling operation line	
06.302	库场作业	storage area operation	
06.303	堆垛	stacking up	
06.304	转垛	stacking area adjustment	
06.305	倒垛	stack transfer	
06.306	集装箱叉车	container fork lift	
06.307	滚上滚下集装箱叉车	roll on/roll off container fork lift	

序　码	汉　文　名	英　文　名	注　释
06.308	侧面集装箱叉车	side container fork lift	
06.309	通用型集装箱叉车	universal container fork lift	
06.310	箱内作业叉车	inside container operation fork lift	
06.311	集装箱吊具	container spreader	
06.312	分拣作业	sorting operation	
06.313	装卸效率	handling efficiency	
06.314	装卸距离	handling distance	
06.315	装卸作业指标	indexes of handling operation	
06.316	货物装卸量	loading/unloading volume	
06.317	货物操作量	tonnage of cargo handled	
06.318	工序吨	unit operation ton	
06.319	机械装卸作业量	machine handling operation volume	
06.320	机械装卸作业比重	share of machine handling operation	
06.321	装卸定额	handling quota	
06.322	装卸工时定额	handling hourly quota	
06.323	装卸工班定额	handling shift quota	
06.324	装卸工班效率	handling shift efficiency	
06.325	装卸工时产量	handling hourly output	
06.326	装卸机械效率指标	handling machinery efficiency index	
06.327	装卸机械生产率	handling machinery productivity	
06.328	装卸工人生产率	handling worker's productivity	
06.329	装卸工作停歇时间	handling intermittence time	
06.330	装卸工作成本	handling operation cost	
06.331	[货物]堆存吨天数	storage ton-days	
06.332	[货物]平均堆存天数	average storage days	
06.333	公路运输管理	highway transport management	
06.334	公路运输计划	highway transport plan	
06.335	重点物资运输计划	priority materials transport plan	
06.336	车辆运行作业计	vehicle operation plan	

序 码	汉 文 名	英 文 名	注 释
	划		
06.337	运量预测	transport volume forecast	
06.338	数量指标	quantitative index	
06.339	质量指标	quality index	
06.340	公路运输量	highway transport volume	
06.341	运量	volume	
06.342	周转量	turnover	
06.343	换算周转量	converted turnover	
06.344	货物平均运距	freight average haul distance	
06.345	旅客平均运距	passenger average haul distance	
06.346	公路运输经济运距	break-even distance for highway transport	
06.347	集装箱化率	containerization ratio	
06.348	总车日	total vehicle-days	
06.349	完好车率	vehicle availability rate	
06.350	完好车日	available vehicle-days	
06.351	工作车日	working vehicle-days	
06.352	工作车班	vehicle working shift	
06.353	停驶车日	vehicle non-working day	
06.354	吨位	rated tonnage	
06.355	客位	rated seat	
06.356	总车吨位日	total vehicle-ton-days	
06.357	总车客位日	total vehicle-seat-days	
06.358	重车行程	loaded mileage	又称"重车里程"。
06.359	空车行程	deadhead mileage	又称"空车里程"。
06.360	出车时间	vehicle line haul hour	
06.361	平均车数	average number of vehicles	
06.362	平均总吨位	average total tonnage	
06.363	平均总客位	average total seats	
06.364	每车平均吨位	average tonnage per vehicle	
06.365	每车平均客位	average seats per vehicle	
06.366	车辆工作率	vehicle working rate	
06.367	车辆停驶率	vehicle non-working rate	
06.368	平均车日行程	average vehicle daily travel	
06.369	营运速度	operation speed	
06.370	行程利用率	operation mileage rate	又称"里程利用率"。
06.371	载重量利用率	truck loading capacity utilization	

序　码	汉文名	英文名	注　释
		rate	
06.372	载客量利用率	bus capacity utilization rate	
06.373	实载率	load factor	
06.374	拖运率	trailer ton-km ratio	
06.375	挂车配比	trailer/tractor ratio	
06.376	单车产量	annual output per vehicle	
06.377	车吨产量	annual output per vehicle tonnage capacity	
06.378	节油率	fuel saving ratio	
06.379	亏油率	fuel deficit ratio	
06.380	公路运输总产值	total highway transport production output value	
06.381	公路运输净产值	net highway transport production output value	
06.382	运输弹性系数	transport elasticity coefficient	
06.383	运输系数	transport coefficient	
06.384	车辆折旧	vehicle depreciation	
06.385	营运费用	operation cost	
06.386	运输管理费	transport administration fee	简称"运管费"。
06.387	车辆事故损失	vehicle accident loss	
06.388	单车核算	unit vehicle accounting	
06.389	车队核算	vehicle fleet accounting	
06.390	营运收入	operation revenue	
06.391	营运支出	operation expenditure	
06.392	全员劳动生产率	all-personnel labour productivity	
06.393	司机及助手劳动生产率	productivity of driver and assistant	
06.394	劳动工资管理	labour wage management	
06.395	行车津贴	operation subsidy	
06.396	物资管理	material management	
06.397	公路运输行政管理	administration of highway transport	
06.398	公路运输行业管理	administration of highway transport industry	
06.399	运输市场	transport market	
06.400	运输市场管理	transport market control	
06.401	开停业管理	entry/exit regulation	

序 码	汉 文 名	英 文 名	注 释
06.402	开业条件	entry conditions	
06.403	开业报批程序	entry application and approval procedure	
06.404	停业报批程序	exit application and approval procedure	
06.405	车辆注册登记	vehicle registration	
06.406	运输服务业	transport service industry	
06.407	经营许可证	business certificate	
06.408	营运证	operation certificate	
06.409	行车路单	waybill	
06.410	营运标志	operation signs	
06.411	违章行为	violation	
06.412	商务纠纷	commercial dispute	
06.413	公路运输法规	highway transport statute	
06.414	汽车运输成本	motor transport cost	
06.415	汽车运输计划成本	motor transport planned cost	
06.416	汽车运输成本分析	motor transport cost analysis	
06.417	汽车运输预测成本	motor transport forecast cost	
06.418	汽车运输成本计算对象	motor transport cost accounting object	
06.419	汽车运输成本计算单位	motor transport cost accounting unit	
06.420	汽车运输成本范围	motor transport cost scope	
06.421	汽车运输成本项目	motor transport cost item	
06.422	汽车运输成本结构	motor transport cost structure	
06.423	汽车运输单车成本	motor transport unit vehicle cost	
06.424	汽车运输车型成本	motor transport cost by vehicle type	
06.425	汽车运输企业成本	motor transport enterprise cost	

序　码	汉　文　名	英　文　名	注　释
06.426	汽车运输车辆大修折旧	motor transport vehicle overhaul depreciation	
06.427	汽车运输装卸成本	motor transport handling cost	
06.428	汽车运输辅助生产成本	motor transport auxiliary production cost	
06.429	汽车运输综合成本	motor transport comprehensive cost	
06.430	汽车运输成本降低率	motor transport cost reduction rate	
06.431	汽车运输成本降低额	motor transport cost reduction amount	
06.432	汽车大修理费用	vehicle overhaul cost	
06.433	旧件修复成本	used parts reconditioning cost	
06.434	轮胎摊提率	tyre depreciation rate	
06.435	汽车运输运营成本	motor transport operation cost	
06.436	汽车运输车辆费用	motor transport vehicle cost	
06.437	装卸直接费用	handling direct cost	
06.438	汽车运输成本控制	motor transport cost control	
06.439	汽车运输边际成本	motor transport marginal cost	
06.440	汽车运输机会成本	motor transport opportunity cost	
06.441	汽车运输沉落成本	motor transport sunk cost	
06.442	运价	tariff, rate	
06.443	汽车运价体系	motor transport tariff system	
06.444	汽车运价构成	formation of motor transport tariff	
06.445	汽车运价率	motor transport tariff rate	
06.446	计价标准	charge standard	
06.447	计费重量	charged weight	
06.448	换算重量	converted weight	
06.449	计费里程	charged mileage	
06.450	运输里程	carriage mileage	

序　码	汉　文　名	英　文　名	注　　释
06.451	包干计费里程	chartered pay mileage	
06.452	重箱里程	loaded container mileage	
06.453	空箱里程	empty container mileage	
06.454	价目	tariff classification	
06.455	基本运价	basic rate	
06.456	运价加成	rate addition	
06.457	运价减成	rate reduction	
06.458	递近递增运价	increasing rate with decreasing distance	
06.459	递远递减运价	decreasing rate with increasing distance	
06.460	理论运价	theoretical rate	
06.461	差别运价	differential rate	
06.462	运价比差	rate ratio	
06.463	汽车运价管理体制	motor transport rate management system	
06.464	零担运价	less-than-truck-load rate	
06.465	整车运价	truck load rate	
06.466	货物分等运价	classified rates for goods	
06.467	特种货物运价	rate for special goods	
06.468	特种车辆运价	rate for special purpose vehicle	
06.469	小型车运价	rate for small vehicle	
06.470	区域运价	zoning rate	
06.471	包车运价	charter rate	
06.472	计时包车运价	charter rate by time	
06.473	计程包车运价	charter rate by distance	
06.474	旅客票价	ticket price	
06.475	全价票	full price ticket	
06.476	半价票	half price ticket	
06.477	普通客票价	ordinary ticket price	
06.478	旅游客车票价	tourist bus ticket price	
06.479	公共客车票价	bus ticket price	
06.480	小型客车票价	ticket price of mini bus	
06.481	集装箱包箱运价	charter rate by number of containers	
06.482	集装箱重箱运价	rate of loaded container	
06.483	集装箱空箱运价	rate of empty container	

序 码	汉 文 名	英 文 名	注 释
06.484	集装箱箱次费	per container charge	
06.485	集装箱中转费	container transit charge	
06.486	行包运价	luggage transport rates	
06.487	免费行包	free-of-charge luggage	
06.488	调车费	dispatch charge	
06.489	延滞费	demurrage	
06.490	装货落空损失费	compensation for failure of loading	
06.491	货物运输杂费	miscellaneous charges for freight transport	
06.492	旅客运输杂费	miscellaneous charges for passenger transport	
06.493	计箱装卸费	handling charge by container number	
06.494	装卸机械计时包用费	chartered handling equipment time based charge	
06.495	装卸机械走行费	travel expense of handling equipment	
06.496	货物装卸费	goods handling charge	
06.497	辅助装卸费	auxiliary handling charge	
06.498	搬移费	moving charge	
06.499	清洗和熏蒸费	cleaning and steaming charge	
06.500	集装箱维修费	container maintenance charge	
06.501	服务手续费	service fee	
06.502	货物站务费	freight terminal charge	
06.503	仓储理货费	storage charge	
06.504	中转劳务包干费	transshipment service package charge	
06.505	承运货物代理费	freight forwarder agency charge	
06.506	车辆站务费	vehicle terminal charge	
06.507	组货服务费	freight sales charge	

07. 筑路机械及运用管理

序 码	汉 文 名	英 文 名	注 释
07.001	推土机	bulldozer	
07.002	履带式推土机	crawler bulldozer	

序　码	汉　文　名	英　文　名	注　释
07.003	轮胎式推土机	wheel bulldozer	
07.004	频爆式推土机	blasting bulldozer	
07.005	水下推土机	underwater bulldozer	
07.006	湿地推土机	swamp bulldozer	
07.007	铲运机	scraper	
07.008	自行式铲运机	motor scraper	
07.009	拖式铲运机	towed scraper	
07.010	升运式铲运机	elevating scraper	
07.011	推土铲运机	scraper-dozer	
07.012	铰接式自卸车	articulated truck	
07.013	平地机	grader	
07.014	自行式平地机	motor grader	
07.015	拖式平地机	towed grader	
07.016	平地升送机	elevation grader	又称"犁扬机"。
07.017	松土机	ripper	
07.018	凿岩机	rock drill, jack hammer	
07.019	内燃凿岩机	motor jack hammer	
07.020	电动凿岩机	electric jack hammer	
07.021	气动凿岩机	pneumatic rock drill	
07.022	气腿式凿岩机	air-rider jack hammer	
07.023	手持式凿岩机	hand-held rock drill	
07.024	液压凿岩机	hydraulic jack hammer	
07.025	冲击式凿岩机	shocking jack hammer	
07.026	旋转式凿岩机	rotary jack hammer	
07.027	冲击旋转式凿岩机	shocking rotary jack hammer	
07.028	凿岩钻车	drill wagon for jack hammer	
07.029	履带式液压凿岩台车	crawler hydraulic bench drill	
07.030	改钻机	drills steel sharpener	
07.031	磨钻机	bits grinders	
07.032	气动岩石破碎机	pneumatic rock breaker	
07.033	液压岩石破碎机	hydraulic rock breaker	
07.034	清岩机	ballast loader	
07.035	铲斗式清岩机	rock shovel	
07.036	蟹耙式清岩机	rake-up ballast remover	
07.037	挖掘机	excavator	

序　码	汉　文　名	英　文　名	注　释
07.038	单斗挖掘机	power shovel, single bucket excavator	
07.039	正铲挖掘机	front shovel	
07.040	反铲挖掘机	backhoe excavator	
07.041	抓铲挖掘机	clam shell excavator	
07.042	拉铲挖掘机	dragline excavator	
07.043	轮胎式单斗挖掘机	wheel type single bucket excavator	
07.044	履带式单斗挖掘机	crawler single bucket excavator	
07.045	机械挖掘机	cable operated excavator	
07.046	液压挖掘机	hydraulic excavator	
07.047	多斗挖掘机	multi-bucket excavator	
07.048	斗轮式挖掘机	bucket wheel excavator	
07.049	伸缩铲挖掘机	telescopic boom excavator	
07.050	挖沟机	trencher	
07.051	链斗式挖沟机	chain bucket trencher	
07.052	斗轮式挖沟机	bucket wheel trencher	
07.053	装载机	loader	
07.054	轮胎式装载机	wheel loader	
07.055	履带式装载机	crawler loader	
07.056	装载挖掘机	loader-excavator	
07.057	开槽机	grooving machine	
07.058	双轮压路机	tandem roller	
07.059	三轮压路机	3-wheel roller	
07.060	轮胎压路机	pneumatic tyred roller	
07.061	振动压路机	vibratory roller	
07.062	凸块压路机	padfoot roller	
07.063	格栅压路机	grid roller	
07.064	拖式羊足碾	towed sheep-foot roller	
07.065	羊足压路机	sheep-foot roller	
07.066	双轮振动压路机	tandem vibratory roller	
07.067	组合式压路机	combination roller	
07.068	振荡压路机	oscillatory roller	
07.069	铰接式压路机	articulated roller	
07.070	拖式振动压路机	towed vibratory roller	
07.071	振动平板夯	vibratory plate compactor	

序　码	汉　文　名	英　文　名	注　释
07.072	振动冲击夯	vibratory tamper	
07.073	冲击夯	shocking rammer	
07.074	爆炸夯	explosion rammer	
07.075	电动冲击夯	electric shocking rammer	
07.076	蛙式夯	electric frog rammer	
07.077	落锤夯	drop weight rammer	
07.078	碎石机	crusher	
07.079	颚式碎石机	jaw crusher	
07.080	转子式碎石机	rotator crusher	
07.081	锤击式碎石机	hammer crusher	
07.082	圆锥式碎石机	cone crusher	
07.083	辊筒式碎石机	roll crusher	
07.084	筛分机	sieving machine	
07.085	振动筛	vibrating screen	
07.086	摇摆筛	swinging screen	
07.087	共振筛	resonance screen	
07.088	滚动筛	drum screen	
07.089	给料机	feeder	
07.090	皮带给料机	belt feeder	
07.091	往复给料机	reciprocating feeder	
07.092	链板给料机	slat feeder	
07.093	圆盘给料机	disk feeder	
07.094	振动给料机	vibrating feeder	
07.095	联合碎石机组	crushing plant	
07.096	移动式联合碎石机组	mobile crushing plant	
07.097	开式联合碎石机组	open type crushing plant	
07.098	闭式联合碎石机组	close type crushing plant	
07.099	材料输送机械	material conveyer	
07.100	带式输送机	belt conveyer	
07.101	刮板输送机	slat conveyer	
07.102	斗式提升机	bucket elevator	
07.103	离心式斗式提升机	centrifugal bucket elevator	
07.104	重力式斗式提升	gravity bucket elevator	

序　码	汉　文　名	英　文　名	注　释
	机		
07.105	转斗式提升机	rotary bucket elevator	
07.106	螺旋输送机	screw conveyer	
07.107	气力输送机	air conveyer	
07.108	稳定土厂拌设备	stabilized soil mixing plant	
07.109	可搬式稳定土厂拌设备	movable stabilized soil mixing plant	
07.110	移动式稳定土厂拌设备	mobile stabilized soil mixing plant	
07.111	强制式稳定土搅拌设备	forced stabilized soil mixing plant	
07.112	自落式稳定土搅拌设备	gravity stabilized soil mixing plant	
07.113	路拌式稳定土搅拌机	soil stabilizer, stabilized road-mixer	
07.114	轮胎式稳定土搅拌机	wheel-type soil stabilizer, wheel-type stabilized soil road-mixer	
07.115	履带式稳定土搅拌机	crawler soil stabilizer, crawler stabilized soil road-mixer	
07.116	超深稳定土搅拌机	super-deep soil stabilizer	
07.117	犁式稳定土搅拌机	blade plough soil stabilizer, blade plough stabilized soil road-mixer	
07.118	转子式稳定土搅拌机	rotary type soil stabilizer, rotary type stabilized soil road mixer	
07.119	一次拌成稳定土搅拌机	single pass soil stabilizer, single pass stabilized soil road mixer	
07.120	粉料撒布机	filler spreader	
07.121	石料摊铺机	aggregate paver	
07.122	稳定土摊铺机	stabilized soil paver	
07.123	路基精整机	subgrade fine trimmer	
07.124	沥青混合料搅拌设备	asphalt mixing plant	
07.125	可搬式沥青混合料搅拌设备	movable asphalt mixing plant	
07.126	移动式沥青混合料搅拌设备	mobile asphalt mixing plant	

序　码	汉　文　名	英　文　名	注　释
07.127	间歇式沥青混合料搅拌设备	batch asphalt mixing plant	
07.128	连续式沥青混合料搅拌设备	continuous asphalt mixing plant	
07.129	滚筒式沥青混合料搅拌设备	drum-mixing asphalt plant	
07.130	除尘装置	dust collector	
07.131	旋风式除尘器	cyclone dust collector	
07.132	布袋除尘器	bag dust collector	
07.133	湿式除尘器	wet dust collector	
07.134	配料给料装置	aggregate feeder units	
07.135	冷料输送机	cold aggregate conveyer	
07.136	干燥筒	drier	
07.137	干燥-搅拌筒	drying-mixing drum	
07.138	热料提升机	hot aggregate elevator	
07.139	热料振动筛	hot aggregate vibrating screen	
07.140	热料仓	hot aggregate bin	
07.141	称重系统	weighing system	
07.142	搅拌器	mixer	
07.143	沥青秤	asphalt weigher	
07.144	沥青保温罐	asphalt tank with heating device	
07.145	粉料仓	filler bin	
07.146	粉料秤	filler weigher	
07.147	粉料计量给料器	filler metering feeder	
07.148	喷燃器	burner	
07.149	鼓风机	blower	
07.150	抽风机	exhauster	
07.151	沥青流量计	asphalt flow meter	
07.152	控制室	control house	
07.153	皮带称重装置	conveyer belt scale	
07.154	热混合料贮仓	hot mix silo	
07.155	混合料刮板升送机	mixture slat elevator	
07.156	混合料斗式升送机	mixture hopper elevator	
07.157	沥青混凝土摊铺机	asphalt paver	

序　码	汉　文　名	英　文　名	注　　释
07.158	轮胎式沥青混凝土摊铺机	wheel-type asphalt paver	
07.159	履带式沥青混凝土摊铺机	crawler-type asphalt paver	
07.160	自行式沥青混凝土摊铺机	self-propelled asphalt paver	
07.161	拖式沥青混合料摊铺机	towed asphalt paver	
07.162	沥青喷洒机	asphalt sprayer	
07.163	石屑撒布机	chip spreader	
07.164	沥青泵	asphalt pump	
07.165	外啮合齿轮沥青泵	outside gear asphalt pump	
07.166	内啮合齿轮沥青泵	inside gear asphalt pump	
07.167	沥青罐车	asphalt truck	
07.168	沥青贮仓	asphalt storage	
07.169	地上沥青贮仓	ground asphalt storage	
07.170	地下沥青贮仓	underground asphalt storage	
07.171	半地下沥青贮仓	semi-underground asphalt storage	
07.172	沥青熔化加热装置	asphalt melting and heating device	
07.173	桶装沥青熔化装置	asphalt barreled melter	
07.174	火管式沥青熔化装置	asphalt firepipemelter	
07.175	蒸气式沥青熔化装置	asphalt steam pipe melter	
07.176	导热油沥青熔化装置	asphalt hot oil melter	
07.177	太阳能沥青熔化装置	asphalt solar energy melter	
07.178	沥青乳化机	bitumen emulsifying machine	
07.179	转子式沥青乳化机	rotary bitumen emulsifying machine	
07.180	叶浆搅拌式沥青乳化机	paddle bitumen emulsifying machine	

序　码	汉　文　名	英　文　名	注　释
07.181	喷嘴式沥青乳化机	nozzle bitumen emulsifying machine	
07.182	沥青乳化设备	bitumen emulsifying plant	
07.183	橡胶沥青装置	rubber asphalt unit	
07.184	泡沫沥青装置	foam asphalt unit	
07.185	聚乙烯沥青装置	polyethylene asphalt unit	
07.186	电热式沥青加热装置	bitumen electrical heating device	
07.187	混凝土搅拌设备	concrete mixing plant	
07.188	强制式混凝土搅拌设备	forced concrete mixing plant	
07.189	自落式混凝土搅拌设备	gravity type concrete mixing plant	
07.190	混凝土搅拌机	concrete mixer	
07.191	混凝土搅拌输送车	truck mixer	
07.192	混凝土摊铺列车	concrete paving train	
07.193	轨模式混凝土摊铺机	rail form concrete paver	
07.194	滑模式混凝土摊铺机	slipform concrete paver	
07.195	混凝土布料机	concrete spreader	
07.196	斗式混凝土布料机	hopper type concrete spreader	
07.197	螺旋式混凝土布料机	screw type concrete spreader	
07.198	刮板式混凝土布料机	slat type concrete spreader	
07.199	混凝土路面整面机	concrete pavement finisher	
07.200	混凝土路面拉毛机	concrete pavement texturing machine	
07.201	混凝土路面切缝机	concrete pavement joint cutting machine	
07.202	混凝土路面清缝机	concrete pavement joint cleaning machine	
07.203	混凝土路面填缝	concrete pavement joint sealing	

序　码	汉　文　名	英　文　名	注　释
	机	machine	
07.204	滑模式混凝土路缘铺筑机	slipform concrete curb paver	
07.205	混凝土输送泵车	concrete pump truck	
07.206	钢丝网铺设机	wire mesh laying machine	
07.207	路面铣削机	pavemill	
07.208	沥青路面加热机	asphalt pavement heater	
07.209	沥青路面火焰加热器	asphalt pavement flame heater	
07.210	沥青路面红外线加热器	asphalt pavement infrared heater	
07.211	沥青路面就地再生机	asphalt pavement *in-situ* recycling machine	
07.212	沥青混合料再生拌和设备	asphalt mixture recycling mixing plant	
07.213	洒水车	sprinkler	
07.214	综合养护车	combined maintenance truck	
07.215	多用养护车	multi-use maintenance truck	
07.216	除雪机	snow remover	
07.217	转子式除雪机	rotary snow remover	
07.218	犁式除雪机	V type snow plough	
07.219	喷气式除雪机	air-jet snow remover	
07.220	螺旋式除雪机	screw snow remover	
07.221	路面清扫机	pavement sweeping machine	
07.222	真空吸尘清扫车	vacuum sweeper	
07.223	旋刷刮板清扫车	rotary broom-slat sweeper	
07.224	抛掷清扫车	abandon sweeper	
07.225	扫砂机	sand sweeping machine	
07.226	回砂机	sand recovering machine	
07.227	撒砂机	sand spraying machine	
07.228	划线机	road marking machine	
07.229	冷漆划线机	cold paint road marking machine	
07.230	热漆划线机	hot paint road marking machine	
07.231	热塑划线机	thermoplastic road marking machine	
07.232	旧线清除机	marking cleaning machine	
07.233	剪草机	mower	

序　码	汉　文　名	英　文　名	注　释
07.234	稀浆封层机	asphalt slurry seal machine	
07.235	地质钻机	exploration drill	
07.236	汽车钻机	drill truck	
07.237	打桩机	pile driver	
07.238	落锤	drop hammer	
07.239	柴油桩锤	diesel pile hammer	
07.240	单作用气动桩锤	single acting pneumatic hammer	
07.241	双作用气动桩锤	double acting pneumatic hammer	
07.242	单作用蒸气桩锤	single acting steam hammer	
07.243	双作用蒸气桩锤	double acting steam hammer	
07.244	桩架	pile frame	
07.245	多功能桩架	multi-purpose pile frame	
07.246	液压打桩机	hydraulic pile driver	
07.247	振动沉拔桩机	vibration pile driver extractor	
07.248	静力压拔桩机	static pile press extractor	
07.249	灌注桩钻孔机	cast-in-place concrete pile rig	
07.250	旋转式钻机	swiveling drill	
07.251	转盘式钻机	rotary table rig	
07.252	反循环钻机	reverse circulation drill	
07.253	正循环钻机	circulation drill	
07.254	潜水钻机	under water rig	
07.255	全套管钻机	all casing drill	
07.256	锤式冲抓斗	hammer grab	
07.257	钻斗	drilling bucket	
07.258	三翼钻头	3-wing bit	
07.259	滚轮式钻头	roller bit	
07.260	泥浆泵	mud pump	
07.261	爪形冲击锤	cross chipper hammer	
07.262	液压压拔套管机	hydraulic casing extractor	
07.263	无钻杆反循环钻机	rodless reverse circulation rig	
07.264	离心式水泵	centrifugal pump	
07.265	膜式水泵	diaphragm pump	
07.266	活塞式水泵	piston pump	
07.267	潜水泵	under water pump	
07.268	深井水泵	deep well pump	
07.269	多级离心泵	multistage centrifugal pump	

序 码	汉 文 名	英 文 名	注 释
07.270	自吸式水泵	self-prime pump	
07.271	浮水设备	floating equipment	
07.272	浮箱	floating box	
07.273	液压桩头破碎器	hydraulic pile head splitter	
07.274	潜水设备	diving equipment	
07.275	灌/压浆设备	grouting equipment	
07.276	架桥设备	bridging equipment	
07.277	混凝土梁滑模浇筑设备	concrete beam slipform device	
07.278	起重设备	hoisting equipment	
07.279	链式起重机	chain hoist, chain block	又称"链滑轮"。
07.280	手摇卷扬机	manual hoist	
07.281	电动卷扬机	electric hoister	
07.282	千斤顶	jack	
07.283	桅杆式起重机	derrick crane	
07.284	履带式起重机	crawler crane	
07.285	轮胎式起重机	wheel crane	
07.286	汽车式起重机	truck crane	
07.287	塔式起重机	tower crane	
07.288	门式起重机	gantry crane	
07.289	缆索起重机	cable crane	
07.290	梁式起重机	beam crane	
07.291	浮式起重机	floating crane	又称"浮吊"。
07.292	顶推设备	incremental launching device	
07.293	真空吸水机械	vacuum water sucker	
07.294	高压油泵	high-pressure oilpump	
07.295	张拉千斤顶	tensioning jack, drawing jack	
07.296	穿束机	strand pulling machine	
07.297	钢筋镦头机	steel-bar header	
07.298	钢丝缠束机	steel wire strand strapping machine	
07.299	卷管机	tendon duct winding machine	卷制预应力筋套管的机械。
07.300	钢筋切断机	steel-bar cutter	
07.301	钢筋调直机	steel-bar straightener	
07.302	钢筋弯曲机	steel-bar bender	
07.303	钢筋对焊机	steel-bar butt welder	

序　码	汉　文　名	英　文　名	注　释
07.304	弧焊机	arc welder	
07.305	点焊机	spot welder	
07.306	空气压缩机	air compressor	
07.307	往复式空气压缩机	reciprocating compressor	
07.308	螺杆式空气压缩机	screw air compressor	
07.309	过顶式装碴机	over-head loader	
07.310	隧道挖掘机	tunnel excavator	
07.311	隧道装载机	tunnel loader	
07.312	隧道掘进机	tunnel boring machine	
07.313	汽车调头转盘	truck turntable	
07.314	喷浆机	concrete injector, cement gun	
07.315	轴流通风机	axial-flow blower	
07.316	起爆器	firing device	
07.317	筑路机械全过程管理	all-life period management of road machine	
07.318	机械运转	machine operation	又称"机械操作"。
07.319	机械技术状况	technical condition of machine	
07.320	机械维护	maintenance of machine	
07.321	例行维护	routine maintenance	
07.322	设备定期维护	equipment periodic maintenance	
07.323	机械故障	failure of machine	
07.324	机械修理	repair of machine	
07.325	筑路机械小修	current repair of road machine	
07.326	筑路机械中修	medium repair of road machine	
07.327	筑路机械大修	major repair of road machine	
07.328	机械完好率	availability rate of machinery	
07.329	机械利用率	operation rate of machinery	
07.330	机械故障率	failure rate of machinery	
07.331	台班使用费	working-day cost	
07.332	基本折旧率	depreciation rate	
07.333	经常维修费	routine repair cost	
07.334	替换设备	spare attachment	
07.335	随机工具	tools attachment	
07.336	附具摊销费	attachment cost	
07.337	维护费	maintenance cost	

序　码	汉　文　名	英　文　名	注　释
07.338	安装拆卸费	mounting and dismounting cost	
07.339	辅助设施费	cost of ancillary facility	
07.340	机械事故	machinery accident	
07.341	安全事故	safety misadventure	
07.342	机械事故处理	accident handling of machinery	
07.343	机械管理责任制	responsibility system of machinery management	
07.344	机械操作证	operation certificate of machinery	
07.345	机械生产定额	production quota of machinery	
07.346	机械技术经济定额	technical economic quota of machinery	
07.347	机械出租	renting of machine	
07.348	机械化程度	level of mechanization	
07.349	设备配套	coordinative composition of equipments	
07.350	机械型谱	model spectrum of machine	
07.351	机械系列	series of machine	
07.352	机械运转记录	operating record of machine	
07.353	机械维修间隔	time interval of machine maintenance and repair	
07.354	机械维修作业项目	items of machine maintenance and repair	
07.355	机械停修台日	suspensive daily shift of machine	
07.356	设备封存	equipment storing up	
07.357	设备更新	equipment replacement	
07.358	设备报废	equipment scrapping	
07.359	设备残值	remanent value of equipment	
07.360	自动操作	automatic operation	
07.361	自动控制	automatic control	
07.362	自动称重	automatic weighing	
07.363	功率自动分配	automatic power apportioning	
07.364	液压比例自控系统	automatic hydraulic proportioning system	
07.365	温度自动控制	automatic temperature control	
07.366	自动找平装置	automatic leveling device	
07.367	液压自动找平装置	hydraulic automatic leveling device	

序 码	汉 文 名	英 文 名	注 释
07.368	电液自动找平装置	electronic-hydraulic automatic leveling device	
07.369	激光自动找平装置	laser automatic leveling device	
07.370	自动划线装置	autoset road marking	
07.371	压实度自动检测装置	automatic compactometer	
07.372	台架试验	bench test	
07.373	现场试验	field test	
07.374	试验室试验	laboratory test	
07.375	使用试验	service test	
07.376	模型试验	model test	
07.377	足尺试验	full-scale test	
07.378	稳定工况试验	steady-state test	
07.379	模拟试验	simulated test	
07.380	实验室模拟试验	laboratory simulation test	
07.381	台架模拟试验	bench simulation test	
07.382	负荷模拟试验	simulated load test	
07.383	环境模拟试验	simulated environment test	
07.384	低温模拟试验	simulated cold climate test	
07.385	高温模拟试验	simulated hot climate test	
07.386	高原气候模拟试验	simulated highland climate test	
07.387	盐蚀模拟试验	salt fog simulated test	
07.388	锈蚀试验	corrosion test	
07.389	风尘环境模拟试验	simulated dust environment test	
07.390	磨损试验	wear test	
07.391	常规试验	normal test	
07.392	加速试验	accelerated test	
07.393	强化试验	intensified test	
07.394	快速试验	prompt test	
07.395	样机试验	prototype test	
07.396	性能试验	performance test	
07.397	可靠性试验	reliability test	
07.398	耐久性试验	durability test	
07.399	研究性试验	research test	

序　码	汉　文　名	英　文　名	注　释
07.400	验证性试验	proving test	
07.401	技术性试验	technical test	
07.402	正式试验	official test	
07.403	验收试验	acceptance test	
07.404	鉴定试验	evaluation test	
07.405	检测性试验	detective test	
07.406	出厂证明试验	certification test	
07.407	抽样试验	sample test	
07.408	用户试验	user test	
07.409	工业性试验	type approval test	
07.410	定置试验	static-state test	
07.411	外形尺寸测定	overall dimensions measurement	
07.412	使用重量测定	operating weight measurement	
07.413	运输重量测定	shipping weight measurement	
07.414	前轴负荷测定	front axle-load measurement	
07.415	质心位置测定	mass center test	
07.416	离地间隙测定	ground clearance measurement	
07.417	操作力测定	operating force measurement	
07.418	操作行程测定	operating distance measurement	
07.419	行驶性能试验	travel performance test	
07.420	最高行驶速度测定	maximum travelling speed measurement	
07.421	最低稳定行驶速度测定	minimum steady travelling speed measurement	
07.422	直线行驶性能试验	straight-line running test	
07.423	加速性能试验	acceleration test	
07.424	制动性能试验	braking ability test	
07.425	牵引试验	drawbar test	
07.426	行驶阻力测定	moving resistance measurement	
07.427	滚动阻力测定	rolling resistance measurement	
07.428	拖动阻力测定	towing resistance measurement	
07.429	滑行试验	coasting test	
07.430	拖动试验	towing test	
07.431	滑行阻力测定	coasting resistance measurement	
07.432	滑行距离测定	coasting distance measurement	
07.433	挂钩牵引力测定	drawbar pull measurement	

序 码	汉 文 名	英 文 名	注 释
07.434	挂钩牵引功率测定	drawbar power measurement	
07.435	挂钩牵引效率测定	drawbar efficiency measurement	
07.436	驱动元件滑转率测定	driving element slipping rate measurement	
07.437	附着系数测定	traction coefficient measurement	
07.438	牵引功率比油耗测定	specific drawbar power fuel consumption measurement	
07.439	作业性能试验	operating performance test	
07.440	爬坡性能试验	climbing ability test	
07.441	转向性能试验	turning test	
07.442	倾翻稳定性试验	overturning stability test	
07.443	生产率测定	productive capacity measurement	
07.444	作业循环时间测定	determination of operating cycle	
07.445	循环作业量测定	determination of operating cycle production quantity	
07.446	每升燃料作业量测定	determination of production quantity per litre fuel consumption	
07.447	驾驶员座位振动测定	vibration measurement of driver seat	
07.448	驾驶室仪表板振动测定	vibration measurement of driver cab instrument board	
07.449	噪声试验	noise level test	
07.450	驾驶室视野试验	test for field of vision of driver cab	
07.451	工作照度试验	operating luminance test	
07.452	平均故障间隔时间	mean time interval between failures	
07.453	平均时间有效率	mean time effective rate	
07.454	当量故障率	equivalent failure rate	
07.455	传动试验	transmission test	
07.456	液压元件试验	hydraulic element test	
07.457	压实机械试验	compaction machine test	
07.458	压实宽度测定	compacting width measurement	
07.459	压实深度测定	compacting depth measurement	
07.460	滚轮静线压力测	drum static linear pressure mea-	

序　码	汉　文　名	英　文　名	注　释
	定	surement	
07.461	振动轮激振力测定	vibration drum exciting force measurement	
07.462	振动轮频率测定	drum vibration frequency measurement	
07.463	振动轮振幅测定	drum vibration amplitude measurement	
07.464	压实速度测定	rolling speed measurement	
07.465	现场压实试验	field compaction test	
07.466	路拌机械试验	road-mixing machine test	
07.467	拌和宽度测定	mixing width measurement	
07.468	拌和深度测定	mixing depth measurement	
07.469	旋转刀具圆周速度测定	rotary tools circular velocity measurement	
07.470	拌和功率测定	mixing power measurement	
07.471	现场拌和试验	field mixing test	
07.472	材料拌和均匀度测定	material mixing uniformity measurement	
07.473	沥青洒布机试验	asphalt distributor test	
07.474	沥青洒布宽度测定	distribution width measurement	
07.475	沥青洒布量测定	asphalt distribution measurement	
07.476	沥青混凝土摊铺机试验	asphalt paver test	
07.477	摊铺宽度测定	paving width measurement	
07.478	摊铺厚度测定	paving depth measurement	
07.479	摊铺作业试验	paving operation test	
07.480	最大摊铺生产率测定	maximum paving capacity measurement	
07.481	摊铺路面平整度试验	paved surface evenness test	
07.482	摊铺材料压实度试验	paved material compaction test	

08. 路用材料及性能

序　码	汉　文　名	英　文　名	注　释
08.001	结合料	binder	
08.002	有机结合料	organic binder	
08.003	沥青	bitumen, asphalt	
08.004	地沥青	asphalt	
08.005	石油沥青	petroleum asphalt, asphalt	
08.006	天然沥青	natural asphalt	
08.007	湖沥青	lake asphalt	
08.008	乳化沥青	emulsified bitumen, asphalt emulsion	又称"沥青乳液"。
08.009	快裂乳化沥青	quick breaking emulsified bitumen	
08.010	中裂乳化沥青	medium breaking emulsified bitumen	
08.011	慢裂乳化沥青	slow breaking emulsified bitumen	
08.012	乳化剂	emulsifying agent	
08.013	阳离子乳化沥青	cationic emulsified bitumen	
08.014	阴离子乳化沥青	anionic emulsified bitumen	
08.015	非离子乳化沥青	non-ionic emulsified bitumen	
08.016	液体沥青	liquid asphalt	
08.017	快凝液体沥青	rapid-curing liquid asphalt	
08.018	中凝液体沥青	medium-curing liquid asphalt	
08.019	慢凝液体沥青	slow-curing liquid asphalt	
08.020	轻制沥青	cutback asphalt	又称"稀释沥青"。
08.021	粘稠沥青	asphalt cement	
08.022	焦油沥青	tar	
08.023	煤沥青	coal tar	
08.024	页岩沥青	shale tar	
08.025	硬沥青	pitch	
08.026	氧化沥青	oxidized asphalt	
08.027	道路沥青	road bitumen	
08.028	橡胶改性沥青	rubber modified asphalt	
08.029	改性沥青	modified asphalt	
08.030	聚合物改性沥青	polymer modified asphalt	
08.031	环氧树脂	epoxy resin	

序　码	汉 文 名	英 文 名	注　释
08.032	重油	heavy oil	
08.033	渣油	residual oil	
08.034	有机聚合物	organic polymer	
08.035	无机结合料	inorganic binder	
08.036	石灰	lime	
08.037	熟石灰	slaked lime	又称"消石灰"。
08.038	生石灰	quick lime	
08.039	石膏	gypsum	
08.040	水泥	cement	
08.041	硅酸盐水泥	portland cement	
08.042	矿渣水泥	slag cement	
08.043	膨胀水泥	expansive cement	
08.044	自应力水泥	self-stressing cement	
08.045	道路水泥	road cement	
08.046	工业废渣	industrial solid waste	
08.047	粉煤灰	fly ash	
08.048	集料	aggregate	又称"骨料"。
08.049	粒料	granular material	
08.050	矿料	mineral aggregate	
08.051	石屑	chips	
08.052	石场弃渣	quarry waste	
08.053	矿粉	mineral powder	
08.054	人造集料	artificial aggregate	
08.055	陶粒	sintered aggregate	
08.056	土的粒组分类	soil classification by grain	
08.057	土的工程分类	engineering classification of soil	
08.058	土的成因分类	soil classification by formation	
08.059	巨粒土	over coarse-grained soil	
08.060	粗粒土	coarse-grained soil	
08.061	砾类土	gravelly soil	
08.062	砂类土	sandy soil	
08.063	细粒土	fine-grained soil	
08.064	特殊土	special soil	
08.065	漂石	boulder	
08.066	块石	angular boulder, block stone	
08.067	卵石	cobble	
08.068	碎石	angular cobble, broken stone,	

序　码	汉　文　名	英　文　名	注　释
		crushed stone	
08.069	片石	rubble	
08.070	料石	ashlar, dressed stone	
08.071	锥形块石	Telford stone	
08.072	亲水性集料	hydrophilic aggregate	
08.073	憎水性集料	hydrophobic aggregate	
08.074	砾	gravel	
08.075	角砾	angular gravel	
08.076	砂砾	sand gravel	
08.077	砂	sand	
08.078	粗砂	coarse sand	
08.079	中砂	medium sand	
08.080	细砂	fine sand	
08.081	土	soil	
08.082	粉土质砂	silty sand	
08.083	粘土质砂	clayey sand	
08.084	含砂低液限粉土	sandy silt of low liquid limit	
08.085	低液限粉土	low liquid limit silt	
08.086	低液限粘土	low liquid limit clay	
08.087	含砂低液限粘土	sandy clay of low liquid limit	
08.088	含砂高液限粘土	sandy clay of high liquid limit	
08.089	高液限粘土	high liquid limit clay	
08.090	混合料	mixture	
08.091	沥青混合料	bituminous mixture	
08.092	沥青混凝土混合料	bituminous concrete mixture	
08.093	沥青碎石混合料	bituminous macadam mixture	
08.094	沥青砂	sand asphalt	
08.095	沥青胶砂	asphaltmastic	曾用名"沥青玛琋脂"。
08.096	沥青稀浆	asphalt slurry	
08.097	再生沥青混合料	reclaimed asphalt mixture	
08.098	水泥砂浆	cement mortar	
08.099	石灰砂浆	lime mortar	
08.100	水泥混凝土混合料	cement concrete mixture	
08.101	水泥混凝土	cement concrete	

序　码	汉文名	英文名	注　释
08.102	粉煤灰水泥混凝土	fly-ash cement concrete	
08.103	早强混凝土	early strength concrete	
08.104	干硬性混凝土	dry concrete	
08.105	贫混凝土	lean concrete	
08.106	轻质混凝土	light-weight concrete	
08.107	高强混凝土	high strength concrete	
08.108	塑性混凝土	plastic concrete	
08.109	引气混凝土	air-entraining concrete	
08.110	变性混凝土	metamorphic concrete	
08.111	纤维混凝土	fiber concrete	
08.112	硅尘混凝土	silica fume concrete	
08.113	浸渍混凝土	impregnated concrete	
08.114	小石子混凝土	micro-aggregate concrete	
08.115	喷射混凝土	shotcrete	
08.116	金属[材料]	metal [material]	
08.117	碳素钢	carbon steel	
08.118	结构钢	structural steel	
08.119	耐候钢	weathering steel	
08.120	钢筋	steel bar	
08.121	螺纹钢筋	deformed bar	又称"变形钢筋"。
08.122	热轧钢筋	hot rolled bar	
08.123	光面钢筋	plain bar	
08.124	冷拉钢筋	cold-stretched steel bar	
08.125	预应力筋	tendon	
08.126	套管	sheath	
08.127	皱纹套管	spirally wound sheath	
08.128	高强度钢丝	high tensile steel wire	
08.129	钢丝束	bundle of steel wire	
08.130	钢铰线	strand	
08.131	冷拔钢丝	cold-drawn wire	
08.132	高强螺栓	high strength bolt	
08.133	大头螺栓	stud bolt	
08.134	金属网	metal mesh	
08.135	钢丝牵索	wire guy	
08.136	波纹钢桥面	corrugated steel deck	
08.137	钢缆索	cable	

序　码	汉　文　名	英　文　名	注　释
08.138	标称直径	nominal diameter	
08.139	索夹	cable clamp	
08.140	外加剂	additive	
08.141	掺加料	admixture	
08.142	沥青改性剂	asphalt modifier	
08.143	抗剥落剂	anti-stripping agent	
08.144	抗老化剂	anti-ager	
08.145	减水剂	water reducing agent	
08.146	加气剂	air entraining agent	
08.147	防冻剂	antifreezing agent	
08.148	早强剂	early strength agent	
08.149	促凝剂	coagulator	
08.150	缓凝剂	retarder	
08.151	稳定剂	stabilizer	
08.152	薄膜养生液	membrane curing solution	
08.153	玻璃纤维	glass fiber	
08.154	金属纤维	metal fiber	
08.155	钢纤维	steel fiber	
08.156	碳纤维	carbon fiber	
08.157	聚乙烯管	polyethylene pipe	又称" PE 管"。
08.158	聚氯乙烯管	polyvinyl chloride pipe	又称"PVC 管"。
08.159	聚四氟乙烯板	polyfluortetraethylene plate	
08.160	丁苯橡胶	buna-S, styrene-butadiene rubber, SBR	
08.161	氯丁橡胶	polymeric chloroprene rubber	
08.162	丁钠橡胶	buna	
08.163	工程塑料	engineering plastic	
08.164	土工织物	geotextile	又称"土工布"。
08.165	无纺土工织物	nonweaven geotextile	
08.166	土工膜	geomembrane	
08.167	土工格栅	geogrid	
08.168	聚丙烯筋带	polypropylene belt	
08.169	加热型塑料	hot laid plastic	
08.170	热塑性塑料	thermoplastic	
08.171	道路标线漆	road mark paint	
08.172	玻璃微珠	micro glass bead	
08.173	反光材料	reflective materials	

序 码	汉 文 名	英 文 名	注 释
08.174	钢材涂料	steel coating	
08.175	材料性能	characteristic of materials	
08.176	孔隙率	porosity	
08.177	孔隙比	void ratio	
08.178	饱和孔隙比	zero air void ratio	
08.179	粒径	grain size	
08.180	颗粒组成	grain composition	
08.181	细度	fineness	
08.182	细度模数	fineness modulus	
08.183	比表面	specific surface	
08.184	筛分	sieve analysis	
08.185	级配	gradation	
08.186	密级配	dense gradation	
08.187	断级配	gap gradation	
08.188	粗级配	coarse gradation	
08.189	细级配	fine gradation	
08.190	开级配	open gradation	
08.191	级配曲线	grading curve	
08.192	最佳级配	optimum gradation	
08.193	含水量	water content	
08.194	天然含水量	natural water content	
08.195	相对含水量	relative water content	
08.196	最佳含水量	optimum water content	
08.197	稠度界限	consistency limit	
08.198	触变性	thixotropy	
08.199	比重	specific gravity	
08.200	相对密实度	relative density	
08.201	干密度	dry density	
08.202	最大干密度	maximum dry density	
08.203	干容量	dry unit weight	
08.204	液限	liquid limit, LL	
08.205	塑限	plastic limit, PL	
08.206	塑性指数	plasticity index, PI	
08.207	缩限	shrinkage limit	
08.208	液性指数	liquidity index	
08.209	毛细水上升高度	capillary water height	
08.210	渗透系数	permeability coefficient	

序 码	汉 文 名	英 文 名	注 释
08.211	固结系数	consolidation coefficient	
08.212	径向固结系数	radial consolidation coefficient	
08.213	无侧限抗压强度	unconfined compression strength	
08.214	酸碱度值	pH value	又称"pH值"。
08.215	自由膨胀率	free swelling rate	
08.216	膨胀量	swelling amount	
08.217	膨胀力	swelling force	
08.218	有机质含量	organic matter content	
08.219	内摩擦系数	internal friction coefficient	
08.220	内粘聚力	internal cohesion	
08.221	内摩擦角	internal friction angle	
08.222	沥青稠度	bitumen consistency	
08.223	针入度	penetration	
08.224	针入度指数	penetration index	
08.225	粘滞度	viscosity	
08.226	动力粘滞度	kinematic viscosity	
08.227	软化点	softening point	
08.228	延度	ductility	
08.229	脆点	brittle point	
08.230	闪点	flash point	
08.231	燃点	burning point	
08.232	溶解度	dissolubility	
08.233	热稳性	thermostability	
08.234	加热损失	heating loss	
08.235	水稳性	water stability	
08.236	浮漂度	floatability	
08.237	粘结力	cohesion	
08.238	粘附力	adhesion	
08.239	含蜡量	paraffin content	
08.240	游离碳含量	free carbon content	
08.241	挥发分含量	volatile matter content	
08.242	灰分含量	ash content	
08.243	酚含量	phenol content	
08.244	组分	constituent	
08.245	马歇尔稳定度	Marshall stability	
08.246	马歇尔劲度	Marshall stiffness	
08.247	流值	flow value	

序　码	汉　文　名	英　文　名	注　释
08.248	油石比	bitumen aggregate ratio	
08.249	含油率	bitumen content	
08.250	压碎值	crushing value	
08.251	磨耗度	abrasiveness	
08.252	石料分级	stone classification	
08.253	石材硬度	hardness of stone	
08.254	石料磨光值	polished stone value	
08.255	泊松比	Poisson's ratio	
08.256	水泥标号	cement mark	
08.257	水泥混凝土标号	cement concrete mark	
08.258	水泥混凝土配合比	proportioning of cement concrete	
08.259	水灰比	water cement ratio	
08.260	砂率	sand ratio	
08.261	和易性	workability	又称"工作度"。
08.262	坍落度	slump	
08.263	硬化	hardening	
08.264	水硬性	hydraulicity	
08.265	气硬性	air hardening	
08.266	离析	segregation	
08.267	收缩	shrinkage	
08.268	蠕变	creep	
08.269	老化	ageing	
08.270	饱和度	degree of saturation	

09．材料与结构试验及设备

序　码	汉　文　名	英　文　名	注　释
09.001	含水量试验	water content test	
09.002	密度试验	density test	
09.003	相对密度试验	relative density test	
09.004	比重试验	specific gravity test	
09.005	颗粒分析试验	grain size test	
09.006	液限试验	liquid limit test	
09.007	塑限试验	plastic limit test	
09.008	天然稠度试验	natural consistency test	

序 码	汉 文 名	英 文 名	注 释
09.009	收缩试验	shrinkage test	
09.010	膨胀试验	swelling test	
09.011	透水度试验	pervious test	
09.012	击实试验	compaction test	
09.013	轻型击实试验	Proctor compaction test	又称"普氏击实试验"。
09.014	重型击实试验	modified compaction test, modified Proctor test	又称"修正击实试验"。
09.015	压实度试验	compactness test	
09.016	固结试验	consolidation test	
09.017	三轴试验	triaxial test	
09.018	动态三轴试验	dynamic triaxial test	
09.019	侧限抗压强度试验	confined compression strength test	
09.020	无侧限抗压强度试验	unconfined compression strength test	
09.021	十字板剪切试验	vane shear test	
09.022	差热分析	differential thermal analysis	
09.023	承载板试验	plate-bearing test	
09.024	加州承载比试验	California bearing ratio test, CBR test	又称"CBR 试验"。
09.025	岩土特性指标试验	geotechnical index property test	
09.026	标准贯入试验	standard penetration test, SPT	
09.027	静力触探试验	static cone penetration test	
09.028	动力触探试验	dynamic sounding	
09.029	弯沉试验	deflection test	
09.030	视比重试验	apparent specific gravity test	
09.031	吸水率试验	water absorptivity test	
09.032	压碎值试验	crushing value test	
09.033	含泥量试验	silt content test	
09.034	有机物含量试验	organic matter content test	
09.035	软颗粒含量试验	soft grain content test	
09.036	磨耗试验	abrasion test	
09.037	石料磨光值试验	polished stone value test	
09.038	剥落试验	stripping test	
09.039	饱水率试验	saturated water content test	

序 码	汉 文 名	英 文 名	注 释
09.040	冻融试验	freezing and thawing test	
09.041	韧度试验	rupture test	
09.042	压缩试验	compression test	
09.043	弹性模量试验	elastic modulus test	
09.044	直接剪切试验	direct shear test	
09.045	干湿试验	wetting and drying test	
09.046	拉力试验	tension test	
09.047	冲击韧度试验	impact toughness test	
09.048	洛氏硬度试验	Rockwell hardness test	
09.049	布氏硬度试验	Brinell hardness test	
09.050	冷弯试验	cold bent test	
09.051	热弯试验	hot bent test	
09.052	疲劳试验	fatigue test	
09.053	坍落度试验	slump test	
09.054	稠度试验	consistency test	
09.055	硬练砂浆强度试验	early-dry mortar strength test	
09.056	软练砂浆强度试验	plastic mortar strength test	
09.057	促凝压蒸试验	accelerated setting autoclave test	
09.058	水泥安定性试验	cement soundness test	
09.059	石灰含量测定法	method for determining the lime content	
09.060	钙电极快速测定法	calcium electric rapid determination method	
09.061	针入度试验	penetration test	
09.062	延度试验	ductility test	
09.063	软化点试验[环球法]	softening point test [ringball method]	
09.064	粘滞度试验	viscosity test	
09.065	闪点试验[开口杯法]	flash point test [open cup method]	
09.066	燃点试验	burning point test	
09.067	加热损失试验	heating loss test	
09.068	溶解度试验	dissolubility test	
09.069	蒸馏试验	distillation test	
09.070	浮漂度试验	floatability test	

序　码	汉　文　名	英　文　名	注　　释
09.071	薄膜加热试验	thin-film heating test	
09.072	脆点试验	brittle point test	
09.073	含蜡量试验	paraffin content test	
09.074	组分试验	constituent test	
09.075	粘结力试验	cohesion test	
09.076	游离碳含量试验	free carbon content test	
09.077	挥发分含量试验	volatile matter content test	
09.078	酚含量试验	phenol content test	
09.079	灰分含量试验	ash content test	
09.080	储存稳定度试验	storage stability test	
09.081	车辙试验	wheel rutting test	
09.082	马歇尔稳定度试验	Marshall stability test	
09.083	铺砂法试验	sand patch test	
09.084	[路面]抗滑试验	anti-skid test	
09.085	混凝土流动性试验	concrete fluidity test	
09.086	钢材验收试验	steel acceptance test	
09.087	钢材外观检验	steel visual inspection	
09.088	钢丝冷拔试验	wire cold-drawn test	
09.089	冷拉率	cold-drawn rate	
09.090	[钢筋]焊接接头强度试验	welded joint strength test	
09.091	桥梁验收荷载试验	bridge acceptance loading test	
09.092	桥梁模型试验	bridge model test	
09.093	桥梁模型风洞试验	bridge model wind tunnel test	
09.094	光弹性试验	photoelastic analysis	
09.095	桥梁病害诊断	bridge defect diagnosis	
09.096	桥梁静载试验	bridge static loading test	
09.097	桥梁动载试验	bridge dynamic loading test	
09.098	挠度横向分布测量	transversal distribution measurement of girder deflection	
09.099	桥梁挠度曲线	bridge deflection curve	
09.100	激光准直挠度测量	laser alignment deflection measurement	

序 码	汉 文 名	英 文 名	注 释
09.101	索力测量	cable force measurement	
09.102	[桥梁]动力响应试验	bridge response to forced vibration	
09.103	行车激振	vibration excited by moving truck	
09.104	跳车激振	vibration excited by truck jumping from threshold on deck	
09.105	放松拉索激振	vibration excited by cutting off holding rope	
09.106	落物激振	vibration excited by dropping weight	
09.107	爆破激振	vibration excited by explosive action	
09.108	桥梁自振频率测量	bridge natural frequency measurement	
09.109	桥梁振型分析	bridge vibration mode analysis	
09.110	桥梁脉动测量	bridge pulsation measurement	
09.111	动力放大过程线	dynamic amplification duration curve	
09.112	峰值响应	peak response	
09.113	正响应区	positive response zone	
09.114	负响应区	negative response zone	
09.115	残留响应区	residual response zone	
09.116	滞后响应区	hysteresis response zone	
09.117	余振自谱分析	residual vibration auto-spectrum analysis	
09.118	冲击系数测定	impact factor evaluation	
09.119	受迫振动互谱分析	forced vibration cross-spectrum analysis	
09.120	无损检测	non-destructive test, NDT	
09.121	混凝土强度超声测量	ultrasonic test of concrete strength	
09.122	混凝土回弹试验	rebound test of concrete	
09.123	裂缝声发射检测	crack detecting by acoustic emission	
09.124	混凝土钻孔内窥镜检查	concrete coring hole inspecting by endoscope	
09.125	梁内导管空隙真	duct void examination by vacuum	

序　码	汉　文　名	英　文　名	注　释
	空加压试验	pressure test	
09.126	混凝土氯化物含量测量	concrete chloride content measurement	
09.127	氯化物含量沿深度分布测量	chloride content depth profile measurement	
09.128	混凝土碳化试验	concrete carbonation test	
09.129	碳化深度酚酞试验	carbonation depth by phenolphthalein test	
09.130	钢筋锈蚀活动性评定	bar corrosion activity evaluation	
09.131	钢筋锈蚀三因素模型	three-factor model of bar corrosion	
09.132	钢筋电阻率测量	bar resistivity measurement	
09.133	钢筋电位测量	bar potential measurement	
09.134	桥基沉降观测	bridge foundation settlement observation	
09.135	试桩	test pile	
09.136	桩轴向荷载试验	axial loading test of pile	
09.137	桩横向荷载试验	lateral loading test of pile	
09.138	桩贯入试验	pile penetration test	
09.139	桩完整性试验	pile integrity test	
09.140	钻孔直径检测	bored hole diameter measurement	
09.141	钻孔垂[直]度检测	bored hole verticality measurement	
09.142	[桩]动测法	dynamic measurement of pile	
09.143	钻孔泥浆试验	boring slurry test	
09.144	工程地质条件分析	engineering geologic condition analysis	
09.145	桥址稳定性评定	bridge site stability evaluation	
09.146	桥基稳定性评定	bridge foundation stability evaluation	
09.147	地震危险性评定	seismic risk evaluation	
09.148	桥址断层活动性评定	fault activity evaluation of bridge site	
09.149	液塑限联合测定仪	liquid-plastic combine tester	
09.150	击实仪	compaction test apparatus	

序　码	汉　文　名	英　文　名	注　释
09.151	核子湿度密度仪	nuclear moisture and density meter	
09.152	核子静态湿度密度仪	nuclear moisture and density static meter	
09.153	核子动态湿度密度仪	nuclear moisture and density dynamic meter	
09.154	贯入仪	penetration test apparatus	
09.155	土圆锥仪	soil cone penetrator	
09.156	固结仪	consolidometer	
09.157	维卡稠度仪	Vicat apparatus	
09.158	承载板	bearing plate	
09.159	路面曲率仪	surface-curvature apparatus	
09.160	路面平整度测定仪	viameter	
09.161	路面透水度测定仪	surface permeameter	
09.162	直剪仪	direct shear apparatus	
09.163	三轴剪切仪	triaxial shear equipment	
09.164	标准筛	standard sieves	
09.165	沥青抽提仪	bitumen extractor	
09.166	砂浆稠度仪	mortar consistency tester	
09.167	坍落度圆锥筒	slump cone	
09.168	冲击韧度试验仪	impact toughness apparatus	
09.169	耐磨硬度试验仪	wear hardness testing apparatus	
09.170	狄法尔磨耗试验机	Deval abrasion testing machine	
09.171	洛杉矶磨耗试验机	Los Angeles abrasion testing machine	
09.172	石料加速磨光仪	accelerated stone polishing tester	
09.173	道路几何数据收集系统	road geometry data acquisition system	
09.174	路面激光测试仪	laser road surface tester	
09.175	路面病害摄影组合仪	photographic road survey group	
09.176	摩擦系数测定仪	friction tester	
09.177	滑溜测量仪	skidometer	
09.178	摆式仪	portable pendulum tester	
09.179	横向力系数测试	sideway force coefficient routine	

序 码	汉 文 名	英 文 名	注 释
	仪	investigation machine, SCRIM	
09.180	颠簸累积式平整仪	bump-integrator roughometer trailer	
09.181	车载式颠簸累积仪	vehicular bump-integrator	
09.182	纵断面分析仪	longitudinal profile analyzer	
09.183	构造深度仪	texture meter	
09.184	手推式构造深度仪	minitexture meter	
09.185	高速构造深度仪	high speed texture meter	
09.186	弯沉仪	deflectometer, Benkelman beam	又称"贝克曼梁"。
09.187	自动弯沉仪	autodeflectometer	
09.188	落锤式弯沉仪	falling weight deflectometer, FWD	
09.189	动力式弯沉仪	dynaflect	
09.190	路面快速加载试验机	accelerated loading facility, ALF	
09.191	试验路	test road	
09.192	试验环道	test loop road	
09.193	长柱试验机	long column testing machine	
09.194	光测弹性试验装置	photoelastic test installation	
09.195	结构振动试验台	shaking table for structure test, vibrostand	
09.196	加载反力架	reaction frame for loading	
09.197	拖车式移动试验车	trailer-typed mobile laboratory vehicle	
09.198	桥梁测试车	bridge testing laboratory vehicle	
09.199	自行式桥梁检测架	self-propelled bridge inspection cradle	
09.200	加载平台	weighted platform	
09.201	激振器	vibration excitor	
09.202	扁千斤顶	flat jack	
09.203	液压稳定器	hydraulic pressure stabilizer	
09.204	应变计	strain gage	
09.205	测力柱	load measurement column	
09.206	压力盒	pressure cell	
09.207	索力测定计	cable tension measurement device	

序　码	汉　文　名	英　文　名	注　　释
09.208	桥面单点加载装置	single point loading device on deck	
09.209	加载水箱	water tank for loading	
09.210	桥面平整度测定仪	travelling-beam testing device for deck surface irregularity	
09.211	回弹仪	rebound tester, Schmidt hammer	
09.212	频率仪	frequency recorder	
09.213	超声脉冲测量仪	ultrasonic pulse velocity measurement device	
09.214	磁性测裂计	magnetic crack detector	
09.215	光纤测裂计	optic fiber sensor for crack monitor	
09.216	电位测量装置	potential test device	
09.217	电位轮	potential test wheel	
09.218	电阻率测量装置	resistivity test device	
09.219	半电池	half cell	
09.220	酸度计	pH meter	
09.221	声发射测量系统	acoustic emission system	
09.222	内窥镜	endoscope	
09.223	基桩病害检测系统	foundation pile diagnosis system	
09.224	打桩分析仪	pile driving analyzer, PDA	
09.225	测斜仪	inclinometer	
09.226	电子水平仪	electrolevel	
09.227	磁感式沉降标	magnetic tell tale	
09.228	孔隙水压力计	pore pressure cell	
09.229	十字板剪力仪	vane-shear apparatus	
09.230	液压静力触探仪	hydraulic static cone penetrometer	
09.231	标贯装置	standard penetration equipment	
09.232	岩心取样器	core sampler	
09.233	薄壳取样器	thin-walled tube sampler	
09.234	钻孔照相机	borehole camera	
09.235	钻孔潜望镜	borehole periscope	
09.236	钢筋锈蚀测定计	bar corrosion activity indicator	
09.237	钢筋保护层测定仪	cover protectometer	
09.238	压电式传感器	piezoelectric transducer	
09.239	拾震器	geophone	

序　码	汉　文　名	英　文　名	注　释
09.240	电阻应变仪	electric resistance strain gauge	
09.241	应变巡检箱	strain scanning unit	
09.242	温度巡检箱	temperature scanning unit	
09.243	可编程序数据采集器	programmable data logger	
09.244	多通道数字记录系统	multi channel digital record system	
09.245	信号处理机	signal processor	
09.246	快速傅里叶变换分析仪	fast Fourier transform analyzer, FFT analyzer	又称"FFT 分析仪"。
09.247	通视能见度检验仪	all-through visibility tester	
09.248	通视浑浊度仪	all-through turbidimeter	

10．公路环境保护

序　码	汉　文　名	英　文　名	注　释
10.001	公路环境	highway environment	
10.002	环境污染	environmental pollution	
10.003	环境损害	environmental hazard	
10.004	公路交通公害	highway traffic nuisance	
10.005	公路施工环境影响	environmental effects in highway construction	
10.006	公路环境保护设计	highway environmental protection desing	
10.007	公路环境影响评价	highway environmental impact evaluation	
10.008	大气环境	air environment	
10.009	光化学烟雾	photochemical smog	
10.010	汽车排放物	vehicle emission	
10.011	施工机械排放物	construction machine emission	
10.012	废气排放标准	exhaust emission standard	
10.013	怠速排放污染物	pollutants of idle speed emission	
10.014	蒸发污染物	evaporative pollutants	
10.015	曲轴箱排放物	crankcase emissions	
10.016	铅粒	particulate lead	

序　码	汉　文　名	英　文　名	注　释
10.017	烟浓度	smoke concentration	
10.018	烟粒	soot	
10.019	柴油机排烟	diesel smoke	
10.020	烟度排放标准	smoke emission standard	
10.021	博什烟度	Bosh smoke unit	
10.022	微粒物	particles	
10.023	总悬浮微粒	total suspended particles	
10.024	一氧化碳监测仪	carbon monoxide detector	
10.025	一氧化碳允许浓度	carbon monoxide allowable concentration	
10.026	正己烷当量	hexane equivalent	
10.027	梅氏曲线	May-curve	
10.028	废气滤清器	exhaust scrubber	
10.029	汽车排放分析仪	vehicle exhaust analyser	
10.030	排气净化器	exhaust purifier	
10.031	隧道排气污染	tunnel's discharge pollution	
10.032	农作物蒙尘	crops covering by dust	
10.033	防尘设施	dustproof facilities	
10.034	水质	water quality	
10.035	水质污染	water pollution	
10.036	工程废水	construction waste water	
10.037	施工废弃物	construction waste	
10.038	消冰剂污染	deicing agent pollution	
10.039	噪声容限	noise margin	
10.040	噪声污染	noise pollution	
10.041	噪声检定值	noise rating number, NRN	
10.042	噪声等级	noise level	
10.043	噪声等值线	noise isoline	
10.044	噪声系数	noise factor, NF	
10.045	背景噪声	background noise	又称"本底噪声"。
10.046	环境噪声标准	environmental noise standard	
10.047	交通噪声	traffic noise	
10.048	交通噪声指数	traffic noise index, TNI	
10.049	交通噪声评价	traffic noise evaluation	
10.050	车辆噪声	vehicle noise	
10.051	车辆允许噪声标准	allowable vehicle noise standard	

序　码	汉　文　名	英　文　名	注　释
10.052	排气噪声	exhaust noise	
10.053	排气消声器	exhaust muffler	
10.054	发动机噪声	motor noise	
10.055	轮胎噪声	tyre noise	
10.056	施工机械噪声	construction machinery noise	
10.057	夜间施工噪声	night construction noise	
10.058	防噪设施	acoustic treatment facilities	
10.059	隔音墙	acoustic barrier	
10.060	隔声道	sound isolate road	
10.061	无噪声打桩技术	noiseless piling technique	
10.062	无噪声路面	noiseless pavement	
10.063	隔声绿化带	sound isolate greenbelt	
10.064	振动	vibration	
10.065	环境振动标准	environmental vibration standard	
10.066	交通振动	traffic vibration	
10.067	施工机械振动	construction machinery vibration	
10.068	爆破振动	blasting vibration	
10.069	减振措施	vibration reducing measure	
10.070	隔振沟	vibration isolate ditch	
10.071	水土保持	water and soil conservation	
10.072	水土流失	erosion loss	
10.073	边坡冲刷防治	slope erosion treatment	
10.074	路基冲刷防治	subgrade erosion protection	
10.075	废方处理	waste bank treatment	
10.076	边坡稳定	slope stabilization	
10.077	植被固坡	planting protect slope	
10.078	植物固沙	sand consolidation by planting	
10.079	生物固沙	sand consolidation with biologic	
10.080	生态	ecology	
10.081	生态环境	ecology environment	
10.082	植物保护	plant protection	
10.083	野生动物保护区	wild animals refuge area	
10.084	保护野生动物标志	wild animals protection sign	
10.085	弃土反耕	waste bank refarming	
10.086	文物保护范围	cultural relic protection scope	
10.087	汽车电磁波干扰	automobile electro-magnetic inter-	

序　码	汉　文　名	英　文　名	注　释
		fere	
10.088	隔离障碍	sovereignty nuisance	
10.089	废车堆积场	auto-wrecking yard	

英 汉 索 引

A

AADT volume 年平均日交通量 03.058

abandon sweeper 抛掷清扫车 07.224

abnormal load 非常荷载 04.480

abnormal noise 异响 05.198

abrasion test 磨耗试验 09.036

abrasiveness 磨耗度 08.251

absolute datum 绝对基面 01.208

absorbed traffic [volume] 吸引交通量 03.126

abutment 桥台 04.380

abutment body 台身 04.381

abutment cap 台帽 04.362

abutment coping 台帽 04.362

abutment pier 制动墩 04.373

accelerated loading facility 路面快速加载试验机 09.190

accelerated setting autoclave test 促凝压蒸试验 09.057

accelerated stone polishing tester 石料加速磨光仪 09.172

accelerated test 加速试验 07.392

accelerating-coasting 加速滑行 05.109

accelerating mode 加速工况 05.122

acceleration-deceleration delay 加减速延误 03.098

acceleration lane 加速车道 02.170

acceleration noise 加速噪声 03.025

acceleration test 加速性能试验 07.423

acceptance 验收 01.146

acceptance certificate 验收证书 01.147

acceptance gap 可插车空档 03.026

acceptance of consignment 承运 06.178

acceptance of luggage consignment 行包承运 06.089

acceptance test 验收试验 07.403

accidental load 偶然荷载 04.479

[accident] black spot 事故易发地点 03.220

accident handling of machinery 机械事故处理 07.342

accident spot map 事故地点图 03.218

accommodation highway 专用公路 02.034

accommodation lane 专用车道 02.182

accompanying parameters of operation process 伴随过程参数 05.168

accumulative equivalent axles 累计当量轴次 02.382

acess adit 服务隧道，*辅助隧道 04.837

acoustic barrier 隔音墙 10.059

acoustic emission system 声发射测量系统 09.221

acoustic treatment facilities 防噪设施 10.058

actions for freight claims 货运理赔 06.226

actions for passenger claims 客运理赔 06.115

active safety 主动安全性 05.142

adaptation lighting 缓和照明，*适应照明 03.456

adaptation of vision 视力适应性 03.527

adaptive traffic signal control system 自适应信号控制系统 03.268

additional stake 加桩 01.192

additional ticket 补充客票 06.098

additive 外加剂 08.140

adhesion 粘附 05.181，粘附力 08.238

ADIS 先进的驾驶员信息系统 03.319

adjustable track trailer 横伸挂车 05.046

adjustable-wheelbase trailer 纵伸挂车 05.045

adjusting crown 调整路拱 02.577

adjusting gradient 调坡 02.576

administration of highway transport 公路运输行政管理 06.397

administration of highway transport industry 公路运输行业管理 06.398

admixture 掺加料 08.141

ADT 平均日交通量 03.064

advanced driver information system 先进的驾驶员信

息系统 03.319

advanced public transportation system 先进的城市
[公共]运输系统 03.320

advanced rural transportation system 先进的长途运
输系统 03.321

advanced traffic management system 先进的交通管
理系统 03.317

advanced vehicle control system 先进的汽车控制系
统，*智能汽车控制系统 03.318

advance exit sign 出口预告标志 03.405

advance mobilization loan 动员费预付贷款 01.134

advance sign 前置标志 03.369

advisory speed 推荐速度 03.072

aerial photogrametry 航空摄影测量 01.232

aerodynamic stability 空气动力稳定性 04.585

aerophoto base line 航摄基线 01.235

aerophoto interpretation 航摄像片判读 01.238

A-framed tower A形索塔 04.324

ageing 老化 08.269

agency 代办站 06.127

agent 代理人 01.104

aggregate 集料，*骨料 08.048

aggregate feeder units 配料给料装置 07.134

aggregate passenger traffic 集结性客流 06.044

aggregate paver 石料摊铺机 07.121

aggregate stabilized soil base 粒料稳定土基层
02.498

aggregate treated earth road 粒料改善土路 02.434

aggregation model 集合模型 03.013

agreement 协议书 01.125

agricultural truck 农用运输汽车 05.014

air compressor 空气压缩机 07.306

air conveyer 气力输送机 07.107

air entraining agent 加气剂 08.146

air-entraining concrete 引气混凝土 08.109

air environment 大气环境 10.008

air hardening 气硬性 08.265

air-jet snow remover 喷气式除雪机 07.219

air-rider jack hammer 气腿式凿岩机 07.022

ALF 路面快速加载试验机 09.190

alignment coordination 线形协调 02.053

alignment elements 线形要素 02.050

alkali-aggregate reaction 碱性集料反应 02.635

all casing drill 全套管钻机 07.255

alligator crack 龟裂 02.603

all-life period management of road machine 筑路机械
全过程管理 07.317

allowable load 容许荷载 04.470

allowable rebound deflection value 容许回弹弯沉值
02.393

allowable stress 容许应力，*许用应力 04.551

allowable stress design of bridge 桥梁容许应力设计
04.505

allowable vehicle noise standard 车辆允许噪声标准
10.051

all-personnel labour productivity 全员劳动生产率
06.392

all red signal 全红信号 03.250

all-round pressurized shield 全气压盾构 04.941

all-through turbidimeter 通视浑浊度仪 09.248

all-through visibility tester 通视能见度检验仪
09.247

alluvial soil 冲积土 01.297

all-weather highway 晴雨通车公路 02.041

alternate system 交互式系统 03.260

alternative line 比较线 02.067

alternative motor fuels 汽车代用燃料 05.160

anchorage 锚碇 04.334，锚具，*工作锚具
04.639

anchorage zone 锚固区 04.651

anchor block 锚碇体 04.335

anchor body 锚碇体 04.335

anchor bolt support 锚杆支护 04.893

anchor cable 锚索 04.329

anchored bulkhead abutment 锚碇板式桥台
04.394

anchored bulkhead retaining wall 锚锭板式挡土墙
02.274

anchor in rock gallery 山洞式锚碇 04.338

anchor pier 锚墩 04.376

anchor pile 锚桩 04.416

anchor plate 锚碇板 04.336

anchor socket 锚座 04.326

anchor span 锚[碇]跨 04.333

ancient drive way 驰道 02.017

angle steel edged joint 角钢镶边接缝 04.306

angular boulder　块石　08.066

angular cobble　碎石　08.068

angular gravel　角砾　08.075

animal corridor　动物通道　03.494

anionic emulsified bitumen　阴离子乳化沥青
08.014

annual average daily traffic volume　年平均日交通量
03.058

annual maximum hourly traffic volume　年最大小时
交通量　03.057

annual mean temperature difference　年均温差
04.590

annual output per vehicle　单车产量　06.376

annual output per vehicle tonnage capacity　车吨产量
06.377

annual thirtieth highest hourly traffic volume　年第三
十位最大小时交通量　03.060

anti-ager　抗老化剂　08.144

anti-collision pier　抗撞墩　04.375

antifreezing agent　防冻剂　08.147

anti-glare panel　防眩板　03.478

anti-glare screen　防眩屏　03.479

anti-knock block　防震挡块　04.379

anti-skid test　[路面]抗滑试验　09.084

anti-skid treatment　防滑处理　02.669

anti-slide pile　抗滑桩　04.432

anti-stripping agent　抗剥落剂　08.143

apparent specific gravity test　视比重试验　09.030

apparent velocity　表观流速　04.056

approach　引道　02.255

approach bridge　引桥　04.008

approach nose　路岛端，＊接近端　02.220

approach span　引桥　04.008

approach to ferry　码头引道　04.956

approach velocity　行近流速　04.057

approximate estimate norm of highway project　公路
工程概算定额　01.078

apron　护坦　04.140

APTS　先进的城市[公共]运输系统　03.320

aqueduct　渡槽　04.824

arbitration　仲裁　01.157

arch　拱　04.278

arch bridge　拱桥　04.215

arch center　拱架　04.686

arch centering　拱架　04.686

arch crown　拱顶　04.342

arch crown block　拱顶石　04.702

arch culvert　拱涵　04.806

arched abutment　拱形桥台　04.389

arched roof in advance of wall tunneling method　先拱
后墙法　04.877

arch rib　拱肋　04.341

arch ring　拱圈，＊拱环　04.340

arch ring stress adjustment　拱圈应力调整　04.681

arch skew block　拱脚斜块　04.703

arch support　拱座　04.343

arch thrust　拱推力　04.498

arc welder　弧焊机　07.304

arterial highway　干线公路　02.035

articulated roller　铰接式压路机　07.069

articulated truck　铰接式自卸车　07.012

artificial aggregate　人造集料　08.054

artificial lighting transition　人工照明过渡　03.463

artificial snow slide　人造雪崩　02.685

ARTS　先进的长途运输系统　03.321

aseismatic design of bridge　桥梁抗震设计　**04.515**

aseismatic stability of bridge　桥梁抗震稳定性
04.516

ash content　灰分含量　08.242

ash content test　灰分含量试验　09.079

ashlar　料石　08.070

ashlar arch bridge　料石拱桥　04.197

asphalt　沥青　08.003，　地沥青　08.004，
石油沥青　08.005

asphalt barreled melter　桶装沥青熔化装置　07.173

asphalt cement　粘稠沥青　08.021

asphalt distribution　喷洒沥青　02.446

asphalt distribution measurement　沥青洒布量测定
07.475

asphalt distributor test　沥青洒布机试验　07.473

asphalt emulsion　乳化沥青，＊沥青乳液　08.008

asphalt firepipemelter　火管式沥青熔化装置
07.174

asphalt flow meter　沥青流量计　07.151

asphalt hot oil melter　导热油沥青熔化装置
07.176

asphaltmastic 沥青胶砂，＊沥青玛琋脂 08.095

asphalt melting and heating device 沥青熔化加热装置 07.172

asphalt mixing plant 沥青混合料搅拌设备 07.124

asphalt mixture recycling mixing plant 沥青混合料再生拌和设备 07.212

asphalt modifier 沥青改性剂 08.142

asphalt overlay 沥青罩面 02.670

asphalt pavement flame heater 沥青路面火焰加热器 07.209

asphalt pavement heater 沥青路面加热机 07.208

asphalt pavement infrared heater 沥青路面红外线加热器 07.210

asphalt pavement *in-situ* recycling machine 沥青路面就地再生机 07.211

asphalt paver 沥青混凝土摊铺机 07.157

asphalt paver test 沥青混凝土摊铺机试验 07.476

asphalt pump 沥青泵 07.164

asphalt slurry 沥青稀浆 08.096

asphalt slurry seal machine 稀浆封层机 07.234

asphalt solar energy melter 太阳能沥青熔化装置 07.177

asphalt sprayer 沥青喷洒机 07.162

asphalt steam pipe melter 蒸气式沥青熔化装置 07.175

asphalt storage 沥青贮仓 07.168

asphalt tank with heating device 沥青保温罐 07.144

asphalt truck 沥青罐车 07.167

asphalt weigher 沥青秤 07.143

assembling of steel bridge 钢桥拼装 04.738

asymptotic stability 渐近稳定性 03.024

at-grade intersection 平面交叉 02.199

ATMS 先进的交通管理系统 03.317

attachment cost 附具摊销费 07.336

autodeflectometer 自动弯沉仪 09.187

automatic compactometer 压实度自动检测装置 07.371

automatic control 自动控制 07.361

automatic electronic ticket system 电子自动售票系统 06.099

automatic hydraulic proportioning system 液压比例自控系统 07.364

automatic leveling device 自动找平装置 07.366

automatic operation 自动操作 07.360

automatic power apportioning 功率自动分配 07.363

automatic temperature control 温度自动控制 07.365

automatic underbody washer 汽车底部自动清洗机 05.364

automatic water spraying fire extinguisher 自动喷水灭火机 04.946

automatic weighing 自动称重 07.362

automobile 汽车 05.004

automobile driving 汽车驾驶 05.086

automobile electro-magnetic interfere 汽车电磁波干扰 10.087

autoset road marking 自动划线装置 07.370

auto-wrecking yard 废车堆积场 10.089

auxiliary handling charge 辅助装卸费 06.497

auxiliary handling operation 辅助装卸作业 06.295

auxiliary lane 附加车道，＊辅助车道 02.179

auxiliary pier 辅助墩 04.377

auxiliary road 辅道 02.032

auxiliary sign 辅助标志 03.358

availability rate of machinery 机械完好率 07.328

available vehicle-days 完好车日 06.350

avalanche 碎落 01.265

AVCS 先进的汽车控制系统，＊智能汽车控制系统 03.318

average costs of vehicle maintenance and repair 汽车维修平均费用 05.316

average daily traffic volume 平均日交通量 03.064

average days during major repair of vehicle 汽车大修平均在修车日 05.314

average days in plant during major repair of vehicle 汽车大修平均在厂车日 05.313

average gradient 平均纵坡 02.115

average man-hours of vehicle maintenance and repair 汽车维修平均工时 05.315

average mileage between vehicle major repairs 汽车大修间隔里程 05.067

average number of vehicles 平均车数 06.361

average seats per vehicle 每车平均客位 06.365

average speed 平均车速 03.074

average storage days [货物]平均堆存天数 06.332

average tonnage per vehicle 每车平均吨位 06.364

average total seats 平均总客位 06.363

average total tonnage 平均总吨位 06.362

average vehicle daily travel 平均车日行程 06.368

avoidance of mixed loading 忌装 06.157

award of contract 中标 01.121

axial-flow blower 轴流通风机 07.315

axial loading test of pile 桩轴向荷载试验 09.136

axle load 轴载 01.090

axle load meter 轴重仪 05.330

axle weight detector 轴重检测器 03.309

azimuth 方位角 01.198

B

back fill 回填土 02.294

back filling behind abutment [桥]台后回填 04.397

back fire 回火 05.212

background noise 背景噪声,＊本底噪声 10.045

back haunch fillet of arch 护拱 04.347

backhoe excavator 反铲挖掘机 07.040

backing 倒车 05.094

backward wave in traffic flow 车流返回波 03.018

back-water 壅水 04.105

back-water at pier 墩前壅水 04.107

back-water height 壅水高度 04.108

back-water in front of bridge 桥前壅水 04.106

bad technical condition of vehicle 汽车不良技术状况 05.164

baffle sill 消能槛 04.139

bag dust collector 布袋除尘器 07.132

Bailey bridge 贝雷桥 04.260

balance weight retaining wall 衡重式挡土墙 02.269

ballast loader 清岩机 07.034

bank erosion 堤岸冲刷 04.112

bank protection 护岸 04.136

bank revetment 护岸 04.136

bar corrosion activity evaluation 钢筋锈蚀活动性评定 09.130

bar corrosion activity indicator 钢筋锈蚀测定计 09.236

bare surface 露骨 02.610

bar potential measurement 钢筋电位测量 09.133

bar resistivity measurement 钢筋电阻率测量 09.132

barrier curb 栏式缘石 02.154

bascule bridge 竖旋桥 04.254

base course 基层 02.354

base line 基线 01.181

basic capacity 基本通行能力 03.164

basic earthquake intensity 地震基本烈度 04.517

basic part repair 基础件修理 05.286

basic rate 基本运价 06.455

basin 盆地 01.227

batch asphalt mixing plant 间歇式沥青混合料搅拌设备 07.127

bateau bridge 舟桥 04.248

battered pile 斜桩 04.415

BBRV anchorage 镦头锚,＊BBRV锚 04.648

beam 梁 04.262

beam bridge 梁桥 04.203

beam crane 梁式起重机 07.290

beam depth 梁高 04.026

bearing 支座 04.444

bearing capacity of pile 单桩承载力 04.596

bearing capacity of subsoil 地基承载力 04.595

bearing pile 支承桩 04.436

bearing plate 承载板 09.158

bearing platform 承台 04.440

bearing stratum 持力层 01.330

bearing stress 支撑应力 04.549

bed course 垫层 02.356

bed material load 推移质 04.123

bedrock 基岩 01.314

behind schedule 晚点 06.072

belt conveyer 带式输送机 07.100

belt feeder 皮带给料机 07.090

belt topographic map 带状地形图 01.215

benched section 台口式断面 02.129

benched subgrade　台口式路基　02.252

bench mark　水准点　01.206

bench simulation test　台架模拟试验　07.381

bench test　台架试验　07.372

bend-ended spur dike　勾头丁坝　04.132

bending moment　力矩，＊弯矩　04.557

benefit from accidents reducing　减少交通事故效益　01.060

benefit from distance shortening　缩短里程效益　01.059

Benkelman beam　弯沉仪，＊贝克曼梁　09.186

bent cap　盖梁，＊帽梁　04.363

berm　护坡道　02.260

bid　投标书　01.109

bid bond　投标担保　01.111

bidding　投标　01.100

bidding documents　招标文件　01.108

bid opening　开标　01.120

bid validity　投标书有效期　01.113

bike lane　自行车车道　02.195

bike-way　自行车道　02.010

bill of lading　货物运单，＊提单　06.181

bill of quantities　工程量清单　01.132

bi-moment　双力矩　04.559

binder　结合料　08.001

binder course　联结层　02.340

biorhythms　生物节律　03.499

bi-purposed bridge　公铁两用桥　04.148

bits grinders　磨钻机　07.031

bitumen　沥青　08.003

bitumen aggregate ratio　油石比　08.248

bitumen consistency　沥青稠度　08.222

bitumen content　含油率　08.249

bitumen electrical heating device　电热式沥青加热装置　07.186

bitumen emulsifying machine　沥青乳化机　07.178

bitumen emulsifying plant　沥青乳化设备　07.182

bitumen extractor　沥青抽提仪　09.165

bituminous concrete mixture　沥青混凝土混合料　08.092

bituminous concrete pavement　沥青混凝土路面　02.416

bituminous macadam mixture　沥青碎石混合料　08.093

bituminous macadam pavement　沥青碎石路面　02.417

bituminous mixture　沥青混合料　08.091

bituminous pavement　沥青路面　02.415

bituminous penetration pavement　沥青贯入式路面　02.418

bituminous surface treatment　沥青表面处治　02.420

black cotton soil　黑棉土　01.282

black-hole effect　黑洞效应　03.469

blade plough soil stabilizer　犁式稳定土搅拌机　07.117

blade plough stabilized soil road-mixer　犁式稳定土搅拌机　07.117

blasting bulldozer　频爆式推土机　07.004

blasting crater　爆破漏斗　02.326

blasting operation　爆破作业　02.325

blasting vibration　爆破振动　10.068

bleeding　泛油　02.611

blind curve　视线不良弯道　02.105

blind ditch　盲沟　02.532

blind drain　盲沟　02.532

block out type safety fence　阻挡式护栏　03.443

block pavement　块料路面　02.424

block stone　块石　08.066

block stone arch bridge　块石拱桥　04.198

blood-alcohol concentration　血中酒精浓度　03.498

blower　鼓风机　07.149

blowing-in shaft　送风竖井　04.916

blowing-out shaft　排风竖井　04.917

blow up　拱胀　02.630

BM　水准点　01.206

boat pole　撑杆　04.969

body and frame measure and correct system　［轿车］车身测量整形机　05.367

body and frame straightener　车身矫正机　05.368

bolted and welded steel bridge　栓焊钢桥　04.192

bolted steel bridge　栓接钢桥　04.191

bolt final twisting by torque method　螺栓扭矩终拧法　04.740

bolt final twisting by twist angle method　螺栓扭角终拧法　04.741

bolting 栓接 04.736

bonded tendon 粘结的预应力筋 04.626

BOQ 工程量清单 01.132

bored hole diameter measurement 钻孔直径检测 09.140

bored hole verticality measurement 钻孔垂[直]度检测 09.141

bored pile 钻孔桩 04.425

borehole camera 钻孔照相机 09.234

borehole periscope 钻孔潜望镜 09.235

boring-and-underreaming method 钻孔扩端法 04.752

boring log 钻孔柱状图 01.311

boring slurry test 钻孔泥浆试验 09.143

boring test 钻探 01.253

borrow earth 借土 02.290

borrow pit 取土坑 02.292

Bosh smoke unit 博什烟度 10.021

bottom-dump trailer 底卸式挂车 05.042

bottom sealing by concreting 混凝土封底 04.772

boulder 漂石 08.065

boundary line of road construction 道路建筑限界 01.347

boundary sign 分界标志 03.361

bowstring arch bridge 系杆拱桥 04.228

box culvert 箱涵 04.804

box girder 箱梁 04.264

box-girder bridge 箱[形]梁桥 04.212

box-girder fence 箱梁型护栏 03.434

box type abutment 箱形桥台 04.388

bracket 托架, *牛腿 04.540

brake distance 制动距离 03.517

brake drum lathe 制动鼓车床 05.356

brake reaction time 制动反应时间 03.516

brake shoe grinder 制动蹄磨床 05.357

brake tester 制动试验台 05.327

braking 制动 05.095

braking ability test 制动性能试验 07.424

braking force 制动力 04.491

braking pier 制动墩 04.373

branch highway 支线公路 02.036

breakdown on the way 抛锚 05.103

break-even distance for highway transport 公路运输经济运距 06.346

breaking-in 磨合 05.298

break-water 防波堤 04.133

brick pavement 砖块路面 02.425

bridge 桥[梁] 04.146

bridge acceptance loading test 桥梁验收荷载试验 09.091

bridge aesthetics 桥梁美学 01.023

bridge approach 桥头引道 04.356

bridge approach to ferry 码头引桥 04.957

bridge aseismatic strengthening 桥梁抗震加固 04.798

bridge collapse 桥梁破坏 04.606

bridge conceptual design 桥梁方案设计 04.462

bridge construction 桥梁施工 04.660

bridge data bank 桥梁数据库 04.782

bridge deck system 桥面系 04.287

bridge deck width 桥面宽度 04.019

bridge defect diagnosis 桥梁病害诊断 09.095

bridge deflection curve 桥梁挠度曲线 09.099

bridge design 桥梁设计 04.461

bridge detail design 桥梁细部设计 04.464

bridge disaster by flood 桥梁水毁 04.795

bridge durability 桥梁耐久性 04.594

bridge dynamic loading test 桥梁动载试验 09.097

bridge end backfilling design 桥头回填设计 04.522

bridge end transition slab 桥头搭板 04.357

bridge engineering 桥梁工程学 01.007

bridge erection by floating method 浮运架桥法 04.674

bridge erection by swinging method 转体架桥法 04.673

bridge erection monitoring 桥梁安装监测 04.775

bridge erection tolerance 桥梁安装容许误差 04.780

bridge evaluation system 桥梁评价系统 04.784

bridge failure 桥梁破损 04.605

bridge fault repairing 桥梁病害整治 04.792

bridge foundation 桥梁基础 04.398

bridge foundation settlement observation 桥基沉降观测 09.134

bridge foundation stability evaluation 桥基稳定性评

定 09.146

bridge head 桥头堡 04.320

bridge in danger 危桥 04.164

bridge inspection 桥梁检查 04.787

bridge inspection category 桥梁检查类别 04.789

bridge inspection cycle 桥梁检查周期 04.790

bridge inspection regulation 桥梁检查规则 04.788

bridge load limit 桥梁限载 01.355

bridge maintenance 桥梁养护 04.781

bridge management system 桥梁管理系统 04.783

bridge model test 桥梁模型试验 09.092

bridge model wind tunnel test 桥梁模型风洞试验 09.093

bridge monitoring system 桥梁监测系统 04.785

bridge name 桥名 04.003

bridge natural frequency measurement 桥梁自振频率测量 09.108

bridge numbering 桥梁编号 04.791

bridge on slope 坡桥 04.161

bridge opening contraction 桥孔压缩 04.104

bridge overall planning 桥梁总体规划 04.001

bridge pulsation measurement 桥梁脉动测量 09.110

bridge pylon 桥塔 04.321

bridge railing 桥梁栏杆 04.319

bridge response to forced vibration [桥梁]动力响应试验 09.102

bridge retrofitting 桥梁翻新 04.799

bridge shallow foundation protection 桥梁浅基防护 04.793

bridge sidewalk 桥侧人行道 04.318

bridge site 桥位 04.004

bridge site selection 桥位选择 04.005

bridge site stability evaluation 桥址稳定性评定 09.145

bridge site survey 桥位勘测 01.302

bridge static loading test 桥梁静载试验 09.096

bridge strengthening 桥梁加固 04.797

bridge structure design 桥梁结构设计 04.463

bridge structure erection control 桥梁结构安装控制 04.776

bridge technical file 桥梁技术档案 04.786

bridge testing laboratory vehicle 桥梁测试车 09.198

bridge toll 过桥费 01.033

bridge tower 桥塔 04.321

bridge type 桥型 04.002

bridge vibration mode analysis 桥梁振型分析 09.109

bridging equipment 架桥设备 07.276

bridle path 驮道 02.013

Brinell hardness test 布氏硬度试验 09.049

brittle point 脆点 08.229

brittle point test 脆点试验 09.072

broken back curve 断背曲线 02.086

broken chain 断链 01.193

broken stone 碎石 08.068

broom-finish 拉毛 02.479

bucket elevator 斗式提升机 07.102

bucket wheel excavator 斗轮式挖掘机 07.048

bucket wheel trencher 斗轮式挖沟机 07.052

buckling 压曲, *压屈 04.566

budgetary norm of highway project 公路工程预算定额 01.079

buffeting 抖振, *击振 04.571

bulb-end anchorage 镦头锚, *BBRV 锚 04.648

bulk goods 散装货物 06.136

bulk tanker trailer 粉状货挂车 05.043

bulldozer 推土机 07.001

bumper 防撞垫, *缓冲垫 04.967

bump-integrator roughometer trailer 颠簸累积式平整仪 09.180

buna 丁钠橡胶 08.162

buna-S 丁苯橡胶 08.160

bundle of steel wire 钢丝束 08.129

buoyancy 浮力 04.496

buried abutment 埋置式桥台 04.391

buried culvert 暗涵 04.803

buried depth of tunnel 隧道埋深 04.849

buried joint 埋式接缝 04.304

burner 喷燃器 07.148

burning 烧伤 05.183

burning point 燃点 08.231

burning point test 燃点试验 09.066

bus 客车 05.015

bus capacity utilization rate 载客量利用率 06.372

bus dispatching 客车调度 06.073

business certificate 经营许可证 06.407

bus run cycle 客车运行周期 06.074

bus running record 客运动态记录单 06.102

bus schedule 客运班期 06.068

bus service quality 旅客运输质量 06.103

bus station 公共汽车站 02.186

bus stop 停靠站 06.126

bus streamlined service 旅客一条龙服务 06.104

bus terminal 公共汽车终点站 02.187

bus ticket price 公共客车票价 06.479

bus transport network 公路客运网 06.062

button 路钮 03.475

buttressed abutment 扶壁式桥台 04.392

butt type expansion installation 对接式伸缩装置 04.300

bypass 绕行公路 02.029

C

cable 钢缆索 08.137

cable clamp 索夹 08.139

cable crane 缆索起重机 07.289

cable force control 索力控制 04.777

cable force measurement 索力测量 09.101

cable guardrail 钢索护栏 03.432

cable operated excavator 机械挖掘机 07.045

cable saddle 索鞍 04.327

cable stayed bridge 斜拉桥 04.235

cable support tower 索塔 04.322

cable tension measurement device 索力测定计 09.207

CAD for bridge 桥梁计算机辅助设计 04.609

CAFE 公司平均燃料经济性 05.154

caisson cutting edge 沉井刃脚 04.404

calcium electric rapid determination method 钙电极快速测定法 09.060

calculated rise of arch 计算矢高 04.028

calculated span 计算跨径 04.015

California bearing ratio 加州承载比 02.413

California bearing ratio test 加州承载比试验, *CBR 试验 09.024

calling for tenders 招标 01.099

camber 预拱度 04.719

cantilever beam bridge 悬臂梁桥 04.206

cantilever erection method 悬臂拼装法 04.662

cantilever retaining wall 悬臂式挡土墙 02.270

cantilever slab 悬臂板 04.286

cantilever traffic sign 悬臂式交通标志 03.371

canyon 峡谷 01.228

capacity index 通行能力指数 03.175

capillary water 毛细水 02.502

capillary water height 毛细水上升高度 08.209

carbonation depth by phenolphthalein test 碳化深度酚酞试验 09.129

carbon fiber 碳纤维 08.156

carbon monoxide allowable concentration 一氧化碳允许浓度 10.025

carbon monoxide detector 一氧化碳监测仪 10.024

carbon steel 碳素钢 08.117

car carrier semitrailer 轿车运载[半挂]车 05.048

card reader 读卡机 03.331

card sender 发卡机 03.330

car ferrying 汽车轮渡 04.950

car-following theory 跟车理论 03.006

car lift 汽车举升机 05.358

car pool lane 合乘车辆车道 03.239

carriage box volume capacity 车厢容积 06.237

carriage mileage 运输里程 06.450

carriageway 车行道, *行车道 02.160

carrier 承运人 06.173

cart road 畜力车道, *大车道 02.011

carving groove 刻槽 02.481

car washer 汽车清洗机 05.362

casing hole-boring method 套管护壁钻孔法 04.747

casting yard 预制场 04.709

cast-in-place balancing cantilever method 平衡悬浇法 04.665

cast-in-place cantilever method 悬臂浇筑法 04.664

cast-in-place concrete pile rig　灌注桩钻孔机
　07.249

cast-in-place method　膺架浇筑法，＊就地浇筑法
　04.668

cast-in-situ pile　就地灌注桩　04.426

cast-in-situ pile by excavation　挖孔桩　04.438

cast on scaffolding method　膺架浇筑法，＊就地浇
　筑法　04.668

catastrophic collapse　毁灭性破坏　04.607

catchment area　汇水面积　04.077

cat eyes　猫眼　03.477

cationic emulsified bitumen　阳离子乳化沥青
　08.013

cattle fence　防牲畜护栏　03.493

catwalk　施工步道，＊猫道　04.705

caution sign　警告标志　03.354

cavitation　穴蚀　05.184

CBR　加州承载比　02.413

CBR test　加州承载比试验，＊CBR试验　09.024

cement　水泥　08.040

cement concrete　水泥混凝土　08.101

cement concrete mark　水泥混凝土标号　08.257

cement concrete mixture　水泥混凝土混合料
　08.100

cement concrete pavement　水泥混凝土路面
　02.451

cement gun　喷浆机　07.314

cement mark　水泥标号　08.256

cement mortar　水泥砂浆　08.098

cement-soil base　水泥土基层　02.494

cement soundness test　水泥安定性试验　09.058

centering by clamped planks in arch shape　夹板拱架
　04.693

centering unloading　拱架卸落　04.695

center island　中心岛　02.218

center island marking　中心岛标线　03.423

center lane　中间车道　02.177

center line　中线　01.187

center line stake　中线桩　01.189

center line survey　中线测量　01.188

central pilot tunneling method　中央导洞法　04.874

centrifugal bucket elevator　离心式斗式提升机
　07.103

centrifugal force　离心力　04.489

centrifugal pump　离心式水泵　07.264

certification test　出厂证明试验　07.406

chain block　链式起重机，＊链滑轮　07.279

chain bucket trencher　链斗式挖沟机　07.051

chain hoist　链式起重机，＊链滑轮　07.279

chalky clay　白垩土　01.292

changeable message sign　可变信息标志　03.365

channel contraction　河槽压缩　04.103

channelization island　导流岛　02.217

channelized intersection　渠化交叉口　02.207

channelized traffic　渠化交通　03.139

channelizing marking　渠化标线　03.422

characteristic of materials　材料性能　08.175

charged mileage　计费里程　06.449

charged weigh　计费重量　06.447

charge standard　计价标准　06.446

charter bill　客运包车票　06.101

chartered bus transport　包车客运　06.059

chartered handling equipment time based charge　装卸
　机械计时包用费　06.494

chartered pay mileage　包干计费里程　06.451

charter rate　包车运价　06.471

charter rate by distance　计程包车运价　06.473

charter rate by number of containers　集装箱包箱运
　价　06.481

charter rate by time　计时包车运价　06.472

chassis dynamometer　底盘测功机　05.326

check load　验算荷载　01.089

children crossing sign　学童过街标志　03.410

chips　石屑　08.051

chip spreader　石屑撒布机　07.163

chloride content depth profile measurement　氯化物含
　量沿深度分布测量　09.127

chute　急流槽　02.535

circular curve　圆曲线　02.082

circulation boring method　正循环钻孔法　04.756

circulation drill　正循环钻机　07.253

city approach highway　城市出入口公路　02.031

city bus　城市客车　05.016

civilian vehicle　民用车辆　05.053

claims　索赔　01.154

clamper　夹具，＊工具锚具　04.640

clam shell excavator 抓铲挖掘机 07.041

classified highway 等级公路 02.024

classified rates for goods 货物分等运价 06.466

clay bound macadam 泥结碎石路面 02.431

clayey sand 粘土质砂 08.083

clay-lime bound macadam 泥灰结碎石路面 02.432

cleaning and steaming charge 清洗和熏蒸费 06.499

clearance 净空 02.125

clearance above bridge deck 桥面净空 04.022

clearance height 净空高度 04.025

clearance limit 极限间隙 05.302

clearance of tunnel 隧道净空 04.851

clearance time 清尾[车]时间 03.256

clearance under span 桥下净空 04.023

clear span 净跨径 04.016

climbing ability test 爬坡性能试验 07.440

climbing lane 爬坡车道 02.172

climbing lane sign 爬坡车道标志 03.393

closed toll system 封闭式收费系统 03.347

close to traffic 封闭交通 01.353

close type crushing plant 闭式联合碎石机组 07.098

closure by jacking and sealing-off crown 千斤顶法封顶 04.685

closure by wedging-in crown 尖拱法封顶, *刹尖 04.684

closure of arch ring 拱圈封顶 04.683

closure of bridge structure 桥梁合拢 04.774

clothoid 回旋线 02.093

clover-leaf interchange 苜蓿叶形立交 02.228

clutch slippage 离合器滑转 05.216

coagulator 促凝剂 08.149

coal tar 煤沥青 08.023

coarse gradation 粗级配 08.188

coarse-grained soil 粗粒土 08.060

coarse joining method 毛接法 02.666

coarse sand 粗砂 08.078

coasting 滑行 05.105

coasting distance measurement 滑行距离测定 07.432

coasting in neutral 脱档滑行 05.106

coasting mode 滑行工况 05.125

coasting resistance measurement 滑行阻力测定 07.431

coasting test 滑行试验 07.429

coasting with clutch disengaged 分离离合器滑行 05.108

coasting with engine off 熄火滑行 05.107

cobble 卵石 08.067

cobble stone pavement 圆石路面 02.428

code number of highway route 公路路线编号 01.339

coefficient of container dead weight 集装箱自重系数 06.239

coefficient of expansion 膨胀系数 04.532

coefficient of lanes 车道系数 02.404

coefficient of loose laying 松铺系数 02.309

coefficient of roughness 糙率 04.064

coefficient of scouring 冲刷系数 04.119

coefficient of wheel tracking transverse distribution 轮迹横向分布系数 02.403

cofferdam 围堰 04.764

cohesion 粘结力 08.237

cohesion test 粘结力试验 09.075

cold aggregate conveyer 冷料输送机 07.135

cold bent test 冷弯试验 09.050

cold breaking-in 冷磨合 05.299

cold-drawn rate 冷拉率 09.089

cold-drawn wire 冷拔钢丝 08.131

cold laid method 冷铺法 02.441

cold mixing method 冷拌法 02.439

cold paint road marking machine 冷漆划线机 07.229

cold-stretched steel bar 冷拉钢筋 08.124

collapse 崩坍 01.264

collapse blasting 抛坍爆破 02.330

collapsed loess 湿陷性黄土 01.281

collector-distributor road 集散道路 02.033

collision prevention of pier and abutment 桥梁墩台防撞 04.794

colonnade foundation 管柱基础 04.406

color adaptability 颜色适应性 03.528

color matching system 调[配]漆机 05.371

column 柱 04.277

column pier 柱式[桥]墩 04.367

combination roller 组合式压路机 07.067

combined highway and railway bridge 公铁两用桥 04.148

combined maintenance truck 综合养护车 07.214

combined stress 组合应力 04.550

commercial dispute 商务纠纷 06.412

commercial vehicle 商用汽车 05.055

commercial vehicle operation management system 商用车辆运行管理系统 03.322

compacting depth measurement 压实深度测定 07.459

compacting factor 压实系数 02.298

compacting width measurement 压实宽度测定 07.458

compaction 压实 02.296

compaction depth 压实厚度 02.308

compaction energy 压实能量 02.306

compaction machine test 压实机械试验 07.457

compaction test 击实试验 09.012

compaction test apparatus 击实仪 09.150

compactness test 压实度试验 09.015

compensation for failure of loading 装货落空损失费 06.490

compensation for loss 按质赔偿 06.118

complete failure 完全故障 05.188

completion of construction 竣工 01.145

component replacement 构件更换 04.800

composite abutment 组合式桥台 04.396

composite beam 组合梁 04.267

composite beam bridge 组合梁桥 04.213

composite deck cable stayed bridge 组合梁式斜拉桥 04.238

composite lining 组合衬砌 04.888

composite longitudinal ventilation 组合式纵向通风 04.912

composite shell-slab and I-beam bridge 微弯板组合梁桥 04.246

composite type concrete pavement 复合式水泥混凝土路面 02.456

compound curve 复曲线 02.083

comprehensive stabilized base 综合稳定基层 02.496

comprehensive traffic management 综合交通管理 03.205

comprehensive transportation system 综合运输系统 03.140

compression gauge 压缩压力计, *气缸压力表 05.344

compression test 压缩试验 09.042

computer aided design for bridge 桥梁计算机辅助设计 04.609

computerized vehicle inspection system 全自动汽车检测系统 05.325

concave vertical curve 凹形竖曲线 02.123

concentrated drainage 集中排水 02.533

concentrated load 集中荷载 04.484

concrete beam slipform device 混凝土梁滑模浇筑设备 07.277

concrete carbonation test 混凝土碳化试验 09.128

concrete chloride content measurement 混凝土氯化物含量测量 09.126

concrete coring hole inspecting by endoscope 混凝土钻孔内窥镜检查 09.124

concrete curing 混凝土养生 04.712

concrete deck cable stayed bridge 混凝土斜拉桥 04.236

concrete fluidity test 混凝土流动性试验 09.085

concrete injector 喷浆机 07.314

concrete mixer 混凝土搅拌机 07.190

concrete mixing plant 混凝土搅拌设备 07.187

concrete pavement finisher 混凝土路面整面机 07.199

concrete pavement joint cleaning machine 混凝土路面清缝机 07.202

concrete pavement joint cutting machine 混凝土路面切缝机 07.201

concrete pavement joint sealing machine 混凝土路面填缝机 07.203

concrete pavement texturing machine 混凝土路面拉毛机 07.200

concrete paving train 混凝土摊铺列车 07.192

concrete pile [钢筋]混凝土桩 04.419

concrete pipe pile 混凝土管桩 04.424

concrete pump truck 混凝土输送泵车 07.205

concrete spreader 混凝土布料机 07.195

concreting　混凝土浇筑　04.710

concreting by pumping　混凝土泵送法　04.711

condition-based maintenance　视情维护　05.258

cone anchorage　锥形锚具，＊弗氏锚　04.641

cone crusher　圆锥式碎石机　07.082

cone penetration test　触探　01.254

confined compression strength test　侧限抗压强度试验　09.019

conical slope　[桥头]锥坡　04.135

connecting rod aligner　连杆校正器　05.351

connecting rod bearing boring machine　连杆轴承镗床　05.352

consigned luggage　托运行包　06.085

consignee　收货人　06.175

consignment　托运　06.177

consignment plan　托运计划　06.176

consignor　托运人　06.172

consistency limit　稠度界限　08.197

consistency test　稠度试验　09.054

consolidated subsoil　加固地基　02.312

consolidation　固结　02.314

consolidation coefficient　固结系数　08.211

consolidation test　固结试验　09.016

consolidometer　固结仪　09.156

constituent　组分　08.244

constituent test　组分试验　09.074

constrained operation　约束运行　03.183

construction contract terms　施工合同条款　01.129

construction drawing design　施工图设计　01.045

construction height of bridge　桥梁建筑高度　04.018

construction joint　施工缝　02.462

construction joint for arch ring　拱圈施工缝　04.682

construction load　施工荷载　04.467

construction machine emission　施工机械排放物　10.011

construction machinery noise　施工机械噪声　10.056

construction machinery vibration　施工机械振动　10.067

construction plan　施工计划　01.097

construction section capacity　施工区通行能力　03.174

construction section traffic management　施工路段交通管理　03.206

construction site management　施工现场管理　01.096

construction specification　施工规范　01.075

construction survey　施工测量　01.182

construction time limit　工期　01.149

construction waste　施工废弃物　10.037

construction waste water　工程废水　10.036

construction water level　施工水位　04.100

contain bag　集装袋　06.246

container carrier　集装箱货车　05.012

container dropside trailer　集装箱栏板挂车　05.035

container flatframe trailer　集装箱专用挂车　05.033

container fork lift　集装箱叉车　06.306

container handling machinery　集装箱装卸机械　06.281

containerization ratio　集装箱化率　06.347

container load capacity utilizing ratio　[集装]箱载重利用率　06.242

container load freight　适箱货　06.267

container lot　[集装]箱位　06.243

container maintenance charge　集装箱维修费　06.500

container platform trailer　集装箱平板挂车　05.034

container specific volume capacity coefficient　[集装]箱容系数，＊集装箱比容　06.240

container spreader　集装箱吊具　06.311

container trailer　集装箱挂车　05.032

container transit charge　集装箱中转费　06.485

container transport　集装箱运输　06.169

container volume capacity　集装箱内容积　06.238

container volume capacity utilizing ratio　[集装]箱容利用率　06.241

contingencies　不可预见费　01.135

continuity model　连续性模型　03.010

continuous arch bridge　连续拱桥　04.222

continuous arch effect　连拱作用　04.591

continuous asphalt mixing plant　连续式沥青混合料搅拌设备　07.128

continuous beam bridge　连续梁桥　04.205

continuous contact of interface　界面连续接触

02.401

continuous deck　连续桥面　04.289

continuous passenger traffic　持续性客流　06.039

continuous reinforced concrete pavement　连续配筋混凝土路面　02.453

continuous rigid frame bridge　连续刚构桥　04.234

continuous slurry wall　地下连续墙　04.443

continuous traffic count station　连续式交通量观测站　03.121

continuous traffic flow　连续性交通流　03.047

contour interval　等高距　01.217

contour line　等高线　01.216

contraband goods　禁运货物　06.154

contract　合同　01.126

contraction crack　收缩裂缝　02.600

contraction joint　缩缝　02.461

contract management　合同管理　01.128

contractor　承包人　01.102

contract transport　合同运输　06.024

control house　控制室　07.152

control point of route　路线控制点　02.051

control stress for prestressing　张拉控制应力　04.729

converging sign　汇流标志　03.382

converging traffic volume　合流交通量　03.185

converted turnover　换算周转量　06.343

converted weight　换算重量　06.448

convex vertical curve　凸形竖曲线　02.122

conveyer belt scale　皮带称重装置　07.153

coordinated control　联动控制　03.299

coordinated traffic signals　联动交通信号　03.258

coordinate method　坐标法　01.200

coordinative composition of equipments　设备配套　07.349

cordon counts　区界交通调查　03.137

cordon traffic survey　区界交通调查　03.137

core　岩心　01.333

core sampler　岩心取样器　09.232

corner break　角裂　02.625

corner steel　角隅钢筋　02.473

corporation average fuel economy　公司平均燃料经济性　05.154

corridor traffic management program　走廊交通管理计划　03.203

corrosion test　锈蚀试验　07.388

corrugated beam barrier　波形梁护栏　03.438

corrugated-metal pipe culvert　波纹管涵　04.811

corrugated steel deck　波纹钢桥面　08.136

corrugation　搓板　02.618

cost/benefit analysis　费用效益分析　01.056

cost of ancillary facility　辅助设施费　07.339

cost quota of vehicle maintenance and repair　汽车维修费用定额　05.065

cost quota of vehicle major repair　汽车大修费用定额　05.066

counter flow transport　对流运输　06.028

counterfort abutment　扶壁式桥台　04.392

counterfort retaining wall　扶臂式挡土墙　02.271

county highway　县公路，＊县道　01.343

coupler　连接器　04.650

coupling and uncoupling of hooks　挂摘钩　06.297

cover protectometer　钢筋保护层测定仪　09.237

crack　裂缝　02.597

crack control　裂缝控制　04.598

crack detecting by acoustic emission　裂缝声发射检测　09.123

crack grouting　裂缝灌注　02.662

cracking load　开裂荷载　04.468

cracking rate　[路面]裂缝率　02.598

cracking ratio　[路面]裂缝度　02.599

crack pattern　裂缝分布图　04.600

crack resistance　抗裂性　04.597

cradle　挂篮　04.708

crankcase emissions　曲轴箱排放物　10.015

cranked platform trailer　凹式平板挂车　05.028

crash bearer　防撞墩　03.446

crawler bulldozer　履带式推土机　07.002

crawler crane　履带式起重机　07.284

crawler hydraulic bench drill　履带式液压凿岩台车　07.029

crawler loader　履带式装载机　07.055

crawler single bucket excavator　履带式单斗挖掘机　07.044

crawler soil stabilizer　履带式稳定土搅拌机　07.115

crawler stabilized soil road-mixer　履带式稳定土搅拌

机 07.115

crawler-type asphalt paver 履带式沥青混凝土摊铺机 07.159

creep 蠕变 08.268

creep loss of concrete 混凝土徐变损失 04.635

critical buckling load 临界压曲荷载 04.500

critical density 临界密度 03.087

critical failure 致命故障 05.190

critical gap 临界空挡 03.027

critical height of subgrade 路基临界高度 02.284

critical load 临界荷载 02.409

critical speed 临界车速 03.068

critical V/C ratio 临界V/C比 03.091

critical wind speed 临界风速 04.587

crops covering by dust 农作物蒙尘 10.032

cross beam 横梁 04.275

cross-buck sign 岔道标志 03.378

cross chipper'hammer 爪形冲击锤 07.261

cross fall 路拱横坡 02.137

crossing [平面]交叉口 02.197

crossing lift 架式举升机 05.360

cross road 十字形交叉 02.202

cross section 横断面 02.124

cross section survey 横断面测量 01.204

cross walk 人行横道 03.486

crown 路拱 02.135

crown curve 路拱曲线 02.136

crown slope 路拱横坡 02.137

cruising mode 等速工况 05.123

crushed stone 碎石 08.068

crusher 碎石机 07.078

crushing plant 联合碎石机组 07.095

crushing value 压碎值 08.250

crushing value test 压碎值试验 09.032

crust 硬表层 01.312

CS 圆缓点 02.100

cultural relic protection scope 文物保护范围 10.086

culture 地物 01.219

culvert 涵[洞] 04.801

culvert aperture 涵洞孔径 04.828

culvert barrel length 涵洞洞身长度 04.827

culvert end lift 涵洞口隆起 04.826

culvert inlet 涵洞进水口 04.819

culvert inlet/outlet apron 涵洞口铺砌 04.823

culvert on steep grade 陡坡涵洞 04.814

culvert outlet 涵洞出水口 04.820

culvert straight end wall 涵洞一字墙 04.817

culvert water drop 涵洞跌水 04.821

culvert wing wall 涵洞八字墙 04.818

curb [路]缘石 02.151

curb marking 缘石标记 03.429

curb parking 路边停车 03.153

current deflecting 水流导向 04.125

current direction 流向 04.065

current-obstruction area 阻水面积 04.053

current repair of road machine 筑路机械小修 07.325

current velocity 流速 04.054

curved bridge 弯桥，*曲线桥 04.160

curve length 曲线长 02.090

curve sign 弯道标志 03.377

curve widening 平曲线加宽 02.132

cushion guardail 缓冲护栏 03.436

cut 挖方 02.289

cut-and-cover method 随挖随盖法 04.867

cut and fill section 半填半挖断面 02.128

cutback asphalt 轻制沥青，*稀释沥青 08.020

cut-fill transition diagram 土方调配图 02.073

cutting 路堑 02.251

cutting ring of shield 盾构切口环 04.943

CVOM 商用车辆运行管理系统 03.322

cycle length 周期长 03.243

cycle track 自行车道 02.010

cyclic dispatching method 循环调度法 06.214

cyclic flashing 周期性闪光 03.468

cyclone dust collector 旋风式除尘器 07.131

cylinder boring machine 镗缸机 05.349

cylinder honing machine 汽缸珩磨机 05.353

cylinder scoring 拉缸 05.215

D

daily dispatched passenger number 旅客日发送量 06.052

damage 损伤 05.186

damage of luggage 行包损坏 06.113

damage rate of goods 货损率 06.231

dampness type 潮湿类型 01.166

daywork labor 计日工 01.142

DDHV 定向设计小时交通量 03.066

dead end highway 断头路 02.043

deadhead mileage 空车行程，＊空车里程 06.359

deadline of bid 投标截止期 01.117

dead load 恒载 04.476

death rate of accident 事故死亡率 03.217

debonded tendon 消除粘结的预应力筋 04.627

debris flow 泥石流 01.270

decelerating mode 减速工况 05.124

deceleration lane 减速车道 02.171

deck 桥面 04.288

deck bridge 上承式桥 04.155

deck drainage system 桥面排水系统 04.290

deck elevation 桥面标高 04.031

deck expansion installation 桥面伸缩装置 04.299

deck expansion joint 桥面伸缩缝 04.298

deck pavement 桥面铺装 04.291

deck profile grade 桥面纵坡 04.032

deck slab 桥面板 04.280

deck transverse slope 桥面横坡 04.033

decreasing rate with increasing distance 递远递减运价 06.459

deep-buried tunnel 深埋隧道 04.838

deep foundation 深基础 04.400

deep well pump 深井水泵 07.268

default 违约 01.150

defect 缺陷 05.185

defects liability period 缺陷责任期 01.155

defects liability release certificate 缺陷责任终止证书 01.156

deflection 挠度 04.556

deflection angle 偏角 01.197

deflection angle method 偏角法 01.201

deflection basin 弯沉盆 02.396

deflection coefficient 弯沉系数 02.394

deflection limit for serviceability 使用状态挠度限值 04.603

deflection test 弯沉试验 09.029

deflection value 弯沉值 02.389

deflectometer 弯沉仪，＊贝克曼梁 09.186

deformed bar 螺纹钢筋，＊变形钢筋 08.121

degree of compaction 压实度 02.297

degree of congestion 拥挤度 03.178

degree of parking saturation 停车饱和度 03.157

degree of prestressing 预应力度 04.611

degree of saturation 饱和度 08.270

dehris 碎落 01.265

deicing 消冰 02.680

deicing agent pollution 消冰剂污染 10.038

delay 延误 03.092

delayed delivery 逾期交付 06.190

delayed pick-up 逾期提货 06.191

delineation line 轮廓标线 03.420

delivery on arrival 到达交付 06.188

Demag joint 戴马格接缝 04.313

demand for parking spaces 停车车位需要 03.147

demurrage 延滞费 06.489

denial of passenger 甩客 06.109

dense gradation 密级配 08.186

dense waterway net region 水网地区 04.050

density of highway network 公路网密度 01.028

density of runs 班次密度 06.067

density test 密度试验 09.002

depersonalized repair method 混装修理法 05.310

depreciation rate 基本折旧率 07.332

depth-span ratio 高跨比 04.029

derrick crane 桅杆式起重机 07.283

desert 沙漠 01.268

desert soil　荒漠土，＊沙漠土　01.289

design capacity　设计通行能力　03.166

design current velocity　设计流速　04.055

design discharge　设计流量　04.069

design elevation　设计高程　01.210

design elevation of subgrade　路基设计高程　02.281

design flood frequency　设计洪水频率　04.072

design hourly traffic volume　设计小时交通量　03.065

design intensity of earthquake　设计烈度　04.518

design load　设计荷载　01.088

design specification　设计规范　01.074

design speed　计算行车速度，＊设计车速　01.081

design traffic volume　设计交通量　01.080

design water level　设计水位　04.098

desired speed　期望车速　03.080

detective test　检测性试验　07.405

determinacy model　确定性模型　03.011

determination of operating cycle　作业循环时间测定　07.444

determination of operating cycle production quantity　循环作业量测定　07.445

determination of production quantity per litre fuel consumption　每升燃料作业量测定　07.446

detonation　爆燃，＊爆震　05.210

detour bridge　便桥　04.181

detour sign　绕行标志　03.400

Deval abrasion testing machine　狄法尔磨耗试验机　09.170

DHV　设计小时交通量　03.065

diagnostic expert system　诊断专家系统　05.233

diagnostic method　诊断方法　05.229

diagnostic norms　诊断规范　05.238

diagnostic parameters　诊断参数　05.237

diagnostic technology　诊断工艺　05.228

diagonal parking　斜列式停车　03.199

diameter of equivalent circle　当量圆直径　02.379

diamond interchange　菱形立交　02.230

diaphragm pump　膜式水泵　07.265

diaphragm wall　地下连续墙　04.443

diesel fuel injection pump tester　柴油机喷油泵试验台　05.345

diesel injection timing tester　柴油机喷油定时测定

器　05.342

diesel pile hammer　柴油桩锤　07.239

diesel smoke　柴油机排烟　10.019

diesel smoke meter　柴油机烟度计　05.336

differential photo　微分法测图　01.241

differential rate　差别运价　06.461

differential settlement　沉降差　04.659

differential thermal analysis　差热分析　09.022

digital terrain model　数字地面模型　01.244

diluvial soil　洪积土　01.299

dip fault　倾斜断层　01.326

direct benefit　直接效益　01.053

direct expense of project　工程直接费　01.062

direct field of vision　直接视野　03.504

directional blasting　定向爆破　02.331

directional carriageway marking　定向行车道标记　03.430

directional design hourly traffic volume　定向设计小时交通量　03.066

directional interchange　定向式立交　02.231

directional sign　指路标志　03.357

direction angle　方向角　01.199

direction distribution　方向分布　03.054

direct shear apparatus　直剪仪　09.162

direct shear test　直接剪切试验　09.044

disaggregation model　非集成模型　03.014

discharge flow cross section　过水断面　04.051

discharge opening of bridge　过水孔径　04.052

disk feeder　圆盘给料机　07.093

dispatch charge　调车费　06.488

dispatching log　调度日志　06.209

dispatching method　调度方法　06.210

dispatching of freight transport　货运调度　06.204

dispatching order　调度命令　06.211

dispersive passenger traffic　分散性客流　06.045

displacement　位移　04.555

displacement pile　排土桩　04.433

dissolubility　溶解度　08.232

dissolubility test　溶解度试验　09.068

distillation test　蒸馏试验　09.069

distortion　扭曲，＊畸变　04.567

distributed load　均布荷载　04.482

distribution width measurement　沥青洒布宽度测定

07.474

disturbed sample　扰动土样　01.336

ditch cleaning out　清理边沟　02.657

diverging traffic volume　分流交通量　03.184

diversion dike　导流堤　04.127

diversion transport　分流运输　06.030

diverted traffic [volume]　转移交通量　03.125

divided carriageway　分隔式车行道　02.161

divided highway sign　分隔行驶公路标志　03.384

dividing ridge　分水岭　04.076

diving equipment　潜水设备　07.274

document of passenger transport　客运单证　06.095

Domke honeycomb joint　多姆克蜂房接缝　04.314

door-to-door transport　门到门运输　06.013

double acting jack　双作用千斤顶　04.731

double acting pneumatic hammer　双作用气动桩锤
　　07.241

double acting steam hammer　双作用蒸气桩锤
　　07.243

double amber lines　双黄线　03.414

double-decked ferry boat　双层渡车船　04.961

double-layer-truss centering　叠桁式拱架　04.692

double-leaf bascule bridge　双叶竖旋桥　04.256

double-leaf swing bridge　双叶平转桥　04.253

double plane cable stayed bridge　双索面斜拉桥
　　04.239

double pylon cable stayed bridge　双塔斜拉桥
　　04.242

double-slope drainage　[路面]双向排水　02.528

double steel comb joint　双梳式接缝　04.310

double trailer train　双挂汽车列车　05.052

double-walled steel cofferdam and bored pile foundation
　　双壁钢围堰钻孔桩基础　04.407

double-walled steel waling　双层钢围囹　04.769

dowel bar　传力杆　02.468

downdrag　下拉荷载，*桩负摩擦力　04.497

dragging　拖滞　05.221

dragline excavator　拉铲挖掘机　07.042

drainage ability　排水能力　02.505

drainage design　排水设计　02.508

drainage ditch　排水沟　02.520

drainage facility　排水设施　02.507

drainage sand mat　排水砂垫层　02.319

drainage system　排水系统　02.506

drainage works　排水工程　02.509

drain opening　泄水口　02.542

drawbar efficiency measurement　挂钩牵引效率测定
　　07.435

drawbar power measurement　挂钩牵引功率测定
　　07.434

drawbar pull measurement　挂钩牵引力测定
　　07.433

drawbar test　牵引试验　07.425

drawing jack　张拉千斤顶　07.295

dressed stone　料石　08.070

dressed stone pavement　细琢石路面　02.426

drier　干燥筒　07.136

drifter　漂浮物　04.121

drift ice　流冰　04.124

drilling bucket　钻斗　07.257

drills steel sharpener　改钻机　07.030

drill truck　汽车钻机　07.236

drill wagon for jack hammer　凿岩钻车　07.028

driven pile　打入桩　04.439

driver behavior pattern　驾驶员行为特性　03.524

[driver] decipherment distance　[司机]识别距离
　　03.513

[driver] judgement time　[司机]判断时间　03.512

[driver] perception-reaction distance　[司机]感觉反
　　应距离　03.515

[driver] perception-reaction time　[司机]感觉反应
　　时间　03.514

driver psychological and physiological reaction　驾驶
　　员心理和生理反应　03.525

[driver] reaction distance　[司机]反应距离
　　03.510

driving ability　驾驶能力　05.110

driving adaptability　驾驶适应性　03.526

driving behavior　驾驶行为　03.523

driving element slipping rate measurement　驱动元件
　　滑转率测定　07.436

driving fatigue　驾驶疲劳　03.520

driving habits　驾驶习惯　03.522

driving information　驾驶信息　03.500

driving in running-in period　走合期驾驶　05.115

driving on snowy and icy road　冰雪路面驾驶

05.114

driving operation ability 驾驶操纵能力 03.521

driving simulator 驾驶模拟器 05.116

driving skill 驾驶技能 05.111

driving technique 驾驶技术 05.112

drop hammer 落锤 07.238

dropside trailer 栏板挂车 05.025

drop weight rammer 落锤夯 07.077

drum-mixing asphalt plant 滚筒式沥青混合料搅拌设备 07.129

drum screen 滚动筛 07.088

drum static linear pressure measurement 滚轮静线压力测定 07.460

drum vibration amplitude measurement 振动轮振幅测定 07.463

drum vibration frequency measurement 振动轮频率测定 07.462

dry concrete 干硬性混凝土 08.104

dry density 干密度 08.201

drying-mixing drum 干燥－搅拌筒 07.137

dry traffic accident 非雨天事故 03.233

dry type 干燥类型 01.164

dry unit weight 干容量 08.203

dual carriageway 双幅车行道 02.162

dual lane 双车道 02.164

dual wheel loading 双轮荷载 02.378

duct grouting 管道灌浆 04.733

ductility 延度 08.228

ductility test 延度试验 09.062

duct void examination by vacuum pressure test 梁内导管空隙真空加压试验 09.125

dummy joint 假缝 02.466

dump trailer 自卸挂车 05.036

dump truck 自卸货车 05.011

durability test 耐久性试验 07.398

duration of tunnel passage 过隧经历时间 04.935

dust collector 除尘装置 07.130

dustproof facilities 防尘设施 10.033

dusty goods 扬尘性货物 06.148

dynaflect 动力式弯沉仪 09.189

dynamic amplification duration curve 动力放大过程线 09.111

dynamic axle weight detector 动态轴重检测器 03.312

dynamic consolidation method 动力固结法 02.315

dynamic deflection 动弯沉 02.390

dynamic diagnosis 动态诊断 05.236

dynamic load 动载 04.474

dynamic measurement of pile [桩]动测法 09.142

dynamic rebound deflection 动态回弹弯沉 02.391

dynamic sounding 动力触探试验 09.028

dynamic triaxial test 动态三轴试验 09.018

dynamic visual acuity 动视觉敏锐度 03.530

Dywidag anchorage system 精轧螺纹锚，＊迪维达格锚固系统 04.649

E

early-dry mortar strength test 硬练砂浆强度试验 09.055

early strength agent 早强剂 08.148

early strength concrete 早强混凝土 08.103

earth cofferdam 土围堰 04.766

earthen centering 土牛拱架 04.694

earth pressure 土压力 04.493

earthquake load 地震荷载，＊地震力 04.488

earthquake magnitude 地震震级 04.519

earth road 土路 02.040

earth shoulder 土路肩 02.143

earth work 土方[工程] 02.287

easiness of vehicle control 汽车操纵轻便性 05.147

EBCR 经济效益费用比 01.047

eccentric compression method 偏心受压法 04.577

ecology 生态 10.080

ecology environment 生态环境 10.081

economical capacity 经济通行能力 03.172

economic benefit cost ratio 经济效益费用比 01.047

economic internal rate of return 经济内部收益率 01.049

economic net present value 经济净现值 01.048

economic span 经济跨径 04.017

economic speed 经济车速 03.079

economy survey 经济调查 01.158

edge failure 啃边 02.619

edge line 边缘线 03.416

edge reinforcement 边缘钢筋 02.472

effective green time 有效绿灯时间 03.247

effectiveness measures of LOS 服务水平量度参数 03.163

effective rolling depth 有效碾压深度 02.303

EIRR 经济内部收益率 01.049

elastic layer system theory 弹性层状体系理论 02.375

elastic modulus test 弹性模量试验 09.043

elastic semi-infinite foundation 弹性半无限地基 02.377

elastic supported beam method 弹性梁支承法 04.581

elastomeric pad bearing 合成橡胶支座 04.447

electrical resistivity survey 电探 01.255

electric curing of concrete 混凝土电热养生 04.714

electric frog rammer 蛙式夯 07.076

electric-heat prestressing 电热[法]张拉 04.720

electric hoister 电动卷扬机 07.281

electric jack hammer 电动凿岩机 07.020

electric resistance strain gauge 电阻应变仪 09.240

electric shocking rammer 电动冲击夯 07.075

electrolevel 电子水平仪 09.226

electronic-hydraulic automatic leveling device 电液自动找平装置 07.368

electronic toll system 电子收费系统 03.349

element 构件，＊杆件 04.533

elevated road 高架道路 02.014

elevating scraper 升运式铲运机 07.010

elevation 高程，＊标高 01.207

elevation grader 平地升送机，＊犁扬机 07.016

embankment 路堤 02.249

embeded length of pile 桩入土长度 04.034

embedment depth of foundation 基础埋置深度 04.035

emergency braking 紧急制动 05.096

emergency fire extinguisher 紧急灭火机 04.945

emergency management system 应急管理系统 03.323

emergency parking strip 应急停车带 02.175

emergency telephone 应急电话 03.295

emergency telephone sign 应急电话标志 03.403

emission analyzer [汽车]排气分析仪 05.335

employer 业主 01.101

empty container mileage 空箱里程 06.453

emulsified bitumen 乳化沥青，＊沥青乳液 08.008

emulsifying agent 乳化剂 08.012

endoscope 内窥镜 09.222

engine analyzer 发动机诊断仪 05.340

engine cleaner 发动机清洗机 05.366

engineering classification of soil 土的工程分类 08.057

engineering geologic condition analysis 工程地质条件分析 09.144

engineering geologic sketch 工程地质略图 01.308

engineering geology 工程地质学 01.009

engineering material 工程材料学 01.013

engineering plastic 工程塑料 08.163

engineering quality control 工程质量管理 01.094

engineering surveying 工程测量学 01.010

engineer's estimate 标底 01.118

engine rebuilding 发动机大修 05.282

engine remanufacture 发动机大修 05.282

enlarged preliminary design 扩大初步设计 01.044

ENPV 经济净现值 01.048

entrance ramp 驶入匝道 02.238

entrance ramp control 入口匝道控制 03.274

entry application and approval procedure 开业报批程序 06.403

entry conditions 开业条件 06.402

entry/exit regulation 开停业管理 06.401

environmental effects in highway construction 公路施工环境影响 10.005

environmental hazard 环境损害 10.003

environmental load 环境荷载 04.481

environmental noise standard 环境噪声标准 10.046

environmental pollution 环境污染 10.002

environmental protection 环境保护 01.017

environmental vibration standard 环境振动标准 10.065

epoxy resin 环氧树脂 08.031

equipment periodic maintenance 设备定期维护 07.322

equipment replacement 设备更新 07.357

equipment scrapping 设备报废 07.358

equipment storing up 设备封存 07.356

equivalent axles 当量轴次 02.381

equivalent failure rate 当量故障率 07.454

equivalent thickness 当量厚度 02.395

equivalent traffic volume 换算交通量, * 当量交通量 03.061

equivalent uniform distributed load 等代均布荷载 04.483

erection by longitudinal pulling method 纵向拖拉法 04.672

erection with cableway 缆索吊装法, * 无支架吊装法 04.661

erection with scaffolding method 膺架式架设法 04.667

ergonomics 工效学 01.018

erosion 冲刷 02.643

erosion loss 水土流失 10.072

escort handling machinery 随车装卸机械 06.280

escort hydraulic arm 随车液压起重臂 06.285

escorting 押运 06.187

estimate index of highway project 公路工程估算指标 01.077

estuarial crossing 港口立交桥 04.153

evaluation of bids 评标 01.119

evaluation test 鉴定试验 07.404

evaporation pond 蒸发池 02.524

evaporative pollutants 蒸发污染物 10.014

examination of dangerous situation 险情检查 02.559

excavation 挖方 02.289

excavator 挖掘机 07.037

excessive dampness type 过湿类型 01.167

excessive fuel consumption 费油 05.207

exhaust emission standard 废气排放标准 10.012

exhauster 抽风机 07.150

exhaust gas analyzer [汽车]排气分析仪 05.335

exhaust muffler 排气消声器 10.053

exhaust noise 排气噪声 10.052

exhaust purifier 排气净化器 10.030

exhaust scrubber 废气滤清器 10.028

existing bridge 旧桥 04.163

existing highway 现有公路 01.029

exit application and approval procedure 停业报批程序 06.404

exit ramp 驶出匝道 02.239

exit ramp control 出口匝道控制 03.273

exit sign 出口标志 03.404

expansion bearing 活动支座 04.446

expansion joint 张缝 02.460

expansion plate 伸缩板 04.545

expansion roller 伸缩滚轴 04.458

expansive cement 膨胀水泥 08.043

expansive soil 膨胀土 01.283

exploration 勘探 01.249

exploration drill 地质钻机 07.235

explosion rammer 爆炸夯 07.074

expressway 汽车专用公路 02.022

extension of time 延期 01.153

exterior beam 外纵梁 04.274

exterior stringer 外纵梁 04.274

external distance 外距 02.091

external prestressing 体外预应力 04.618

extrados 拱背[线] 04.345

F

fabricated and detachable steel bridge 装拆式钢桥 04.259

fabricated bridge 装配式桥 04.258

fabrication yard 预制场 04.709

factory and mine road 厂矿道路 02.005

fagot 柴捆 04.142

fagot dam 柴捆坝 04.143

failure branch chart 故障树 05.224

failure distribution 故障分布 05.223

failure load 破坏荷载 04.471

failure load by buckling 屈服破坏荷载 04.501

failure mechanism 故障机理 05.196

failure mode　故障模式　05.197

failure of machine　机械故障　07.323

failure rate　故障率　05.222

failure rate of machinery　机械故障率　07.330

failure tree　故障树　05.224

fall　崩坍　01.264

falling weight deflectometer　落锤式弯沉仪　09.188

falsework　脚手架，＊膺架，＊工作架　04.704

fan-type stay cable　扇形拉索　04.331

fast Fourier transform analyzer　快速傅里叶变换分析仪，＊FFT 分析仪　09.246

fast lane　内侧车道，＊快车道　02.176

fatigue design　疲劳设计　04.521

fatigue life of pavement　路面疲劳寿命　02.373

fatigue limit　疲劳限度　04.601

fatigue load　疲劳荷载　04.469

fatigue test　疲劳试验　09.052

fault　断层　01.324

fault activity evaluation of bridge site　桥址断层活动性评定　09.148

faulting of slab ends　板端错台　02.627

fault strike　断层走向　01.325

FCL　整箱货　06.268

feasibility study　可行性研究　01.040

feeder　给料机　07.089

feeder highway　支线公路　02.036

ferry　渡口　04.948

ferry boat　渡轮　04.958

ferry crossing　渡口　04.948

ferry house　渡口管理所　04.972

ferry management　渡口管理　04.970

ferry safety regulation　渡口安全规则　04.971

ferry toll　船渡费　01.035

ferry wharf　渡运码头　04.953

FFT analyzer　快速傅里叶变换分析仪，＊FFT 分析仪　09.246

fiber concrete　纤维混凝土　08.111

fiber concrete pavement　纤维混凝土路面　02.454

field compaction test　现场压实试验　07.465

field location　野外定线　02.070

field mixing test　现场拌和试验　07.471

field of front vision　前方视野　03.502

field of lateral vision　侧向视野　03.503

field of vision　视野　03.501

field test　现场试验　07.373

fill　填方　02.288

filled spandrel arch bridge　实腹拱桥　04.217

filler bin　粉料仓　07.145

filler metering feeder　粉料计量给料器　07.147

filler spreader　粉料撒布机　07.120

filler weigher　粉料秤　07.146

final account of project　工程决算　01.148

final survey　竣工测量　01.183

fine gradation　细级配　08.189

fine-grained soil　细粒土　08.063

fineness　细度　08.181

fineness modulus　细度模数　08.182

fine sand　细砂　08.080

fine-weather highway　晴通雨阻公路　02.042

finishing　抹光　02.476

firing device　起爆器　07.316

first class highway　一级公路　02.021

fissured clay　裂隙粘土　01.290

fixed bearing　固定支座　04.445

fixed handling machinery　固定式装卸机械　06.278

fixed line operation　定线运行　06.077

fixed-message sign　固定信息标志　03.366

fixed ticket　定站客票　06.096

fixed-time control　定时信号控制　03.272

flange　翼缘，＊法兰　04.536

flared intersection　拓宽路口式交叉口　02.208

flare wing-walled abutment　八字形桥台　04.385

flashing signal　闪光信号　03.254

flash point　闪点　08.230

flash point test [open cup method]　闪点试验[开口杯法]　09.065

flat jack　扁千斤顶　09.202

flat joint　平缝　02.464

flat trailer　平板挂车　05.026

flexible pavement　柔性路面　02.414

flexible pier　柔性墩　04.372

flexible safety fence　柔性护栏　03.440

flexural-tensile strain　弯拉应变　02.407

flexural-tensile strength　弯拉强度　02.408

flexural-tensile stress　弯拉应力　02.406

floatability　浮漂度　08.236

floatability test 浮漂度试验 09.070

floating box 浮箱 07.272

floating bridge 浮桥 04.249

floating caisson 浮运沉井 04.742

floating crane 浮式起重机，＊浮吊 07.291

floating equipment 浮水设备 07.271

flood 洪水 02.501

flood damage 水毁 02.645

flood discharge 洪水流量 04.067

flooded area 泛滥区 04.049

flood frequency 洪水频率 04.071

flood land 河滩 04.088

flood land line water level 平滩水位 04.102

flood level duration curve 洪水位过程线 04.074

flood relief bridge 排涝桥 04.177

flood run-off 洪水径流 04.081

flood survey 洪水调查 01.172

flow discharge 流量 04.066

flow duration curve 流量过程线 04.070

flowing water pressure 流水压力 04.495

flow method for vehicle maintenance 汽车维护流水作业法 05.266

flow method for vehicle repair 汽车修理流水作业法 05.306

flow value 流值 08.247

flush curb 平缘石 02.152

flutter 颤振 04.569

fluttering 振抖 05.208

fly ash 粉煤灰 08.047

fly-ash cement concrete 粉煤灰水泥混凝土 08.102

flyover 跨线桥 04.151

foam asphalt unit 泡沫沥青装置 07.184

fog detector 雾检测器 03.311

fold 褶皱 01.327

foot brake pedal pressure gauge [制动]踏板力计 05.348

forced concrete mixing plant 强制式混凝土搅拌设备 07.188

forced idling mode 强制急速工况 05.121

forced stabilized soil mixing plant 强制式稳定土搅拌设备 07.111

forced traffic flow 强制性交通流 03.044

forced vibration cross-spectrum analysis 受迫振动互谱分析 09.119

ford 过水路面 02.543

fording drive 涉水驾驶 05.113

forest road 林区道路 02.006

for-hire transport 营业性运输 06.019

for-hire vehicle 营运汽车 05.056

formation of motor transport tariff 汽车运价构成 06.444

form correction repair 整形修理 05.284

formwork 模板 04.716

foundation engineering 基础工程学 01.011

foundation pile diagnosis system 基桩病害检测系统 09.223

foundation pit well point drainage method 基坑井点排水法 04.773

foundation settlement 基础沉降 04.658

four-way stop 四向停车 03.190

fracture zone 破碎带 01.328

fragile goods 易碎货物 06.151

free bound 行车碾压 02.448

free carbon content 游离碳含量 08.240

free carbon content test 游离碳含量试验 09.076

free flow speed 自由流车速 03.073

free-of-charge luggage 免费行包 06.487

free swelling rate 自由膨胀率 08.215

free traffic flow 自由交通流 03.041

freeway 高速公路 02.020

freeway mainline control 高速公路主线控制 03.277

freeway surveillance and control 高速公路监控 03.276

freezing and thawing test 冻融试验 09.040

freight average haul distance 货物平均运距 06.344

freight bill 货票 06.183

freight claims 货运索赔 06.225

freight damage 货损 06.219

freight flow 货流 06.199

freight forwarder agency charge 承运货物代理费 06.505

freight hub terminal 货运枢纽站 06.250

freight information center 货运信息中心 06.251

freight list 货物清单 06.182

freight lot 货位 06.259

freight production balance 货运生产平衡 06.194

freight quality indicator 货运质量指标 06.229

freight sales charge 组货服务费 06.507

freight shortage 货差 06.218

freight source 货源 06.192

freight source survey 货源调查 06.193

freight terminal 货运站 06.248

freight terminal charge 货物站务费 06.502

freight traffic diagram 货流图 06.200

freight traffic direction 货物流向 06.202

freight traffic time 货物流时 06.203

freight traffic volume 货物流量 06.201

freight transport 货物运输 06.017

freight transport business 货运业务 06.171

freight transport quality 货运质量 06.216

freight volume fluctuation coefficient 运量波动系数 06.196

frequency of freight quality accidents 货运质量事故频率 06.230

frequency of vehicle current repair 汽车小修频率 05.068

frequency recorder 频率仪 09.212

fresh and living goods 鲜活货物 06.145

Freyssinet cone anchorage 锥形锚具，*弗氏锚 04.641

friction pile 摩擦桩 04.435

friction tester 摩擦系数测定仪 09.176

front axle-load measurement 前轴负荷测定 07.414

front-mooring ferry boat 端靠式渡船 04.963

front shovel 正铲挖掘机 07.039

front wall 前墙，*胸墙 04.382

frost boiling 翻浆 02.614

frost damage 冻害 02.647

frost depth 冰冻深度 01.169

frost heaving 冻胀 02.648

frost region 冰冻地区 01.168

frost thawing 冻融 02.649

frozen soil 冻土 01.271

fuel consumption gauge 燃油流量计 05.346

fuel deficit ratio 亏油率 06.379

fuel saving device 汽车节能装置 05.161

fuel saving ratio 节油率 06.378

full circulation operation 大循环运行 06.075

full container load 整箱货 06.268

full control of access 全部控制进入，*全封闭 02.235

full cut section 全挖式断面 02.130

full depth asphalt pavement 全厚式沥青路面 02.421

full depth resurfacing 全厚翻修 02.671

full-face tunneling method 全断面开挖法 04.872

full load mode 全负荷工况 05.126

full price ticket 全价票 06.475

full-scale test 足尺试验 07.377

full span wooden bent centering 排架式满布木拱架 04.689

full span wooden inclined strut centering 斜撑式满布木拱架 04.690

full trailer 全挂车 05.023

full trailer train 全挂汽车列车 05.050

fully-actuated signal control 全感应式信号控制 03.265

fully prestressed concrete 全预应力混凝土 04.612

FWD 落锤式弯沉仪 09.188

G

gabion 石笼 04.144

galloping 驰振 04.570

gangboard 跳板 04.968

gang maintenance 道班养护 02.587

gantry crane 门式起重机 07.288

gap gradation 断级配 08.187

gas emission in tunnel 隧道瓦斯泄出 04.930

gas explosion in tunnel 隧道瓦斯爆炸 04.929

gate effect 门口效应 03.511

gate type automatic car washer 汽车门式自动清洗机 05.363

gear shifting 换档 05.089

general conditions of contract 合同通用条件 01.130

general goods 普通货物 06.134

general inspection and diagnosis 综合诊断 05.234

general scour at bridge opening 桥下一般冲刷 04.114

geographic information system 地理信息系统 01.245

geogrid 土工格栅 08.167

geologic legend 地质图例 01.310

geologic logging 地质测井 01.309

geologic profile 地质剖面图 01.307

geologic survey 地质调查 01.306

geomembrane 土工膜 08.166

geophone 拾震器 09.239

geophysical prospecting 物探 01.256

geotechnical index property test 岩土特性指标试验 09.025

geotextile 土工织物，＊土工布 08.164

girder 梁 04.262

girder bridge 梁桥 04.203

girder depth 梁高 04.026

glass fiber 玻璃纤维 08.153

glass fiber reinforced plastic bridge 玻璃钢桥 04.202

glass fiber reinforced plastic fence 玻璃钢护栏 03.442

global positioning system 全球定位系统 01.243

glued timber bridge 胶合木桥 04.200

gobi 戈壁 01.269

goods handling 货物装卸 06.273

goods handling charge 货物装卸费 06.496

goods mark 货物标志 06.160

goods packing 货物包装 06.158

good technical condition of vehicle 汽车完好技术状况 05.163

gore area 分道角区 02.211

GPS 全球定位系统 01.243

gradation 级配 08.185

grade change point 变坡点 02.119

grade compensation 纵坡折减 02.113

graded aggregate base 级配集料基层 02.490

graded aggregate pavement 级配路面 02.433

grade length limit 坡长限制 02.117

grader 平地机 07.013

grade separation 立体交叉，＊立交 02.222

grade separation without ramps 分离式立交，＊非互通式立交 02.223

grading curve 级配曲线 08.191

gradual failure 渐变故障 05.193

grain composition 颗粒组成 08.180

grain size 粒径 08.179

grain size test 颗粒分析试验 09.005

grand bridge 特大桥 04.167

granular material 粒料 08.049

graphic dispatching method 图上作业法 06.212

grass planting 植草 02.655

grated inlet 栅式进水口 02.538

gravel 砾 08.074

gravelly soil 砾类土 08.061

gravity abutment 重力式桥台 04.390

gravity anchor 重力式锚碇 04.337

gravity bucket elevator 重力式斗式提升机 07.104

gravity pier 重力式[桥]墩 04.364

gravity retaining wall 重力式挡土墙 02.268

gravity stabilized soil mixing plant 自落式稳定土搅拌设备 07.112

gravity type concrete mixing plant 自落式混凝土搅拌设备 07.189

great bridge 大桥 04.168

green belt 绿化带 02.156

greening mileage 绿化里程 02.550

green ratio 绿信比 03.248

green time 绿灯时间 03.245

green wave band 绿波带 03.249

grey control system 灰色控制系统 03.016

grey system theory 灰色系统理论 03.015

grid roller 格栅压路机 07.063

grillage analogy 格构比拟 04.610

grooving machine 开槽机 07.057

ground asphalt storage 地上沥青贮仓 07.169

ground clearance measurement 离地间隙测定 07.416

ground control point survey 地面控制点测量 01.234

ground elevation 地面高程 01.211

ground stereo-photogrametry 地面立体摄影测量 01.233

ground water 地下水 01.257

ground water level 地下水位 01.258

grouted curtain 灌浆帷幕 04.442

grouted cut-off wall 灌浆截水墙 04.441

grouting equipment 灌/压浆设备 07.275

growth rate method 增长率法 03.120

guard fence 防护栅 03.490

guard post 护柱 03.450

guard wall 护墙 02.262

guide sign 指路标志 03.357

gully 集水槽 02.534

gully bed neighboring to culvert 涵洞沟床 04.822

gusset plate 结点板，＊节点板 04.541

gutter 天沟 02.518

gutter drainage 明渠排水 02.521

Guyon-Massonet method GM 法 04.583

gypsum 石膏 08.039

H

hair-like crack 发裂 02.604

hairpin curve 发针形曲线 02.088

half cell 半电池 09.219

half price ticket 半价票 06.476

half-through bridge 中承式桥 04.156

half tunnel 半山洞 02.253

hammer crusher 锤击式碎石机 07.081

hammer grab 锤式冲抓斗 07.256

hand-held rock drill 手持式凿岩机 07.023

handling charge by container number 计箱装卸费 06.493

handling direct cost 装卸直接费用 06.437

handling distance 装卸距离 06.314

handling efficiency 装卸效率 06.313

handling hourly output 装卸工时产量 06.325

handling hourly quota 装卸工时定额 06.322

handling intermittence time 装卸工作停歇时间 06.329

handling machinery 装卸机械 06.277

handling machinery efficiency index 装卸机械效率指标 06.326

handling machinery productivity 装卸机械生产率 06.327

handling mechanization scheme 装卸机械化方案 06.288

handling mechanization system 装卸机械化系统 06.287

handling mode 装卸方式 06.274

handling operation cost 装卸工作成本 06.330

handling operation line 装卸作业线 06.301

handling procedure 装卸工序 06.293

handling quality standard 装卸质量标准 06.294

handling quota 装卸定额 06.321

handling shift efficiency 装卸工班效率 06.324

handling shift quota 装卸工班定额 06.323

handling technological process 装卸工艺流程 06.290

handling technology 装卸工艺 06.286

handling technology design 装卸工艺设计 06.289

handling worker's productivity 装卸工人生产率 06.328

hand luggage 自理行包 06.086

hardening 硬化 08.263

hardness of stone 石材硬度 08.253

hardpan 硬盘 01.313

hardrock 硬岩 01.315

hard shoulder 硬路肩 02.144

harp-type stay cable 坚琴形拉索 04.330

haunch 加劲腋，＊梁腋 04.539

hazardous goods 危险货物 06.140

heading 导洞 04.846

heading and bench tunneling method 导洞与层阶法 04.879

head light tester ［汽车］前大灯试验仪 05.334

head-on collision 迎面碰撞 03.227

head wall abutment 一字形桥台 04.386

heating loss 加热损失 08.234

heating loss test 加热损失试验 09.067

heave force of foundation 基础隆胀力 04.499

heavy and bulky goods 长大笨重货物 06.143

heavy haul trailer 重型平板挂车 05.029

heavy oil 重油 08.032

heavy tamping method 重锤夯实法 02.300

heavy traffic 重交通 02.384

height of cut and fill at center stake 中桩填挖高度 01.212

herringbone drainage system 人字形排水系统 02.531

hexane equivalent 正己烷当量 10.026

hidden peril of accident 交通事故隐患 03.234

hidden work acceptance 隐蔽工程验收 01.143

high class pavement 高级路面 02.334

highest water level 最高水位 04.092

high liquid limit clay 高液限粘土 08.089

high mast lighting 高杆照明 03.457

high occupancy vehicle lane 多乘员车辆车道 03.238

high platform 高站台 06.255

high-pressure oilpump 高压油泵 07.294

high-rise platform pile foundation 高承台桩基础 04.411

high speed texture meter 高速构造深度仪 09.185

high strength bolt 高强螺栓 08.132

high strength concrete 高强混凝土 08.107

high tensile steel wire 高强度钢丝 08.128

high water level bridge 高水位桥 04.172

highway 公路 02.002

highway administration 公路路政管理 01.338

highway aesthetics 公路美学 01.022

highway and transportation 公路交通 01.001

highway and transportation planning 公路交通规划 03.108

highway appearance 路容 02.560

highway appurtenance 公路附属设施 01.360

highway boundary frame 公路限界架 01.363

highway bridge 公路桥 04.147

highway capacity 公路通行能力 03.159

highway capital construction procedure 公路基本建设程序 01.038

highway condition 路况 02.552

highway condition inspection 路况检查 02.555

highway condition investigation 路况调查 02.553

highway condition patrol 路况巡视 02.556

highway condition registration 路况登记 02.554

highway construction administration 公路建设管理 01.024

highway construction cost 公路造价 01.061

highway construction fund 公路建设基金 01.031

highway construction management 公路施工管理 01.093

highway container transfer terminal 公路集装箱中转站 06.253

highway data bank 公路数据库 02.565

highway disaster 公路灾害 02.584

highway distress 公路病害 02.596

highway drainage 公路排水 02.499

highway engineering 公路工程学 01.003

highway engineering supervision 公路工程监理 01.098

highway environment 公路环境 10.001

highway environmental impact evaluation 公路环境影响评价 10.007

highway environmental protection desing 公路环境保护设计 10.006

highway esthetic design 公路美学设计 02.061

highway ferry 公路渡口 04.949

highway ferry station 轮渡站 04.951

highway function 公路功能 01.086

highway functional design 公路功能设计 02.060

highway geometric design 公路几何设计 02.048

highway greening 公路绿化 02.695

highway improvement 公路改善 02.574

highway intermediate maintenance 公路中修 02.571

highway intersection 公路交叉 02.196

highway landscape 公路景观 02.692

highway landscape design 公路景观设计 02.062

highway load standard 公路荷载标准 01.087

highway maintenance 公路养护 02.544

highway maintenance fee 养路费 01.036

highway major maintenance 公路大修 02.572

highway mileage 公路里程 02.547

highway natural zoning 公路自然区划 01.092

highway network 公路网 01.026

highway network planning 公路网规划 01.027

highway passenger transport 公路旅客运输

06.034

highway perspective view 公路透视图 02.055

highway planting 公路绿化 02.695

highway property 路产 01.348

highway-railway grade separation 公路铁路立交 02.246

highway reconnaissance and survey regulations 公路勘测规程 01.072

highway relocation 公路改线 02.579

highway residual value 公路残值 01.068

highway routine maintenance 公路小修[保养] 02.570

highway service facilities 公路服务设施 01.362

highway standardized and beautified project 公路标准化美化工程，*GBM工程 02.691

highway technical administration 公路技术管理 01.069

highway technical file 公路技术档案 02.564

highway technical standard 公路工程技术标准 01.070

highway traffic mode 公路交通方式 03.109

highway traffic nuisance 公路交通公害 10.004

highway traffic supervision 公路交通监理 01.359

highway transport 公路运输 06.002

highway transport management 公路运输管理 06.333

highway transport plan 公路运输计划 06.334

highway transport statute 公路运输法规 06.413

highway transport vehicle 公路运输车辆 05.001

highway transport vehicle management 公路运输车辆管理 05.058

highway transport volume 公路运输量 06.340

highway tunnel 公路隧道 04.832

highway user's benefit 公路使用者效益 01.058

highway user's cost 公路使用者费用 01.057

hilling terrain 重丘区 01.223

hill side line 山坡线 02.065

hinge connected beam method 铰接梁法 04.579

hingless arch bridge 无铰拱桥 04.218

historic flood level 历史洪水位 01.173

hoisting capacity 起重量 06.291

hoisting equipment 起重设备 07.278

hoisting fixture 吊货工夹具 06.284

hollow pier 空心[桥]墩 04.366

hollow slab bridge 空心板桥 04.208

hopper type concrete spreader 斗式混凝土布料机 07.196

horizontal alignment 平面线形 02.074

horizontal clearance 净宽 02.127

horizontal curve 平曲线 02.076

horizontal handling machinery 水平搬运机械 06.282

hot aggregate bin 热料仓 07.140

hot aggregate elevator 热料提升机 07.138

hot aggregate vibrating screen 热料振动筛 07.139

hot bent test 热弯试验 09.051

hot breaking-in 热磨合 05.300

hot laid method 热铺法 02.440

hot laid plastic 加热型塑料 08.169

hot mixing method 热拌法 02.438

hot mix silo 热混合料贮仓 07.154

hot paint road marking machine 热漆划线机 07.230

hot rolled bar 热轧钢筋 08.122

humidity absorbing goods 吸湿性货物 06.150

30HV 年第三十位最大小时交通量 03.060

HWL 最高水位 04.092

hydraulic automatic leveling device 液压自动找平装置 07.367

hydraulic casing extractor 液压压拔套管机 07.262

hydraulic computation 水力计算 04.043

hydraulic cushion guardrail 液压缓冲护栏 03.444

hydraulic drop 跌水 02.536

hydraulic element test 液压元件试验 07.456

hydraulic excavator 液压挖掘机 07.046

hydraulicity 水硬性 08.264

hydraulic jack hammer 液压凿岩机 07.024

hydraulic jack prestressing 油压千斤顶张拉 04.721

hydraulic pile driver 液压打桩机 07.246

hydraulic pile head splitter 液压桩头破碎器 07.273

hydraulic pressure stabilizer 液压稳定器 09.203

hydraulic radius 水力半径 04.063

hydraulic rock breaker 液压岩石破碎机 07.033

hydraulic static cone penetrometer 液压静力触探仪 09.230

hydraulic structure 水工构造物 04.047

hydrodynamics theory 流体动力学理论 03.005

hydrogeological survey 水文地质调查 01.170

hydrogeology 水文地质学 01.008

hydrographic survey 水道测量 04.037

hydrologic analysis 水文分析 04.042

hydrologic data 水文资料 04.040

hydrologic observation at bridge site 桥址水文观测 04.044

hydrologic station 水文站 04.045

hydrologic survey 水文测量 04.039

hydrophilic aggregate 亲水性集料 08.072

hydrophobic aggregate 憎水性集料 08.073

hydroplaning phenomena 水滑现象 02.372

hysteresis response zone 滞后响应区 09.116

I

ICB 国际竞争性招标 01.114

ice apron 破冰体 04.378

ice cone 冰锥 01.274

ice hummock 冰丘 01.273

ice jam 冰障 02.651

ice pressure 冰压力 04.494

idling mode 怠速工况 05.120

illuminated sign 灯光照明标志 03.409

imaginary intersection point 虚交点 01.195

immersed tunneling method 沉埋法 04.869

impact factor 冲击系数，*动力系数 04.530

impact factor evaluation 冲击系数测定 09.118

impact force 冲击力 04.490

impact toughness apparatus 冲击韧度试验仪 09.168

impact toughness test 冲击韧度试验 09.047

impregnated concrete 浸渍混凝土 08.113

inclined shaft 斜井 04.844

inclinometer 测斜仪 09.225

increasing rate with decreasing distance 递近递增运价 06.458

incremental launching construction by multipoint jacking 多点同步顶推 04.670

incremental launching device 顶推设备 07.292

incremental launching method 顶推法 04.669

increment of traffic [volume] 增长交通量 03.123

indexes of handling operation 装卸作业指标 06.315

indirect benefit 间接效益 01.054

indirect expense of project 工程间接费 01.063

indirect field of vision 间接视野 03.505

individual pile 单桩 04.413

induced traffic [volume] 诱增交通量 03.124

induction sign 诱导标志 03.362

industrial solid waste 工业废渣 08.046

inertial type brake tester 惯性式制动试验台 05.329

influence area 影响面积 04.565

influence line 影响线 04.564

information of freight source 货源信息 06.198

infrared oven stand 红外线烘干装置 05.370

infra-red vehicle detector 红外线车辆检测器 03.291

initial maintenance 初期养护 02.567

inland container depot 内陆还箱站 06.266

inlet 进水口，*雨水口 02.537

inlet submerged culvert 半压力式涵洞 04.809

inlet unsubmerged culvert 无压力涵洞 04.810

inorganic binder 无机结合料 08.035

inquiry form survey 调查表调查 06.132

inside container operation fork lift 箱内作业叉车 06.310

inside gear asphalt pump 内啮合齿轮沥青泵 07.166

inspection and classification of parts 零件检验分类 05.296

inspection and repair only as necessary 视情检修 05.290

inspection of goods 验货 06.184

installing pile shaft by excavation 挖孔灌桩法 04.753

instruction to bidder 投标人须知 01.110

instrument of vehicle maintenance and repair 汽车维修工具 05.322

instrument station 测站 01.179

insurance of passenger unexpected injury 旅客意外伤害保险 06.111

insured transport 保价运输 06.170

integral lining 整体式衬砌 04.886

intelligent vehicle highway system 智能车路系统 03.316

intensified test 强化试验 07.393

intercepting ditch 截水沟 02.519

interchange 互通式立交 02.226

interchange toll station 互通式立交收费站 03.340

intercity bus 长途客车 05.017

interface drainage 界面排水 02.529

interior beam 内纵梁 04.273

interior stringer 内纵梁 04.273

intermediate class pavement 中级路面，*过渡式路面 02.336

intermittent passenger traffic 阵发性客流 06.040

intermittent traffic count station 间歇式交通量观测站 03.122

intermodal freight terminal 货运联运站 06.252

intermodal transport 联合运输，*联运 06.011

internal cohesion 内粘聚力 08.220

internal friction angle 内摩擦角 08.221

internal friction coefficient 内摩擦系数 08.219

international competitive bidding 国际竞争性招标 01.114

international roughness index 国际平整度指数 02.702

international transport 国际运输，*出入境运输 06.015

inter-regional passenger traffic 跨区客流，*区间客流 06.037

interrupted traffic flow 中断性交通流 03.048

interruption of firing 断火 05.214

intersection [平面]交叉口 02.197

intersection angle 交叉角 02.198

intersection capacity 交叉口通行能力 03.168

intersection delay 交叉口延误 03.095

intersection point 交点 01.194

intersection with widened corners 加宽转角式交叉口 02.209

interstitial water 缝隙水 01.260

interurban road 城间道路 02.004

interval between diagnosis 诊断周期 05.239

interval between vehicle maintenances 汽车维护周期 05.269

intrados 拱腹[线] 04.344

intra-regional passenger traffic 区内客流 06.038

intra-zone traffic 区内交通 03.113

inundated district 淹水地区 04.048

inverted filter 反滤层 02.530

inverted siphon culvert 倒虹吸涵 04.813

inverted Y tower 倒Y形索塔 04.325

invitation to bid 投标邀请书 01.112

IP 交点 01.194

IRI 国际平整度指数 02.702

IROAN 视情检修 05.290

I[-shaped] beam 工形梁 04.268

isolated signal control 点控制 03.297

items of machine maintenance and repair 机械维修作业项目 07.354

IVCS 先进的汽车控制系统，*智能汽车控制系统 03.318

IVHS 智能车路系统 03.316

J

jack 千斤顶 07.282

jack hammer 凿岩机 07.018

jacking-in method of culvert or subsurface bridge 桥涵顶入法 04.675

jack knifing [列车]折叠 05.104

jam density 阻塞密度 03.086

jaw crusher 颚式碎石机 07.079

JM12 anchorage JM12型锚具 04.647

joint 接缝 02.457

joint failure 接缝破损 02.623

joint filler 填缝料 02.470

joint filler failure 填缝料破损 02.624

joint filling 填缝 02.663

joint-filling material 岩缝填充物 01.318

joint plate 填缝板 02.471

joint sealing materials 填缝材料 04.302

joint ventures bidder 联合投标 01.116

joint with sealing plate 填缝板接缝 04.303

journey speed 行程车速 03.076

journey time 行程时间 03.082

junction [平面]交叉口 02.197

K

kaolin 高岭土 01.288

karst 喀斯特，＊岩溶 01.261

keyed joint 楔形缝 04.305

key pile 主桩 04.417

kilometer stone 里程碑 02.194

kinematic viscosity 动力粘滞度 08.226

knuckle bearing 铰式支座 04.460

L

laboratory simulation test 实验室模拟试验 07.380

laboratory test 试验室试验 07.374

labour wage management 劳动工资管理 06.394

lack of power 乏力 05.205

lagging 支拱板条 04.698

lake asphalt 湖沥青 08.007

laminated rubber bearing 板式橡胶支座 04.448

lamps and lanterns 照明灯具 03.471

landscape planting 景观栽植 02.694

landslide 滑坡 01.263，坍方 02.642

land-use forecast model 土地利用预测模型 03.111

lane 车道 02.159

lane balance 车道平衡 03.186

lane capacity 车道通行能力 03.169

lane distribution 车道分布 03.055

lane line 车道分界线 03.415

lane load 车道荷载 04.485

lane occupancy ratio 车道占有率 03.100

lane shift 车道偏移 03.188

lane toll machine 车道收费机 03.325

lapped steel plate joint 搭接钢板接缝 04.308

laser alignment deflection measurement 激光准直挠度测量 09.100

laser automatic leveling device 激光自动找平装置 07.369

laser road surface tester 路面激光测试仪 09.174

lateral bracing 横系杆 04.543

lateral clearance 侧向余宽 02.157

lateral clear distance of horizontal curve 平曲线横净距 02.081

lateral loading test of pile 桩横向荷载试验 09.137

laterally intermittent illuminant 横向间断光源 03.467

launching nose 导梁 04.707

layer spread method 层铺法 02.435

laying arch by rings 分环砌拱 04.678

laying arch by sections 分段砌拱 04.677

laying arch by sections and rings 分段分环结合砌拱 04.679

laying arch by transverse overhanging method 横向悬砌拱法 04.680

laying arch continuously 连续砌拱 04.676

LCB 国内竞争性招标 01.115

LCL 拼箱货 06.269

lead-rubber bearing 铅橡胶支座 04.451

leakage 漏浆 02.485，泄漏 05.199

lean concrete 贫混凝土 08.105

lean concrete base 贫混凝土基层 02.487

left turn guide line 左转弯导向线 03.418

left turn ramp 左转弯匝道 02.242

3-leg interchange 喇叭形立交 02.233

length of tunnel 隧道长度 04.850

less than container load 拼箱货 06.269

less-than-truck-load rate 零担运价 06.464

less-than-truck-load terminal 零担货运站 06.249

less-than-truck-load transport 零担[货物]运输 06.166

levee 河堤 04.134

leveling course 整平层 02.361·

leveling survey 水准测量 01.205

level of mechanization 机械化程度 07.348

level of service 服务水平 03.162

lever principle method 杠杆原理法 04.576

lift bridge 升降桥 04.257

light control system 光度控制系统 03.470

light goods 轻浮货物 06.142

lighting 照明 03.455

lighting adaptation section 照明适应段 03.462

lighting facilities 照明设备 03.472

lighting standard 照明灯柱 03.458

lighting transition 照明过渡 03.460

lighting transition section 照明过渡段 03.461

light luggage 轻浮行包 06.083

light signal 灯光信号 03.241

light traffic 轻交通 02.386

light-weight concrete 轻质混凝土 08.106

lime 石灰 08.036

lime-flyash-soil base 石灰粉煤灰土基层 02.495

lime mortar 石灰砂浆 08.099

lime pile 石灰桩 02.322

lime-soil base 石灰土基层 02.493

limited compensation 限额赔偿 06.119

limited prestressed concrete 有限预应力混凝土 04.614

limiting technical condition of vehicle 汽车极限技术状况 05.169

limit of crack opening 裂缝宽度限值 04.599

limit state design of bridge 桥梁极限状态设计 04.506

linear pressure 线压力 02.304

line control 线控制 03.298

line location 定线 02.069

linked control 线控制 03.298

linking-up line 联络线 02.044

lip curb 缓坡缘石 02.155

liquid asphalt 液体沥青 08.016

liquidated damages 延期违约偿金 01.151

liquidity index 液性指数 08.208

liquid limit 液限 08.204

liquid limit test 液限试验 09.006

liquid-plastic combine tester 液塑限联合测定仪 09.149

lithofacies 岩相 01.319

live load 活载 04.475

livestock trailer 牲畜家禽挂车 05.044

LL 液限 08.204

load 荷载 04.465

load carrying capacity of bridge 桥梁承载能力 04.593

loaded container mileage 重箱里程 06.452

loaded mileage 重车行程，*重车里程 06.358

load effect 荷载效应 04.503

loader 装载机 07.053

loader-excavator 装载挖掘机 07.056

load factor 荷载系数 04.531，实载率 06.373

load factor design method of bridge 桥梁荷载系数设计法 04.509

load factor of intersection 交叉口负荷系数 03.253

loading and unloading quality conformity rate 装卸质量合格率 06.235

loading berm 反压护道 02.261

loading combinations 荷载组合 04.502

loading island 乘车安全岛 02.188

loading network 装货网络 06.245

loading/unloading lot 车位 06.258

loading/unloading volume 货物装卸量 06.316

load measurement column 测力柱 09.205

loan repay ability 贷款偿还能力 01.051

local competitive bidding 国内竞争性招标 01.115

local highway 地方公路 02.037

local scour around pier 桥墩局部冲刷 04.115

local stability 局部稳定性 03.023

local traffic 境内交通 03.112

local trip 境内出行 03.133

location survey 定测 01.177

loess 黄土 01.280

logistic diagnosis 逻辑诊断 05.232

long column testing machine 长柱试验机 09.193

longitudinal continuous band illuminant 纵向连续带光源 03.466

longitudinal crack　纵向裂缝　02.605

longitudinal dike　顺坝　04.128

longitudinal drainage　纵向排水　02.514

longitudinal gradient　纵坡　02.112

longitudinal joint　纵缝　02.458

longitudinal profile analyzer　纵断面分析仪　09.182

longitudinal ventilation　纵向通风　04.909

long span bridge　大跨径桥　04.169

long tunnel　长隧道　04.854

loop ramp　环形匝道　02.244

loop vehicle detector　环形线圈式车辆检测器
　03.288

loose laying depth　松铺厚度　02.307

loosening　松动　05.202

loosening blasting　松动爆破　02.328

loose protection course　松散保护层　02.347

LOS　服务水平　03.162

Los Angeles abrasion testing machine　洛杉矶磨耗试
　验机　09.171

loss due to anchorage deformation　锚具变形损失
　04.631

loss due to anchorage temperature difference　锚具温
　差损失　04.632

loss due to bending of concrete member　杆件弯曲损
　失　04.638

loss due to concrete member shortening　杆件缩短损
　失　04.637

loss due to concrete shrinkage　混凝土收缩损失
　04.633

loss due to duct friction　管道摩擦损失　04.630

loss due to elastic compression of concrete　混凝土弹

性压缩损失　04.634

loss due to tendon relaxation　钢筋预应力松弛损失
　04.636

loss in grade　坡度损失　02.116

loss of luggage　行包丢失　06.112

loss of prestress　预应力损失　04.629

low-bed trailer　低平板挂车　05.027

low class pavement　低级路面　02.337

low clearance sign　低净空标志　03.399

lower retaining wall　下挡墙　02.277

lowest water level　最低水位　04.093

low freight lot　低货位　06.256

low liquid limit clay　低液限粘土　08.086

low liquid limit silt　低液限粉土　08.085

low water level　枯水位　04.097

low water level bridge　低水位桥　04.174

luggage　行包　06.080

luggage acceptance　行包受理　06.088

luggage compensation rate　行包赔偿率　06.125

luggage consignment　行包托运　06.087

luggage delivery　行包交付　06.091

luggage label　行包标签　06.100

luggage lot　行包货位　06.129

luggage opening for inspection　开启查验　06.092

luggage storage service　行包保管　06.090

luggages transport regularity rate　行包正运率
　06.124

luggage transport rates　行包运价　06.486

luminous sign　发光标志　03.406

LWL　最低水位　04.093

M

macadam base　碎石基层　02.489

machine handling operation volume　机械装卸作业量
　06.319

machine operation　机械运转，＊机械操作　07.318

machinery accident　机械事故　07.340

macro-traffic model　宏观交通模型　03.009

MADT volume　月平均日交通量　03.059

magnetic card toll machine　磁卡收费机　03.327

magnetic card toll pass ticket　磁卡通行券　03.328

magnetic crack detector　磁性测裂计　09.214

magnetic tell tale　磁感式沉降标　09.227

magnetic vehicle detector　磁感应式车辆检测器
　03.289

main beam　主梁　04.263

main bearing boring machine　主轴承镗床　05.350

main bridge　主桥　04.007

main channel　主槽　04.086

main girder　主梁　04.263

mainline toll station 主线收费站 03.339

main span 主跨 04.010

maintenance after failure 事后维护 05.259

maintenance cost 养护费用 01.065, 维护费 07.337

maintenance counterproposal 养护对策 01.357

maintenance division 养路总段 01.349

maintenance gang 养路道班 01.351

maintenance gang house 道班房 02.588

maintenance management 养护管理 02.545

maintenance mileage 养护里程 02.549

maintenance of machine 机械维护 07.320

maintenance period 大中修周期 02.573

maintenance project 养护工程 02.546

maintenance quality 养护质量 02.686

maintenance section 养路段 01.350

major failure 严重故障 05.191

major maintenance cost 大修工程费 01.066

major repair of road machine 筑路机械大修 07.327

maladjustment 失调 05.203

mandatory sign 指示标志 03.356

manhole 检查井 02.540

man-machine processing system 人机调节系统 03.495

manual handling 人力装卸 06.275

manual hoist 手摇卷扬机 07.280

manually operated ferry boat 人力渡车船 04.960

map display 地图板显示 03.300

marginal strip 路缘带 02.150

marking cleaning machine 旧线清除机 07.232

Marshall stability 马歇尔稳定度 08.245

Marshall stability test 马歇尔稳定试验 09.082

Marshall stiffness 马歇尔劲度 08.246

masonry arch bridge 圬工拱桥 04.194

masonry bridge 圬工桥 04.193

mass center test 质心位置测定 07.415

mass diagram 土方累积图 02.072

mass goods 大宗货物 06.153

master controller 主控制机 03.304

match casting 镶合浇筑 04.715

material conveyer 材料输送机械 07.099

material management 物资管理 06.396

material mixing uniformity measurement 材料拌和均匀度测定 07.472

materials field investigation 料场调查 01.162

materials test methods 材料试验法 01.073

mattress 柴排 04.141

Mauer expansion installation 毛尔伸缩装置 04.315

maximum dry density 最大干密度 08.202

maximum gathered passenger number 旅客最高聚集人数 06.053

maximum paving capacity measurement 最大摊铺生产率测定 07.480

maximum safe speed 最大安全速度 03.286

maximum service flow rate 最大服务流率 03.177

maximum service volume 最大服务交通量 03.176

maximum speed 最高车速 03.078

maximum superelevation rate 最大超高率 02.140

maximum travelling speed measurement 最高行驶速度测定 07.420

May-curve 梅氏曲线 10.027

MC 曲线中点 02.103

McConnell's curve 麦康奈尔螺旋线 02.095

mean grade of vehicle technical condition 车辆平均技术等级 05.075

mean highest water level 平均最高水位 04.094

mean lowest water level 平均最低水位 04.095

mean time effective rate 平均时间有效率 07.453

mean time interval between failures 平均故障间隔时间 07.452

mean water level 常水位 04.096

measurement and payment 计量与支付 01.136

mechanical handling 机械装卸 06.276

mechanical ventilation 机械通风 04.908

median 中间带 02.147

median barrier 中央分隔带护栏 03.445

median dampness type 中湿类型 01.165

median drainage 中间带排水 02.513

median opening 中央分隔带开口 02.149

median separator 中央分隔带 02.148

medium breaking emulsified bitumen 中裂乳化沥青 08.010

medium bridge 中桥 04.170

medium-curing liquid asphalt 中凝液体沥青

08.018

medium repair of road machine 筑路机械中修 07.326

medium sand 中砂 08.079

medium traffic 中等交通 02.385

medium tunnel 中长隧道 04.855

medium water level bridge 中水位桥 04.173

meeting 会车 05.100

melt sinking 溶陷 02.650

member 构件，*杆件 04.533

member support 构件支撑 04.891

membrane curing 薄膜养生 02.484

membrane curing solution 薄膜养生液 08.152

merging end 汇合端，*合流端 02.221

metal fiber 金属纤维 08.154

metal [material] 金属[材料] 08.116

metal mesh 金属网 08.134

metal rollers with neoprene plate joint 金属滚轴带氯丁橡胶板接缝 04.312

metamorphic concrete 变性混凝土 08.110

meteorologic data 气象资料 04.041

meteorologic station 气象站 04.046

method for determining the lime content 石灰含量测定法 09.059

micro-aggregate concrete 小石子混凝土 08.114

micro glass bead 玻璃微珠 08.172

micro-traffic model 微观交通模型 03.008

mid point of curve 曲线中点 02.103

mileage open to traffic 通车里程 02.548

mileage post 里程标 02.193

mile stone 里程碑 02.194

military bridge 军用桥 04.152

milling groove 铣槽 02.482

millisecond delay blasting 毫秒爆破 02.329

mineral aggregate 矿料 08.050

mineral powder 矿粉 08.053

mine tunneling method 矿山法，*钻爆法 04.868

minimum height of fill 最小填土高度 02.282

minimum length of horizontal curve 平曲线最小长度 02.080

minimum length of vertical curve 竖曲线最小长度 02.121

minimum radius of horizontal curve 平曲线最小半径

02.077

minimum safe lateral clearance 侧向最小安全间距 05.101

minimum stated-speed sign 最低限速标志 03.388

minimum steady travelling speed measurement 最低稳定行驶速度测定 07.421

minimum turning radius [汽车]最小转弯半径 02.079

mini-pile 微型桩，*小型桩 04.429

minitexture meter 手推式构造深度仪 09.184

minor failure 一般故障 05.192

miscellaneous charges for freight transport 货物运输杂费 06.491

miscellaneous charges for passenger transport 旅客运输杂费 06.492

misfiring 缺火 05.213

misloading 错装 06.220

misshipment 错运 06.222

missing a bus 漏乘 06.107

mixed loading 混装 06.156

mixed stuffing 拼箱 06.263

mixed toll system 混合式收费系统 03.348

mixed traffic 混合交通 03.002

mixed traffic highway 一般公路 02.023

mixer 搅拌器 07.142

mixing 拌和 02.443

mixing depth measurement 拌和深度测定 07.468

mixing power measurement 拌和功率测定 07.470

mixing width measurement 拌和宽度测定 07.467

mixture 混合料 08.090

mixture hopper elevator 混合料斗式升送机 07.156

mixture slat elevator 混合料刮板升送机 07.155

mobile asphalt mixing plant 移动式沥青混合料搅拌设备 07.126

mobile crushing plant 移动式联合碎石机组 07.096

mobile handling machinery 移动式装卸机械 06.279

mobile stabilized soil mixing plant 移动式稳定土厂拌设备 07.110

model spectrum of machine 机械型谱 07.350

model test 模型试验 07.376

modified asphalt　改性沥青　08.029

modified compaction test　重型击实试验，＊修正击实试验　09.014

modified Proctor test　重型击实试验，＊修正击实试验　09.014

modulus of deformation　形变模量　02.397

modulus of elasticity　弹性模量　02.398

modulus of resilience　回弹模量　02.399

modulus ratio　模量比　02.400

moist curing　湿法养生　02.483

mollisoil　软土　01.279

moment　力矩，＊弯矩　04.557

moment distribution method　力矩分配法，＊弯矩分配法　04.560

moment envelope　力矩包络线　04.561

montmorillonite　蒙脱土，＊蒙脱石　01.291

mortar consistency tester　砂浆稠度仪　09.166

mosaic pavement　嵌花式路面　02.427

motor coach sleeper　卧铺客车　05.018

motor grader　自行式平地机　07.014

motor jack hammer　内燃凿岩机　07.019

motor noise　发动机噪声　10.054

motor scraper　自行式铲运机　07.008

motor transport auxiliary production cost　汽车运输辅助生产成本　06.428

motor transport comprehensive cost　汽车运输综合成本　06.429

motor transport cost　汽车运输成本　06.414

motor transport cost accounting object　汽车运输成本计算对象　06.418

motor transport cost accounting unit　汽车运输成本计算单位　06.419

motor transport cost analysis　汽车运输成本分析　06.416

motor transport cost by vehicle type　汽车运输车型成本　06.424

motor transport cost control　汽车运输成本控制　06.438

motor transport cost item　汽车运输成本项目　06.421

motor transport cost reduction amount　汽车运输成本降低额　06.431

motor transport cost reduction rate　汽车运输成本降低率　06.430

motor transport cost scope　汽车运输成本范围　06.420

motor transport cost structure　汽车运输成本结构　06.422

motor transport enterprise cost　汽车运输企业成本　06.425

motor transport forecast cost　汽车运输预测成本　06.417

motor transport handling cost　汽车运输装卸成本　06.427

motor transport handling station　汽车货运装卸站　06.292

motor transport marginal cost　汽车运输边际成本　06.439

motor transport operation cost　汽车运输运营成本　06.435

motor transport opportunity cost　汽车运输机会成本　06.440

motor transport planned cost　汽车运输计划成本　06.415

motor transport rate management system　汽车运价管理体制　06.463

motor transport sunk cost　汽车运输沉落成本　06.441

motor transport tariff rate　汽车运价率　06.445

motor transport tariff system　汽车运价体系　06.443

motor transport unit vehicle cost　汽车运输单车成本　06.423

motor transport vehicle cost　汽车运输车辆费用　06.436

motor transport vehicle overhaul depreciation　汽车运输车辆大修折旧　06.426

[motor] truck　载货汽车，＊货车　05.005

motor vehicle　汽车　05.004

motor vehicle transport　汽车运输　06.005

motorway　汽车专用公路　02.022

mountain terrain　山岭区　01.224

mounting and dismounting cost　安装拆卸费　07.338

mouthly average daily traffic volume　月平均日交通量　03.059

movable asphalt mixing plant　可搬式沥青混合料搅拌设备　07.125

movable bearing　活动支座　04.446

movable bridge　开启桥，＊活动桥　04.250

movable stabilized soil mixing plant　可搬式稳定土厂拌设备　07.109

moving charge　搬移费　06.498

moving observer method　随车观测法　03.102

moving resistance measurement　行驶阻力测定　07.426

mower　剪草机　07.233

100m stake　百米桩　01.190

muck　腐殖土　01.287

mud　淤泥　01.276

mud avalanche　泥石流　01.270

mud pump　泥浆泵　07.260

multi-bucket excavator　多斗挖掘机　07.047

multi cell box girder　多室箱梁　04.266

multi channel digital record system　多通道数字记录系统　09.244

multi-function pontoon　多用浮箱　04.966

multi-lane　多车道　02.165

multi-level interchange　多层立交　02.227

multiple-channel queue　多路通道排队　03.031

multiple collision　多次碰撞　03.226

multiplex traffic accident prediction method　多角度交通事故预测法　03.236

multi-purpose pile frame　多功能桩架　07.245

multi pylon cable stayed bridge　多塔斜拉桥　04.243

multistage centrifugal pump　多级离心泵　07.269

multi-use maintenance truck　多用养护车　07.215

multiway stop control　多路停车控制　03.269

mushroom-type tunneling method　蘑菇形开挖法　04.878

N

naked goods　裸装货物　06.137

narrow bridge sign　窄桥标志　03.383

national defence highway　国防公路　01.345

national main trunk line　国道主干线　01.340

national trunk highway　国家干线公路，＊国道　01.341

NATM　新奥法　04.871

natural asphalt　天然沥青　08.006

natural consistency test　天然稠度试验　09.008

natural scour　天然演变冲刷　04.110

natural scour of channel　河槽天然冲刷　04.111

natural subsoil　天然地基　02.311

natural ventilation　自然通风　04.907

natural water content　天然含水量　08.194

navigable clearance　通航净空　04.024

navigable span　通航桥孔　04.011

navigable water level　通航水位　04.099

NDT　无损检测　09.120

near-side lane　外侧车道　02.178

negative friction of pile　下拉荷载，＊桩负摩擦力　04.497

negative response zone　负响应区　09.114

neglected loading　漏装　06.221

neolian soil　风积土　01.296

nerve function measurement　神经机能测定　03.496

net highway transport production output value　公路运输净产值　06.381

net-shaped crack　网裂　02.602

New Austrian Tunneling Method　新奥法　04.871

New Jersey safety barrier　新泽西护栏　03.449

NF　噪声系数　10.044

night construction noise　夜间施工噪声　10.057

no entry sign　禁止驶入标志　03.394

noise factor　噪声系数　10.044

noise isoline　噪声等值线　10.043

noiseless pavement　无噪声路面　10.062

noiseless piling technique　无噪声打桩技术　10.061

noise level　噪声等级　10.042

noise level test　噪声试验　07.449

noise margin　噪声容限　10.039

noise pollution　噪声污染　10.040

noise rating number　噪声检定值　10.041

no left turn　禁止左转弯　03.195

no leftturn sign 禁止左转弯标志 03.391

nominal diameter 标称直径 08.138

non-bonded tendon 无粘结的预应力筋 04.628

non-destructive test 无损检测 09.120

non-displacement pile 非排土桩 04.434

non-ionic emulsified bitumen 非离子乳化沥青 08.015

non-navigable span 非通航桥孔 04.012

non-power-driven vehicle 非机动车辆 05.003

non-power-driven vehicle transport 非机动车运输 06.004

non-pressure culvert 无压力涵洞 04.810

non-prestressed reinforcement 非预应力钢筋 04.625

non-scheduled run 非定班运行 06.070

non-scouring velocity 止冲流速 04.059

non-signalized crossing 无信号交叉口 02.215

nonweaven geotextile 无纺土工织物 08.165

nonweaving traffic flow 非交织交通流 03.046

non-woven fabrics expansion installation 无纺布伸缩装置 04.316

no right turn sign 禁止右转弯标志 03.390

norm 定额 01.076

normal luggage 一般行包 06.081

normal test 常规试验 07.391

normal wear 正常磨损 05.174

notification of award 中标通知书 01.122

no U-turn sign 禁止调头标志 03.389

nozzle bitumen emulsifying machine 喷嘴式沥青乳化机 07.181

nozzle tester 喷嘴试验器 05.343

NRN 噪声检定值 10.041

nuclear moisture and density dynamic meter 核子动态湿度密度仪 09.153

nuclear moisture and density meter 核子湿度密度仪 09.151

nuclear moisture and density static meter 核子静态湿度密度仪 09.152

number of runs 班次 06.064

nursery 苗圃 02.590

NWL 通航水位 04.099

O

object marking 立面标记 03.427

observation point 测点 01.180

OD study 起迄点调查，＊OD 调查 01.161

official test 正式试验 07.402

off-road parking 路外停车 03.154

offset 相位差，＊[信号]时差 03.251

offset intersection 错位交叉 02.205

offset pattern 绿灯时差型式 03.255

oil pox 油包 02.612

oil pumping 窜油 05.209

one dimension collision 一维碰撞 03.222

one-way loading transport 单程运输 06.026

one-way ramp 单向匝道 02.240

one-way road 单行路 02.166

one-way sign 单行线标志 03.397

one-way slab 单向板 04.284

on-line control 联机控制 03.271

on schedule 正点 06.071

on-site dispatching 现场调度 06.207

on vehicle survey 随车调查 06.133

open caisson foundation 沉井基础 04.403

open culvert 明涵 04.802

open cut foundation 明挖基础 04.401

open cut method 明挖法 04.865

open cut tunnel 明洞 04.845

open face blasting 多面临空爆破 02.332

open gradation 开级配 08.190

open spandrel arch bridge 空腹拱桥 04.216

open toll system 开放式收费系统 03.346

open to traffic 开放交通，＊通车 01.352

open type crushing plant 开式联合碎石机组 07.097

operating distance measurement 操作行程测定 07.418

operating force measurement 操作力测定 07.417

operating luminance test 工作照度试验 07.451

operating performance test 作业性能试验 07.439

operating record of machine 机械运转记录 07.352

operating speed 运行车速 03.077

operating weight measurement 使用重量测定 07.412

operational delay 行驶延误 03.093

operational route map 营运线路图 06.131

operation area 营运范围 06.010

operation certificate 营运证 06.408

operation certificate of machinery 机械操作证 07.344

operation cost 营运费用 06.385

operation dispatching 执行调度 06.206

operation expenditure 营运支出 06.391

operation manner 营运方式 06.009

operation mileage 营运里程 06.066

operation mileage rate 行程利用率, * 里程利用率 06.370

operation of vehicle maintenance 汽车维护作业 05.262

operation rate of machinery 机械利用率 07.329

operation revenue 营运收入 06.390

operation route 营运线路 06.055

operation signs 营运标志 06.410

operation speed 营运速度 06.369

operation status of truck 货车动态 06.215

operation subsidy 行车津贴 06.395

optic fiber sensor for crack monitor 光纤测裂计 09.215

optic fiber sign 光纤标志 03.408

optimization of traffic line and network 交通线网优化 03.040

optimum density 最佳密度 03.085

optimum design of bridge 桥梁优化设计 04.511

optimum gradation 最佳级配 08.192

optimum speed 最佳速度 03.083

optimum water content 最佳含水量 08.196

ordinary highway 一般公路 02.023

ordinary ticket price 普通客票价 06.477

organic binder 有机结合料 08.002

organic matter content 有机质含量 08.218

organic matter content test 有机物含量试验 09.034

organic polymer 有机聚合物 08.034

organic soil 有机[质]土 01.286

oriental cone anchorage OVM 锚具 04.646

original piece receipt and delivery 原件收付 06.093

origin-destination study 起迄点调查, * OD 调查 01.161

orthotropic slab 正交异性板 04.281

oscillatory roller 振荡压路机 07.068

outer separator 外侧分隔带 02.146

outlet 出水口 02.539

outlet submerged culvert 压力式涵洞 04.808

out of control 失控 05.204

outside gear asphalt pump 外啮合齿轮沥青泵 07.165

overall dimensions measurement 外形尺寸测定 07.411

over coarse-grained soil 巨粒土 08.059

over-distance stacking operation 超远作业 06.299

over-head loader 过顶式装碴机 07.309

overhead traffic sign 门式交通标志 03.370

overheating 过热 05.200

over-height stacking operation 超高作业 06.300

overlay 罩面 02.351

over-limit pollution 污染超限 05.206

overpass 上跨式立交 02.224

oversize or overweight goods 超限货物 06.144

overtaking 超车 05.099

overtaking lane 超车车道 02.181

overtaking prohibited sign 禁止超车标志 03.392

overtaking sight distance 超车视距 02.108

overturn angle tester 倾斜角度试验机 05.331

overturning stability test 倾翻稳定性试验 07.442

own account transport 非营业性运输, * 自用运输 06.020

owner 业主 01.101

oxidized asphalt 氧化沥青 08.026

P

package indicative mark 包装储运标志 06.163

package mark for hazardous goods 危险货物包装标志 06.164

package run 套班运行 06.078

packed goods 包装货物 06.138

packing for transport 运输包装 06.159

pad bearing 垫块支座 04.456

paddle bitumen emulsifying machine 叶桨搅拌式沥青乳化机 07.180

padfoot roller 凸块压路机 07.062

pallet [货物]托盘 06.244

panel 节间 04.534

panimatric photo 综合法测图 01.239

paper location 纸上定线 02.071

paraffin content 含蜡量 08.239

paraffin content test 含蜡量试验 09.073

parallel parking 纵列式停车 03.198

parameter diagnosis 参数诊断 05.230

parameters of operation process 工作过程参数 05.167

parameters of vehicle technical condition 汽车技术状况参数 05.166

parent soil 原状土 01.293

park and drive 存车搭乘 03.107

park and ride 存车换乘 03.106

parking 停车场 02.185, 停车，*存车，*泊车 03.105

parking accumulation 停车车辆总数 03.149

parking area 停车场 02.185

parking ban sign 禁止停放标志 03.398

parking bay 港湾式停车处 02.184

parking deficiency 停车车位短缺 03.150

parking demand 需要停车次数 03.148

parking duration 停车持续时间 03.145

parking fee 停车费 03.158

parking index 车辆停放指数 03.201

parking lane 停车车道 02.174

parking lift 举升泊车机 05.361

parking lot sign 停车场标志 03.367

parking management program 停车管理计划 03.156

parking meter 停车计时器 03.341

parking plan 停车计划 03.155

parking prohibited 禁止停车 03.197

parking rate 停车率 03.144

parking regulation 停车规则 03.211

parking set 停车车位 03.143

parking space limit marking 停放车位标线 03.424

parking supply 停车车位供应量 03.151

parking surplus 停车车位过剩 03.152

parking turnover 停车周转次数 03.146

park-way 风景区道路 02.007

partial circulation operation 小循环运行 06.076

partial clover-leaf interchange 半苜蓿叶形立交 02.229

partial control of access 部分控制进入，*半封闭 02.236

partial failure 局部故障 05.189

partial load mode 部分负荷工况 05.127

partially prestressed concrete 部分预应力混凝土 04.613

partial pressure culvert 半压力式涵洞 04.809

particles 微粒物 10.022

particulate lead 铅粒 10.016

partite transport 分运 06.180

parts repair 零件修理 05.285

parts washer 零件清洗机 05.365

pass 垭口 01.230

passage vehicle detector 通过型车辆检测器 03.293

passenger 旅客，*乘客 06.035

passenger average haul distance 旅客平均运距 06.345

passenger claims 客运索赔 06.114

passenger flow 客流 06.036

passenger number fluctuation coefficient 旅客波动系

数 06.054

passenger source 客源 06.050

passengers transport regularity rate 旅客正运率 06.123

passenger terminal work load 客运站站务工作量 06.051

passenger traffic 客流 06.036

passenger traffic diagram 客流图 06.049

passenger traffic flow direction 旅客流向 06.048

passenger traffic time 旅客流时 06.047

passenger traffic volume 旅客流量 06.046

passenger transport 旅客运输 06.018

passenger unexpected injury 旅客意外伤害 06.110

passing bay 错车道 02.173

passing bay in tunnel 错车洞 04.842

passing bay sign 让车道标志 03.395

passing bridge with reduced speed 减速过桥 01.356

passing lane 超车车道 02.181

passive safety 被动安全性 05.143

patching 补坑 02.664

path 小道 02.009

patrol maintenance 巡回养护 02.568

paved highway 有路面公路 02.039

paved material compaction test 摊铺材料压实度试验 07.482

paved ratio 路面铺装率 02.551

paved surface evenness test 摊铺路面平整度试验 07.481

pavement 路面 02.333

pavement characteristics 路面特性 02.363

pavement condition index 路面状况指数 02.700

pavement damage ratio 路面破损率 02.688

pavement depression 路面沉陷 02.632

pavement design 路面设计 02.362

pavement drainage 路面排水 02.512

pavement engineering 路面工程学 01.006

pavement evaluation model 路面评价模型 02.701

pavement investigation 路面调查 02.561

pavement lettering marking 路面文字标记 03.428

pavement management system 路面管理系统 02.698

pavement monitor and evaluation system 路面检测评价系统 02.699

pavement operating quality index 路面使用质量指标 02.687

pavement periodical investigation 路面定期调查 02.563

pavement random investigation 路面随机调查 02.562

pavement recapping 路面翻修 02.580

pavement recycling 路面再生 02.673

pavement skid resistance condition 路面抗滑性能 02.703

pavement strength 路面强度 02.387

pavement strengthening 路面补强 02.581

pavement strengthening design 路面补强设计 02.411

pavement structural composition design 路面结构组合设计 02.412

pavement structure 路面结构 02.338

pavement sweeping machine 路面清扫机 07.221

pavemill 路面铣削机 07.207

paving 摊铺 02.444

paving depth measurement 摊铺厚度测定 07.478

paving operation test 摊铺作业试验 07.479

paving width measurement 摊铺宽度测定 07.477

PC 直圆点 02.096

PCC 复曲线点 02.102

PDA 打桩分析仪 09.224

peak flow discharge 洪峰流量 04.068

peak hour factor 高峰小时系数 03.056

peak response 峰值响应 09.112

pear-ended spur dike 梨形丁坝 04.131

peat 泥炭 01.278

pedestrian accidents 撞人事故 03.219

pedestrian barrier 行人护栏 03.489

pedestrian bridge 人行桥 04.149

pedestrian control 行人管制 03.207

pedestrian crossing devices 行人横穿设施 03.485

pedestrian crossing sign 行人过街标志 03.375

pedestrian crossing signal lamp 行人过街信号灯 03.306

pedestrian detector 行人检测器 03.285

pedestrian load 人群荷载 04.486

pedestrian overpass 人行天桥 03.488

pedestrian push-button signal controller 行人按钮信号控制机 03.305

pedestrian safety devices 行人安全设施 03.484

pedestrian signal phase 行人信号相位 03.257

pedestrian trip 徒步出行 03.132

pedestrian underpass 人行地道 03.487

peeled resurfacing 剥层翻修 02.672

penetration 针入度 08.223

penetration index 针入度指数 08.224

penetration macadam with coated chips 上拌下贯式路面 02.419

penetration method 贯入法 02.442

penetration test 针入度试验 09.061

penetration test apparatus 贯入仪 09.154

percentage of coring 岩心采取率 01.334

per container charge 集装箱箱次费 06.484

percussion and grabbing method 冲抓法 04.755

performance bond 履约担保 01.124

performance test 性能试验 07.396

periodical inspection 定期检查 02.558

periodical maintenance 定期养护 02.569

period of limitation for freight claims 货运索赔时限 06.228

period of limitation for passenger claims 客运索赔时限 06.117

perishable goods 易腐货物 06.146

permafrost 多年冻土, *永冻土 01.272

permanent bridge 永久性桥 04.178

permanent load 永久荷载 04.477

permeability coefficient 渗透系数 08.210

permeable embankment 透水路堤 02.250

permissible clearance 允许间隙 05.303

permissible wear 允许磨损 05.176

personalized repair method 就车修理法 05.311

perspective examination 透视图检验 02.056

pervious test 透水度试验 09.011

pervious wearing course 透水磨耗层 02.345

petroleum asphalt 石油沥青 08.005

phase 相位 03.244

phase offset 相位差, *[信号]时差 03.251

phenol content 酚含量 08.243

phenol content test 酚含量试验 09.078

PHF 高峰小时系数 03.056

pH meter 酸度计 09.220

photochemical smog 光化学烟雾 10.009

photoelastic analysis 光弹性试验 09.094

photoelastic test installation 光测弹性试验装置 09.194

photogrametry 摄影测量 01.231

photographic map 影像地图 01.236

photographic road survey group 路面病害摄影组合仪 09.175

photo-index 像片索引图, *镶辑复照图 01.237

photo-mosaic 像片镶嵌图 01.242

pH value 酸碱度值, *pH值 08.214

physical separator 实物分隔带 03.447

physiological fatigue 生理疲劳 03.519

physiological function measurement 生理机能测定 03.497

PI 塑性指数 08.206

pick-up empty container 提取空箱 06.264

piece goods 件货 06.139

piece luggage 计件行包 06.082

pier 桥墩 04.359

pier body 墩身 04.360

pier break-water 桥墩分水尖 04.138

pier cap 墩帽 04.361

pier coping 墩帽 04.361

pier-foundation 墩式基础 04.408

piezoelectric transducer 压电式传感器 09.238

pile 桩 04.412

pile-bent foundation 排架桩基础 04.410

pile-bent pier 排架桩墩 04.370

pile by percussion drill method 冲孔成桩法 04.754

pile cap 承台 04.440

pile driver 打桩机 07.237

pile driving 打桩 04.759

pile driving analyzer 打桩分析仪 09.224

pile driving method by hammer 锤击沉桩法 04.760

pile extracting 拔桩 04.763

pile foundation 桩基础 04.409

pile frame 桩架 07.244

pile group 群桩, *桩群 04.414

pile integrity test 桩完整性试验 09.139

pile jacking-in method 静压沉桩法 04.758

pile jetting method 射水沉桩法 04.762

pile penetration test 桩贯入试验 09.138

pile-plank retaining wall 柱板式挡土墙 02.272

pile vibrosinking method 震动沉桩法 04.761

pilot 导洞 04.846

pilot-tunneling method 导洞法 04.873

pin-connected plywood girder bridge 钉板梁桥 04.201

pipe culvert 管涵 04.805

pipe drainage 管道排水 02.522

pipe guardrail 管式护栏 03.433

piping 管涌 02.644

piston action of moving vehicles 车辆活塞作用 04.923

piston pump 活塞式水泵 07.266

piston slap 敲缸 05.211

pitch 硬沥青 08.025

pitching 铺砌 02.449

pitching pavement 弹街路面 02.429

pit test 坑探 01.252

pitting 点蚀 05.180

PL 塑限 08.205

place name sign 地名标志 03.360

plain bar 光面钢筋 08.123

plain stage of slope 边坡平台 02.263

plain terrain 平原区 01.221

plane structure for bridge 桥梁平面结构 04.523

planned dispatching 计划调度 06.205

planned transport 计划运输 06.023

planning assignment 计划任务书 01.041

plan of bridge site 桥位平面图 01.303

plan sketch 平面示意图 02.059

plan survey 平面测量 01.186

planting mileage 绿化里程 02.550

planting protect slope 植被固坡 10.077

plant-mixing method 厂拌法 02.437

plant protection 植物保护 10.082

plastic concrete 塑性混凝土 08.108

plasticity index 塑性指数 08.206

plastic limit 塑限 08.205

plastic limit test 塑限试验 09.007

plastic mortar strength test 软练砂浆强度试验 09.056

plate bearing 平板支座 04.454

plate-bearing test 承载板试验 09.023

platform trailer 平板挂车 05.026

platoon 车队 03.001

PMS 路面管理系统 02.698

pneumatic caisson foundation 气压沉箱基础 04.405

pneumatic rock breaker 气动岩石破碎机 07.032

pneumatic rock drill 气动凿岩机 07.021

pneumatic tyred roller 轮胎压路机 07.060

point of compound curve 复曲线点 02.102

point of curve 直圆点 02.096

point of curve to spiral 圆缓点 02.100

point of spiral to curve 缓圆点 02.099

point of spiral to tangent 缓直点 02.101

point of tangent 圆直点 02.097

point of tangent to spiral 直缓点 02.098

Poisson's ratio 泊松比 08.255

pole trailer 长货挂车 05.040

police patrol 警察巡逻 03.282

polished stone value 石料磨光值 08.254

polished stone value test 石料磨光值试验 09.037

polishing 磨光 02.621

pollutants of idle speed emission 怠速排放污染物 10.013

polyethylene asphalt unit 聚乙烯沥青装置 07.185

polyethylene pipe 聚乙烯管, * PE 管 08.157

polyfluortetraethylene plate 聚四氟乙烯板 08.159

polymeric chloroprene rubber 氯丁橡胶 08.161

polymer modified asphalt 聚合物改性沥青 08.030

polypropylene belt 聚丙烯筋带 08.168

polytetrafluoroethylene bearing 聚四氟乙烯支座 04.450

polyvinyl chloride pipe 聚氯乙烯管, * PVC 管 08.158

pontoon bridge 舟桥 04.248

pore pressure cell 孔隙水压力计 09.228

porosity 孔隙率 08.176

portable pendulum tester 摆式仪 09.178

portal clearance 隧道口净空 04.852

portal frame 门桥 04.965

portal framed tower 门形索塔 04.323

portland cement 硅酸盐水泥 08.041

positive response zone　正响应区　09.113

possible capacity　可能通行能力　03.165

post-assessment　后评价　01.046

post delineator　柱式轮廓标　03.452

post lift　柱式举升机　05.359

post road〔Tang Dynasty〕　驿道　02.018

post-tensioning method　后张法　04.723

post-tensioning strand group anchorage　YM 锚具　04.644

post traffic sign　立柱式交通标志　03.372

potential test device　电位测量装置　09.216

potential test wheel　电位轮　09.217

pot holes　坑槽　02.617

pot rubber bearing　盆式橡胶支座　04.449

power-driven vehicle　机动车辆　05.002

power-driven vehicle transport　机动车运输　06.003

powered ferry boat　机动渡车船　04.959

power shovel　单斗挖掘机　07.038

precast balancing cantilever method　平衡悬拼法　04.663

precast-boxed cofferdam　套箱围堰　04.767

precast lining　拼装式衬砌　04.887

precast pile　预制桩　04.418

precious belongings　贵重行包　06.084

pre-departure operation　发运　06.179

prefab trailer　预制件挂车　05.041

pre-feasibility study　预可行性研究　01.039

preference for domestic bidders　国内投标人优惠　01.123

preheating　预热　05.087

preliminary design　初步设计　01.042

preliminary survey　初测　01.176

preloading centering　拱架预压　04.697

preloading method　预压法　02.316

presence vehicle detector　存在型车辆检测器　03.292

present service index　现有服务指数　02.374

pressure cell　压力盒　09.206

pressure culvert　压力式涵洞　04.808

prestressed concrete bridge　预应力混凝土桥　04.184

prestressed concrete pile　预应力混凝土桩　04.420

prestressing bed　预应力台座　04.730

prestressing by cold-drawn rate　冷拉率控制张拉　04.727

prestressing by subsequent bond　复传力法预应力　04.620

prestressing under stress control　应力控制张拉　04.726

prestressing with bond　体内预应力　04.617

prestressing without tendon by pre-bending　预弯法预应力　04.619

prestressing with subsequent compression and tension　压拉双作用预应力　04.621

pretensioning method　先张法　04.722

pretimed controller　定时信号控制机　03.303

prevention and cure against highway disaster　公路灾害防治　02.585

preventive maintenance　预防性养护　02.566, 预防维护　05.257

preventive repair　预防性修理　04.796

price adjustment　价格调整　01.138

primary collision　一次碰撞　03.224

prime coat　透层　02.343

priority goods　重点货物　06.152

priority lane　优先通行车道　03.189

priority materials transport plan　重点物资运输计划　06.335

priority phase offset　优先时差　03.252

private vehicle　自用车辆　05.057

probabilistic limit state design method of bridge　桥梁概率极限状态设计法　04.510

Proctor compaction test　轻型击实试验，＊普氏击实试验　09.013

production program of vehicle maintenance　汽车维护生产纲领　05.268

production program of vehicle repair　汽车修理生产纲领　05.312

production quota of machinery　机械生产定额　07.345

productive capacity measurement　生产率测定　07.443

productivity of driver and assistant　司机及助手劳动生产率　06.393

profile　纵断面　02.111

profile of bridge axis　桥轴线纵剖面　01.305

profile survey 纵断面测量 01.203

programmable data logger 可编程序数据采集器 09.243

progressive analysis for bridge erection 顺序拼装分析，*正向分析 04.778

prohibitory sign 禁令标志 03.355

prompt test 快速试验 07.394

proportioning of cement concrete 水泥混凝土配合比 08.258

protection course 保护层 02.346

prototype test 样机试验 07.395

provincial trunk highway 省干线公路，*省道 01.342

proving test 验证性试验 07.400

provisional sums 暂定金额 01.139

PSI 现有服务指数 02.374

PSRC 路面抗滑性能 02.703

PT 圆直点 02.097

PTFE bearing 聚四氟乙烯支座 04.450

pulling to one side 跑偏 05.219

pumping drainage 泵站排水 02.526

p-y curve method p-y 曲线法 04.655

Q

quadri-hinge-pipe culvert 四铰管涵 04.812

quality accident 质量事故 06.217

quality index 质量指标 06.339

quality level of highway maintenance 公路养护质量等级 02.689

quality management of passenger transport 客运质量管理 06.105

quantitative index 数量指标 06.338

quarry waste 石场弃渣 08.052

quasi-orthotropic slab method 准正交各向异性板法 04.582

queue rule 排队规则 03.029

queue theory 排队理论 03.028

quick breaking emulsified bitumen 快裂乳化沥青 08.009

quick lime 生石灰 08.038

quick sand 流砂 01.266

quota 定额 01.076

R

rabbet 企口缝 02.463

radar speedometer 雷达测速器 03.294

radial consolidation coefficient 径向固结系数 08.212

radial highway 辐射式公路 02.027

radioactive goods 放射性货物 06.141

rail form concrete paver 轨模式混凝土摊铺机 07.193

railway crossing 铁路平交道口 02.245

railway crossing sign 铁路道口标志 03.412

rainstorm intensity 暴雨强度 04.080

rainstorm run-off 暴雨径流 04.079

raised separator 突起分隔带 03.448

raked pile 斜桩 04.415

rake-up ballast remover 蟹耙式清岩机 07.036

ramp 匝道 02.237

ramp bridge 匝道桥，*坡道桥 04.162

ramp capacity 匝道通行能力 03.170

ramp integrated system control 匝道集成系统控制 03.275

ramp junction 匝道连接处 03.180

ramp marking 匝道标线 03.426

ramp metering 匝道交通调节 03.191

random failure 随机故障 05.195

rapid-curing liquid asphalt 快凝液体沥青 08.017

rapid transit system 快速公共交通系统 03.141

rate 运价 06.442

rate addition 运价加成 06.456

rated compensation 定额赔偿 06.223

rated passenger number 客车定员 06.063

rated seat 客位 06.355

rated tonnage 吨位 06.354

rate for small vehicle 小型车运价 06.469

rate for special goods 特种货物运价 06.467

rate for special purpose vehicle 特种车辆运价 06.468

rate of empty container 集装箱空箱运价 06.483

rate of freight compensation 赔偿率 06.233

rate of good level highway 好路率 02.690

rate of loaded container 集装箱重箱运价 06.482

rate of on-schedule bus runs 客车正班率 06.120

rate of on-time departure 发车正点率 06.122

rate of stops at scheduled bus stops 定站停靠率 06.121

rate of timely freight 货运及时率 06.234

rate ratio 运价比差 06.462

rate reduction 运价减成 06.457

rating 定额 01.076

RCCP 碾压混凝土路面 02.455

reaction frame for loading 加载反力架 09.196

reaction type brake tester 反力式制动试验台 05.328

real-time traffic control 实时交通控制 03.263

reaming-after-boring method 先钻后扩法 04.751

rear-end collision 头尾相撞 03.228

rebound deflection 回弹弯沉 02.388

rebound tester 回弹仪 09.211

rebound test of concrete 混凝土回弹试验 09.122

reciprocating compressor 往复式空气压缩机 07.307

reciprocating feeder 往复给料机 07.091

reclaimed asphalt mixture 再生沥青混合料 08.097

reclaim of compensation 收回赔偿 06.224

reconditioning 修复 05.287

reconnaissance 踏勘 01.175

reconstructed bridge 改建桥 04.165

recreation area 休息娱乐区 02.191

rectangular beam 矩形梁 04.271

red clay 红粘土 01.284

red time 红灯时间 03.246

reference stake 护桩 01.191

reflecting button 反光路钮 03.476

reflecting marking 反光标识 03.474

reflecting sign 反光标志 03.407

reflection crack 反射裂缝 02.601

reflective materials 反光材料 08.173

reflux of freight 重复运输 06.029

refrigerated trailer 冷藏挂车 05.039

refrigerated vehicle 冷藏货车 05.009

refuge island 安全岛 02.219

regular service route 班车线路 06.056

regulating structure 调治构造物 04.126

regulatory traffic sign 管制标志 03.373

rehabilitation work 修复工程 02.583

reinforced concrete bridge 钢筋混凝土桥 04.183

reinforced concrete fence 钢筋混凝土护栏 03.441

reinforced concrete pavement 钢筋混凝土路面 02.452

reinforced earth retaining wall 加筋土挡土墙 02.275

reject of freight claims 货运拒赔 06.227

reject of passenger claims 客运拒赔 06.116

relative density 相对密实度 08.200

relative density test 相对密度试验 09.003

relative elevation 相对高程 01.209

relative water content 相对含水量 08.195

relay transport 接力运输 06.012

reliability datum period 可靠度基准期 04.513

reliability of bridge structure 桥梁结构可靠度 04.512

reliability test 可靠性试验 07.397

remanent value of equipment 设备残值 07.359

remote sensing of engineering geology 工程地质遥感测量 01.247

removing landslide 清除坍方 02.674

renting of machine 机械出租 07.347

repair after failure 事后修理 05.278

repair before failure 事前修理 05.277

repair of machine 机械修理 07.324

repair on technical conditions 视情修理 05.289

repair size 修理尺寸 05.301

repayment period of investment 投资回收期 01.050

replacement of earth 换土 02.295

replacement unit 周转总成 05.309

research test 研究性试验 07.399

reserve capacity 储备通行能力 03.173

reserve fund of project 工程预备费 01.064

reserve settlement 预留沉落量 02.310

resident supervising engineer　驻地监理工程师　01.107

residual oil　渣油　08.033

residual response zone　残留响应区　09.115

residual vibration auto-spectrum analysis　余振自谱分析　09.117

residue oil pavement　渣油路面　02.422

resistivity test device　电阻率测量装置　09.218

resonance screen　共振筛　07.087

respiratory goods　呼吸性货物　06.149

responsibility system of machinery management　机械管理责任制　07.343

rest area　休息区　02.189

rest area sign　休息区标志　03.401

restrain block　防震挡块　04.379

restricted goods　限运货物　06.155

resultant gradient　合成坡度　02.118

retaining wall　挡土墙　02.265

retarder　缓凝剂　08.150

retention money　保留金　01.137

retrogressive analysis for bridge erection　逆向拆除分析，*反向分析　04.779

return empty container　回送空箱　06.265

return factor　回程系数　06.195

return period of flood　洪水重现期　04.073

return rate of vehicle major repair　汽车大修返修率　05.317

reverse circulation boring method　反循环钻孔法　04.757

reverse circulation drill　反循环钻机　07.252

reverse curve　反向曲线　02.085

reverse loop　回头曲线　02.087

reversible center lane line　可变向中心车道线　03.421

reversible lane　可变向车道　02.180

ribbed arch bridge　肋拱桥　04.225

ribbed slab bridge　肋板桥　04.209

ridge　山脊　01.226

ridge crossing line　越岭线　02.066

ridge line　山脊线　02.064

right-angled intersection　正交叉　02.200

right bridge　正交桥　04.158

right-of-way　公路用地　01.346

right turn guide line　右转弯导向线　03.419

right turn ramp　右转弯匝道　02.243

rigid connected beam method　刚接梁法　04.580

rigid cross beam method　刚性横梁法　04.578

rigid frame bridge　刚构桥　04.230

rigid-framed arch bridge　刚架拱桥　04.223

rigidity　刚度　04.554

rigid pavement　刚性路面　02.450

ring road　环形道路　02.028

ripper　松土机　07.017

riprap protection　抛石防护　04.145

rise of arch　矢高　04.027

rise-span ratio　矢跨比　04.030

river bank　河岸　04.089

river basin　流域　04.075

river bed　河床　04.084

river bed gradient　河床比降　04.061

river bed paving　河床铺砌　04.137

river channel　河槽　04.085

river-crossing bridge　跨河桥　04.150

river survey　河道调查　04.038

riveted steel bridge　铆接钢桥　04.189

riveting　铆接　04.735

road　道路　02.001

road bed　路床　02.359

road bitumen　道路沥青　08.027

roadblock　路障　02.592

road cement　道路水泥　08.045

road economical analysis　道路经济分析　01.055

road engineering　道路工程学　01.002

road geometry data acquisition system　道路几何数据收集系统　09.173

road hypnosis　道路催眠状态　03.518

road inspection　验道　06.186

road machinery　筑路机械　01.016

road marking machine　划线机　07.228

road markings　道路标识　03.413

road mark paint　道路标线漆　08.171

road-mixing machine test　路拌机械试验　07.466

road-mixing method　路拌法　02.436

road parking lot　路上停车点　03.200

road reflecting mirror　道路反光镜　03.473

road section capacity　路段通行能力　03.167

roadside green belt　路旁绿化带　02.696

roadside material terrace　堆料台　02.593

roadside radio broadcasting　路侧广播　03.296

roadside trees　行道树　02.589

road toll　过路费　01.032

road toll system　道路收费系统　03.332

road transport　道路运输　06.001

road trough　路槽　02.360

roadway　路幅　02.158

rock blasting　石方爆破　02.324

rock drill　凿岩机　07.018

rocker bearing　摆式支座　04.453

rock shovel　铲斗式清岩机　07.035

rock stratum　岩层　01.317

Rockwell hardness test　洛氏硬度试验　09.048

rodless reverse circulation rig　无钻杆反循环钻机　07.263

roll crusher　辊筒式碎石机　07.083

rolled shape steel reinforced concrete bridge　钢骨混凝土桥　04.185

roller bearing　滚轴支座　04.452

roller bit　滚轮式钻头　07.259

rolling　碾压　02.301

rolling compacted concrete pavement　碾压混凝土路面　02.455

rolling groove　压槽　02.480

rolling resistance measurement　滚动阻力测定　07.427

rolling slurry process　滚浆法　02.667

rolling speed　碾压速度　02.305

rolling speed measurement　压实速度测定　07.464

rolling terrain　微丘区　01.222

roll on/roll off container fork lift　滚上滚下集装箱叉车　06.307

roll on/roll off transport　滚装运输　06.033

rotary bitumen emulsifying machine　转子式沥青乳化机　07.179

rotary broom-slat sweeper　旋刷刮板清扫车　07.223

rotary bucket elevator　转斗式提升机　07.105

rotary interchange　环形立交　02.234

rotary intersection　环形交叉　02.206

rotary jack hammer　旋转式凿岩机　07.026

rotary snow remover　转子式除雪机　07.217

rotary table rig　转盘式钻机　07.251

rotary tools circular velocity measurement　旋转刀具圆周速度测定　07.469

rotary type soil stabilizer　转子式稳定土搅拌机　07.118

rotary type stabilized soil road mixer　转子式稳定土搅拌机　07.118

rotator crusher　转子式碎石机　07.080

rough tunnel　未衬砌隧道　04.858

roundabout　环形交叉　02.206

roundabout transport　迂回运输　06.027

round trip passenger traffic　往返性客流　06.042

route　路线　02.045

route alignment　路线线形　02.047

route computer aided design　路线计算机辅助设计　02.054

route continuity　路线连续性　02.052

route design　路线设计　02.046

route development　展线　02.068

route guidance system　道路引导系统　03.284

route optimum design　路线优化设计　02.058

route plan　路线平面图　01.185

route selection　选线　02.049

route survey　路线测量　01.184

route three-dimensional space design　路线三维空间设计　02.057

routine dispatching　值班调度　06.208

routine inspection　日常检查　02.557

routine maintenance　例行维护　07.321

routine repair cost　经常维修费　07.333

royal road [Qin Dynasty]　驰道　02.017

rubber asphalt unit　橡胶沥青装置　07.183

rubber joint　橡胶接缝　04.307

rubber modified asphalt　橡胶改性沥青　08.028

rubble　片石　08.069

running-in　走合　05.297

running-in maintenance　走合维护　05.256

running speed　行驶速度　03.075

running time　行驶时间　03.081

run-off　径流　04.078

run-off area　径流面积　04.082

run-off coefficient　径流系数　04.083

run-off-road accident 驶出路外事故 03.230

rupture strength 抗折强度 02.405

rupture test 韧度试验 09.041

rural road 乡村道路 02.008

rush-repair work 抢修工程 02.582

rut 车辙 02.616

S

safety factor 安全系数 04.529

safety fence 防护栅 03.490

safety marked coat 标志服 02.591

safety measure 安全措施 03.213

safety misadventure 安全事故 07.341

salivary flow ice 涎流冰 01.275

salt corrosion 盐侵蚀 02.637

salt fog simulated test 盐蚀模拟试验 07.387

salt heaving 盐胀 02.638

salty soil 盐渍土 01.285

same direction adjacent curve 同向曲线 02.084

sample test 抽样试验 07.407

sand 砂 08.077

sand asphalt 沥青砂 08.094

sand bag well 袋装砂井 02.317

sand barrier 沙障 02.676

sand brooming 匀砂 02.661

sand consolidation by planting 植物固沙 10.078

sand consolidation with biologic 生物固沙 10.079

sand cylinder for centering unloading 卸架砂筒 04.700

sand dune 沙丘 01.267

sand gravel 砂砾 08.076

sand hazard 沙害 02.653

sanding 铺砂 02.659

sand liquefaction 砂土液化 01.301

sand patch test 铺砂法试验 09.083

sand pile 砂桩 02.321

sand protection dike 防沙坝 02.679

sand protection facilities 防沙设施 02.675

sand protection green 防沙林 02.677

sand ratio 砂率 08.260

sand recovering machine 回砂机 07.226

sand spraying machine 撒砂机 07.227

sand sweeping 回砂 02.660

sand sweeping machine 扫砂机 07.225

sand well 砂井 02.323

sandy clay of high liquid limit 含砂高液限粘土 08.088

sandy clay of low liquid limit 含砂低液限粘土 08.087

sandy silt of low liquid limit 含砂低液限粉土 08.084

sandy soil 砂类土 08.062

satellite remote sensing 卫星遥感测量 01.246

saturated water content test 饱水率试验 09.039

saturation volume 饱和流量 03.160

saturation volume rate 饱和流率 03.161

sawn joint 锯缝 02.469

saw-tooth type wharf 锯齿形码头 04.954

SBR 丁苯橡胶 08.160

SC 缓圆点 02.099

scaffold 脚手架,＊膺架,＊工作架 04.704

scaling 脱皮 02.620

scheduled bus transport 班车客运 06.057

scheduled preventive maintenance and repair system 计划预防维修制度 05.244

scheduled repair 计划修理 05.288

scheduled run 定班运行 06.069

schistosity 片理 01.323

Schmidt hammer 回弹仪 09.211

school sign 学校标志 03.411

scoring 刮伤 05.179

scouring 冲刷 02.643

scouring at bridge foundation 桥基冲刷 04.113

scouring depth 冲刷深度 04.118

scour without sediment motion 清水冲刷 04.116

scour with sediment motion 浑水冲刷 04.117

scraper 铲运机 07.007

scraper-dozer 推土铲运机 07.011

scratching 擦伤 05.178

screw air compressor 螺杆式空气压缩机 07.308

screw conveyer 螺旋输送机 07.106

screw snow remover 螺旋式除雪机 07.220

screw type concrete spreader 螺旋式混凝土布料机 07.197

SCRIM 横向力系数测试仪 09.179

seal coat 封层 02.350

sealing-off and covering anchorage 封锚 04.734

seasonal freight source 季节性货源 06.197

seasonal maintenance 季节性维护 05.255

seasonal passenger traffic 季节性客流 06.041

seating steel bridge on supports 钢桥就位 04.739

secondary collision 二次碰撞 03.225

sedimentation basin 沉淀池 02.525

sedimentation joint 沉积节理 01.321

sediment-moving incipient velocity 泥沙起动流速 04.058

seepage water 渗透水 02.503

seepage well 渗井 02.523

segmental precast erection method 节段预制拼装法 04.671

segregation 离析 08.266

segregation of traffic 交通分隔 03.204

seismic acceleration 地震加速度 04.520

seismic force 地震荷载, ＊地震力 04.488

seismic hazard 震害 02.654

seismic region 地震区 01.300

seismic risk analysis 地震危险性分析 04.514

seismic risk evaluation 地震危险性评定 09.147

seizure 咬粘 05.182

self-anchored post tensioning prestressing 后张自锚法预应力 04.622

self-climbing formwork 自升式模板 04.717

self-loading container trailer 集装箱自装卸挂车 05.047

self-owned terminal 自办站 06.128

self-prime pump 自吸式水泵 07.270

self-propelled asphalt paver 自行式沥青混凝土摊铺机 07.160

self-propelled bridge inspection cradle 自行式桥梁检测架 09.199

self-stressing cement 自应力水泥 08.044

semi-actuated signal control 半感应式信号控制 03.266

semi-directional interchange 半定向式立交 02.232

semi-permanent bridge 半永久性桥 04.179

semi-rigid base 半刚性基层 02.491

semi-shield 半盾构 04.942

semi-trailer 半挂车 05.024

semi-trailer train 半挂汽车列车 05.051

semi-transverse ventilation 半横向通风 04.911

semi-underground asphalt storage 半地下沥青贮仓 07.171

separated subgrade 分离式路基 02.254

separation fence 隔离栅 03.491

separation net 隔离网 03.492

separator 分隔带 02.145

serial number of bus run 车次 06.065

series of machine 机械系列 07.351

serviceability limit state of bridge 桥梁使用能力极限状态 04.508

service area 服务区 02.190

service area sign 服务区标志 03.402

service fee 服务手续费 06.501

service load 实用荷载 04.466

service patrol 服务巡逻 03.283

service test 使用试验 07.375

service tunnel 服务隧道, ＊辅助隧道 04.837

setting out 放样 01.141

settlement [桥基]沉降 02.631

settlement joint in culvert 涵口沉降缝 04.825

shaft 竖井 04.843

shaft pier 柱式[桥]墩 04.367

shaft test 井探 01.251

shaking table for structure test 结构振动试验台 09.195

shale tar 页岩沥青 08.024

shallow-buried tunnel 浅埋隧道 04.839

shallow foundation 浅基础 04.399

shaping 整形 02.475

share of machine handling operation 机械装卸作业比重 06.320

shear 剪力 04.558

shear distribution 剪力分布 04.562

shear key 剪力键 04.542

shear lag 剪力滞后, ＊剪滞 04.563

sheath 套管 08.126

sheep-foot roller 羊足压路机 07.065

sheet drainage 排水板法 02.318

sheet pile 板桩 04.428

sheet pile cofferdam 板桩围堰 04.765

shell bridge 扁壳桥 04.229

shelter belt 防护林带 02.697

shield method 盾构法 04.870

ship collision force 船舶撞击力 04.492

shipper 发货人 06.174

shipping mark 收发货标志 06.161

shipping weight measurement 运输重量测定 07.413

shoal 沙洲 04.087

shocking jack hammer 冲击式凿岩机 07.025

shocking rammer 冲击夯 07.073

shocking rotary jack hammer 冲击旋转式凿岩机 07.027

shortage rate of goods 货差率 06.232

short tunnel 短隧道 04.856

shotcrete 喷射混凝土 08.115

shotcrete and rock bolt support 喷锚支护 04.892

shotcrete lining 喷射混凝土衬砌 04.889

shoulder 路肩 02.142

shoulder gap 路肩缺口 02.640

shrinkage 收缩 08.267

shrinkage limit 缩限 08.207

shrinkage test 收缩试验 09.009

side collision 侧向碰撞 03.221

side container fork lift 侧面集装箱叉车 06.308

side ditch 边沟 02.517

side heading method 侧壁导洞法 04.880

side-mooring ferry boat 侧靠式渡船 04.964

side parking 路边停车 03.153

side skidding 侧滑 05.220

side slip tester [汽车]侧滑试验台 05.332

side slope 边坡 02.256

sideway force coefficient routine investigation machine 横向力系数测试仪 09.179

sieve analysis 筛分 08.184

sieving machine 筛分机 07.084

sight distance 视距 02.106

sight distance of intersection 交叉口视距, *路口视距 02.212

sight line 视线 02.104

sight obstruction 视线障碍物 02.109

sightseeing stand 观景台 02.192

sight triangle 视距三角形 02.213

sign 标志 03.352

signal ahead sign 前方信号标志 03.381

signal control 信号控制 03.240

signal controller 信号控制器 03.301

signal cycle 信号周期 03.242

signalized crossing 信号控制交叉口 02.214

signal processor 信号处理机 09.245

sign legibility 标志视认性 03.359

silica fume concrete 硅尘混凝土 08.112

silt 淤泥 01.276

silt content test 含泥量试验 09.033

silting 淤积 04.120

silty sand 粉土质砂 08.082

simple-supported beam bridge 简支梁桥 04.204

simulated cold climate test 低温模拟试验 07.384

simulated dust environment test 风尘环境模拟试验 07.389

simulated environment test 环境模拟试验 07.383

simulated highland climate test 高原气候模拟试验 07.386

simulated hot climate test 高温模拟试验 07.385

simulated load test 负荷模拟试验 07.382

simulated test 模拟试验 07.379

simultaneous traffic accident prediction method 同步交通事故预测法 03.237

single acting pneumatic hammer 单作用气动桩锤 07.240

single acting steam hammer 单作用蒸气桩锤 07.242

single bucket excavator 单斗挖掘机 07.038

single cell box girder 单室箱梁 04.265

single-channel queue 单路通道排队 03.030

single-columned pier 单柱式[桥]墩 04.368

single direction thrusted pier 单向推力墩 04.374

single-hinged arch bridge 单铰拱桥 04.221

single lane 单车道 02.163

single-leaf bascule bridge 单叶竖旋桥 04.255

single-leaf swing bridge 单叶平转桥 04.252

single pass soil stabilizer 一次拌成稳定土搅拌机 07.119

single pass stabilized soil road mixer 一次拌成稳定

土搅拌机 07.119

single pile 单桩 04.413

single plane cable stayed bridge 单索面斜拉桥 04.240

single point loading device on deck 桥面单点加载装置 09.208

single pylon cable stayed bridge 独塔斜拉桥 04.241

single shaft pier 单柱式[桥]墩 04.368

single-slope drainage [路面]单向排水 02.527

sinking open caisson by injected air curtain 空气幕法下沉沉井 04.746

sinking open caisson by slurry coating 泥浆套法下沉沉井 04.745

sinking open caisson by undrained dredging 不排水下沉沉井 04.744

sinking open caisson on built island 筑岛沉井 04.743

sintered aggregate 陶粒 08.055

site inspection 现场考察 01.133

site stability evaluation 场地稳定性评价 01.337

skeleton reinforced concrete bridge 劲性骨架混凝土桥 04.186

skew angle 斜交角 04.036

skew bridge 斜交桥 04.159

skew culvert 斜交涵洞 04.815

skew intersection 斜交叉 02.201

skidometer 滑溜测量仪 09.177

skid resistant course 防滑层 02.349

skid sign 路滑标志 03.387

slab 板 04.279

slab bridge 板桥 04.207

slab culvert 盖板涵 04.807

slab pumping 路面板唧泥 02.634

slab-rib arch bridge 板肋拱桥 04.226

slab rupture 板体断裂 02.629

slab staggering 错台 02.626

slab stress due to thermal warping 板体温度翘曲应力 02.410

slab warping 板体翘曲 02.628

slab with haunched ribs 肋腋板 04.283

slag base 矿渣基层 02.497

slag cement 矿渣水泥 08.042

slaked lime 熟石灰，＊消石灰 08.037

slant legged rigid frame bridge 斜腿刚构桥 04.233

slat conveyer 刮板输送机 07.101

slat feeder 链板给料机 07.092

slat type concrete spreader 刮板式混凝土布料机 07.198

sliding contact of interface 界面滑动接触 02.402

sliding expansion bearing 活动伸缩缝支座 04.457

slipform concrete curb paver 滑模式混凝土路缘铺筑机 07.204

slipform concrete paver 滑模式混凝土摊铺机 07.194

slip-form concreting 滑模浇筑混凝土 04.718

slope erosion 边坡冲蚀 02.639

slope erosion treatment 边坡冲刷防治 10.073

slope landscape 坡面景观 02.693

slope protection 护坡 02.259

slope sign 坡道标志 03.385

slope stabilization 边坡稳定 10.076

slope trimming 边坡修整 02.658

slope wash soil 坡积土，＊堆积土 01.298

slow breaking emulsified bitumen 慢裂乳化沥青 08.011

slow-curing liquid asphalt 慢凝液体沥青 08.019

slowing down 减速 05.091

slow vehicle mixed rate 慢速车混入率 03.179

slump 坍落度 08.262

slump cone 坍落度圆锥筒 09.167

slump test 坍落度试验 09.053

slurry hole-boring method 泥浆护壁钻孔法 04.748

slurry penetration 灌浆 02.447

slurry seal 稀浆封层 02.423

small bridge 小桥 04.171

smoke concentration 烟浓度 10.017

smoke emission standard 烟度排放标准 10.020

snow [deposit] markerpost 积雪标杆 03.480

snow-fence 防雪栅 02.684

snow hazard 雪害 02.646

snow protection facilities 防雪设施 02.682

snow remover 除雪机 07.216

snow removing 除雪 02.681

snow slide protection 雪崩防治 02.683

socketed pile　嵌岩桩　04.437

socketing pile in prebored hole　预钻插桩法　04.750

sodding　铺草皮　02.656

soffit　拱腹[线]　04.344

softening point　软化点　08.227

softening point test [ringball method]　软化点试验
[环球法]　09.063

soft grain content test　软颗粒含量试验　09.035

soft ground　软弱地基　02.313

softrock　软岩　01.316

soft soil　软土　01.279

soil　土　08.081

soil classification by formation　土的成因分类
08.058

soil classification by grain　土的粒组分类　08.056

soil cone penetrator　土圆锥仪　09.155

soil stabilizer　路拌式稳定土搅拌机　07.113

soil-structure interaction　土－结构共同作用
04.592

soil survey　土质调查　01.248

soldier pile　支挡桩　04.431

solid pier　实体[桥]墩　04.365

soot　烟粒　10.018

sorting operation　分拣作业　06.312

sounding　触探　01.254

sound isolate greenbelt　隔声绿化带　10.063

sound isolate road　隔声道　10.060

sound level meter　声级计　05.347

sovereignty nuisance　隔离障碍　10.088

space headway　车头间距　03.088

space mean speed　空间平均速度　03.071

space structure for bridge　桥梁空间结构　04.524

span　跨径　04.009

span by span construction method　逐跨施工法
04.666

spandrel arch　腹拱　04.352

spandrel column　拱上立柱　04.351

spandrel cross wall　拱上横墙　04.350

spandrel structure　拱上结构　04.348

spandrel wall　拱上侧墙　04.349

span length　跨径长　04.014

spare attachment　替换设备　07.334

special conditions of contract　合同专用条件
01.131

special goods　特种货物　06.135

special purpose vehicle　专用汽车　05.013

special soil　特殊土　08.064

special trailer　特种挂车　05.030

special truck　特种载货汽车　05.006

special truck transport　特种车辆货物运输　06.167

specialty vehicle repair shop　汽车专项修理厂
05.320

specific drawbar power fuel consumption measurement
牵引功率比油耗测定　07.438

specific gravity　比重　08.199

specific gravity test　比重试验　09.004

specific surface　比表面　08.183

speed　车速　03.067

speed change lane　变速车道　02.169

speeding up　加速　05.090

speed limit　车速限制　03.187

speedometer tester　[汽车]速度[表]试验台
05.333

spherical bearing　球形支座　04.455

spillway bridge　溢洪桥　04.176

spiral curve　螺旋线　02.094

spirally wound sheath　皱纹套管　08.127

split　绿信比　03.248

spoiler　扰流板　04.355

spontaneous out-of-gear　脱档　05.217

spot control　点控制　03.297

spot speed　点速度　03.069

spot welder　点焊机　07.305

spray booth　喷漆房　05.369

spread foundation　扩大基础　04.402

spreading　撒布　02.445

springing　弹簧现象　02.615,　起拱点　04.346

sprinkler　洒水车　07.213

SPT　标准贯入试验　09.026

spur dike　丁坝　04.129

ST　缓直点　02.101

stability　稳定性　04.553

stability of surrounding rock　围岩稳定　04.861

stabilized material base　稳定材料基层　02.492

stabilized protection course　稳定保护层　02.348

stabilized road-mixer　路拌式稳定土搅拌机

07.113

stabilized soil mixing plant　稳定土厂拌设备
　　07.108

stabilized soil paver　稳定土摊铺机　07.122

stabilize hooks　稳钩　06.298

stabilizer　稳定剂　08.151

stable traffic flow　稳定交通流　03.042

stacking area adjustment　转垛　06.304

stacking up　堆垛　06.303

stack transfer　倒垛　06.305

stage construction　分期修建　01.030

stage for heaping debris　碎落台　02.264

staggered intersection　错位交叉　02.205

standard axle load　标准轴载　02.380

standard design　标准设计　01.071

standard design of bridge　桥梁标准设计　04.006

standard highway　等级公路　02.024

standard penetration equipment　标贯装置　09.231

standard penetration test　标准贯入试验　09.026

standard sieves　标准筛　09.164

standard truck loading　标准车辆荷载　04.504

standing　暂停　03.103 ·

starting　起步　05.088

starting delay　起动延误　03.094

starting of construction　开工　01.140

starting wave in traffic flow　车流起动波　03.021

stated-speed sign　限速标志　03.386

statically determinate bridge structure　静定桥梁结构
　　04.525

statically indeterminate bridge structure　超静定桥梁
　　结构　04.526

static axle weight detector　静态轴重检测器
　　03.313

static cone penetration test　静力触探试验　09.027

static load　静载　04.473

static pile press extractor　静力压拔桩机　07.248

static-state test　定置试验　07.410

static visual acuity　静视觉敏锐度　03.531

static water head hole-boring method　静水护壁钻孔
　　法　04.749

statistical diagnosis　统计诊断　05.231

status monitoring maintenance　状况监测维护
　　05.253

stay cable　斜索·　04.328

steady-state test　稳定工况试验　07.378

steam curing　蒸汽养生　02.486

steam curing of concrete　混凝土蒸汽养生　04.713

steel acceptance test　钢材验收试验　09.086

steel arch centering　拱式钢拱架　04.688

steel-arched support　钢拱支撑　04.895

steel bar　钢筋　08.120

steel-bar bender　钢筋弯曲机　07.302

steel-bar butt welder　钢筋对焊机　07.303

steel-bar corrosion　钢筋锈蚀　04.608

steel-bar cutter　钢筋切断机　07.300

steel-bar header　钢筋镦头机　07.297

steel-bar straightener　钢筋调直机　07.301

steel beam centering　梁式钢拱架　04.687

steel bridge　钢桥　04.188

[steel] cable splay chamber　缆索散开室　04.339

steel coating　钢材涂料　08.174

steel comb joint　梳式接缝　04.309

steel deck cable stayed bridge　钢斜拉桥　04.237

steel fiber　钢纤维　08.155

steel guardrail　钢板护栏　03.437

steel pile　钢桩　04.422

steel pipe-encased concrete bridge　钢管混凝土桥
　　04.187

steel pipe pile　钢管桩　04.423

steel plate girder bridge　钢板梁桥　04.210

steel visual inspection　钢材外观检验　09.087

steel waling　钢围图　04.768

steel wire strand strapping machine　钢丝缠束机
　　07.298

steep mountain　峻岭　01.225

stepped culvert　阶梯式涵洞　04.816

stiffener　加劲杆，*加劲肋　04.537

stiffening girder　加劲梁　04.538

stiffness　刚度　04.554

stiff safety fence　刚性护栏　03.439

stochastic model　随机性模型　03.012

stock yard　堆货场　06.257

stone arch bridge　石拱桥　04.196

stone bridge　石桥　04.195

stone classification　石料分级　08.252

stop ahead sign　前方停车标志　03.379

stop delay　停车延误　03.097

stop line　停车线　03.425

stop line delay　停车线延误　03.096

stopping　停止　03.104

stopping sight distance　停车视距　02.107

stopping wave in traffic flow　车流停驶波　03.019

stop sign　停车标志　03.374

storage area operation　库场作业　06.302

storage area utilization ratio　仓库面积使用率　06.260

storage charge　仓储理货费　06.503

storage length　停储长度　03.202

storage stability test　储存稳定度试验　09.080

storage ton-days　[货物]堆存吨天数　06.331

stowing machinery　堆垛机械　06.283

straight line　直线　02.075

straight-line running test　直线行驶性能试验　07.422

straight line type wharf　直线形码头　04.955

straight road [Qin Dynasty]　直道　02.016

strain gage　应变计　09.204

strain scanning unit　应变巡检箱　09.241

strand　钢铰线　08.130

strand pulling machine　穿束机　07.296

strand tapered anchorage　夹片式锚具　04.642

stratification　层理　01.322

stratified soil　层状土　01.332

stratum　地层　01.329

strengthened neoprene plate joint　加强氯丁橡胶板缝　04.311

strengthening course　补强层　02.352

stress absorbing membrane　应力吸收薄膜　02.341

stress concentration　应力集中　04.546

stress corrosion　应力腐蚀　04.547

stress dispersal beneath footing　基底应力扩散　04.657

stressing sequence　张拉程序　04.725

stress wave equation method　应力波方程法　04.656

stripping of aggregate　集料剥落　02.609

stripping test　剥落试验　09.038

structural integrity　结构完整性　04.528

structural steel　结构钢　08.118

structure ductility　结构延展性　04.604

structure joint　构造节理　01.320

structure system transform　结构体系转换　04.527

stud bolt　大头螺栓　08.133

stuffing　装箱　06.262

styrene-butadiene rubber　丁苯橡胶　08.160

subaqueous tunnel　水底隧道　04.836

subbase　底基层　02.355

subcontracting　分包　01.127

subcontractor　分包人　01.103

subgrade　路基　02.247

subgrade drainage　路基排水　02.511

subgrade engineering　路基工程　02.248

subgrade erosion protection　路基冲刷防治　10.074

subgrade fine trimmer　路基精整机　07.123

subgrade moisture content　路基含水量　02.283

subgrade protection　路基防护　02.285

subgrade stability　路基稳定性　02.278

subgrade strength　路基强度　02.279

subgrade strengthening　路基加固　02.286

sub-high class pavement　次高级路面　02.335

submerged spur dike　淹没丁坝　04.130

submersible bridge　漫水桥　04.175

subsidence　[路基]沉降　02.633

subsoil　底土　01.295

subsoil bearing capacity　土基承载能力　02.280

subsoil drainage　地下排水　02.510

subsoil moistness classification　土基干湿类型　01.163

subsoil reaction modulus c method for laterally loaded pile　桩基 c 值法　04.654

subsoil reaction modulus method for laterally loaded pile　横向承载桩地基系数法　04.652

subsoil reaction modulus m method for laterally loaded pile　桩基 m 值法　04.653

substandard highway　等外公路　02.025

substructure　下部结构　04.358

sudden failure　突然故障　05.194

suitable brilliance in tunnel　隧道适宜亮度　03.465

sulphur corrosion　硫侵蚀　02.636

super-deep soil stabilizer　超深稳定土搅拌机　07.116

superelevation　超高　02.138

superelevation run-off 超高缓和段 02.141

superelevation slope 超高横坡度 02.139

super long tunnel 特长隧道 04.853

superstructure 上部结构 04.261

supervising engineer 监理工程师 01.105

supervising engineer's representative 监理工程师代表 01.106

supported type abutment 支撑式桥台，*轻型桥台 04.395

supporting course 承重层 02.353

supporting ring of shield 盾构支承环 04.944

support type expansion installation 支承式伸缩装置 04.301

surcharge preloading 超载预压 02.320

surface course 面层 02.339

surface-curvature apparatus 路面曲率仪 09.159

surface evenness 表面平整度 02.364

surface friction coefficient 表面摩擦系数 02.366

surface gathered water 路面积水 02.594

surface loosening 松散 02.608

surface macro texture 表面宏观构造 02.369

surface mega texture 表面巨观构造 02.371

surface micro texture 表面微观构造 02.370

surface permeameter 路面透水度测定仪 09.161

surface pockmark 麻面 02.607

surface roughness 表面粗糙度 02.365

surface skidding 路面滑溜 02.595

surface texture 表面构造，*表面纹理 02.367

surface texture depth 表面构造深度，*纹理深度 02.368

surface water 地表水 02.500

surface weeping of concrete 混凝土表面泌水 02.477

surrounding rock 围岩 04.859

surrounding rock classification 围岩分类 04.863

surrounding rock pressure 围岩压力 04.860

surrounding rock self-supporting capacity 围岩自承能力 04.862

survey 测量 01.174

susceptivity analysis 敏感性分析 01.052

suspended beam 挂梁 04.272

suspended load 悬移质 04.122

suspended ribbon bridge 悬带桥 04.245

suspender 吊杆 04.332

suspension bridge 悬索桥，*吊桥 04.244

suspensive daily shift of machine 机械停修台日 07.355

swamp 沼泽 01.277

swamp bulldozer 湿地推土机 07.006

swelling amount 膨胀量 08.216

swelling force 膨胀力 08.217

swelling soil 膨胀土 01.283

swelling test 膨胀试验 09.010

swing bridge 平转桥，*平旋桥 04.251

swinging screen 摇摆筛 07.086

switch back curve 回头曲线 02.087

swiveling drill 旋转式钻机 07.250

synchronous system 同步式系统 03.259

T

table dispatching method 表上作业法 06.213

T-abutment T形桥台 04.387

tack coat 粘层 02.342

taking wrong bus 错乘 06.108

tallying of cargoes 理货 06.185

talus 岩堆 01.262

tamping 夯实 02.299

tandem roller 双轮压路机 07.058

tandem vibratory roller 双轮振动压路机 07.066

tangent length 切线长 02.089

tangent offset method 切线支距法 01.202

tank trailer 液罐挂车 05.037

tank truck 液罐货车 05.010

tapered section 渐变段 02.210

tar 焦油沥青 08.022

tariff 运价 06.442

tariff classification 价目 06.454

taxi 出租汽车 05.019

taxi transport 出租汽车客运 06.061

T-beam bridge T[形]梁桥 04.211

technical checking 技术检验 05.294

technical condition of machine 机械技术状况

07.319

technical condition of vehicle 汽车技术状况
07.162

technical design 技术设计 01.043

technical economic quota of machinery 机械技术经
济定额 07.346

technical reformation of existing road 旧路技术改造
02.578

technical service 技术服务 05.252

technical test 技术性试验 07.401

techno-economic rating for vehicle operation 汽车技
术经济定额 05.062

technological equipment of vehicle maintenance and
repair 汽车维修工艺设备 05.321

telescopic boom excavator 伸缩铲挖掘机 07.049

Telford base 块石基层 02.488

Telford stone 锥形块石 08.071

temperature difference 温差 04.589

temperature scanning unit 温度巡检箱 09.242

temporary bridge 临时性桥 04.180

temporary bridge for construction 施工便桥
04.182

temporary road 便道 02.026

temporary works 临时工程 01.144

tender 投标书 01.109

tendon 预应力筋 08.125

tendon duct winding machine 卷管机 07.299

tension-compression bearing 拉压支座 04.459

tensioning jack 张拉千斤顶 07.295

tensioning procedure 张拉程序 04.725

tension test 拉力试验 09.046

terminal and yard facilities of motor transport 汽车运
输站场设施 06.247

termination of contract 终止合同 01.152

terrace 台地 01.229

test for field of vision of driver cab 驾驶室视野试验
07.450

test loop road 试验环道 09.192

test pile 试桩 09.135

test road 试验路 09.191

test run loading for bridge 桥梁试运行荷载
04.472

texture meter 构造深度仪 09.183

theoretical deflection coefficient 理论弯沉系数
02.392

theoretical rate 理论运价 06.460

thermal gradient 温度梯度 04.588

thermal insulated trailer 保温挂车 05.038

thermal insulated vehicle 保温货车 05.008

thermal insulation course 隔温层 02.358

thermal stress 热应力，*温差应力 04.548

thermoplastic 热塑性塑料 08.170

thermoplastic road marking machine 热塑划线机
07.231

thermostability 热稳性 08.233

the silk road ［古］丝绸之路 02.019

thin-film heating test 薄膜加热试验 09.071

thin-walled tube sampler 薄壳取样器 09.233

thixotropy 触变性 08.198

thread bar anchorage 精轧螺纹锚，*迪维达格锚固
系统 04.649

three-dimension expansion installation with sealing
strip 密封条三维伸缩装置 04.317

three-dimension prestressing 三向预应力 04.624

three-factor model of bar corrosion 钢筋锈蚀三因素
模型 09.131

three-hinged arch bridge 三铰拱桥 04.219

through bridge 下承式桥 04.157

through bus transport 直达客运 06.058

through highway 过境公路 02.030

through traffic 过境交通 03.114

through transport 直达运输 06.014

throwing blasting 抛掷爆破 02.327

ticket price 旅客票价 06.474

ticket price of mini bus 小型客车票价 06.480

tidal level 潮水位 04.101

tie bar 拉杆 02.467

tied arch bridge 系杆拱桥 04.228

tie rod anchored retaining wall 锚杆式挡土墙
02.273

timber bridge 木桥 04.199

timbering support 木支撑 04.894

timber pile 木桩 04.421

time effect treatment 时效处理 04.728

[time] headway 车头时距 03.089

time interval of machine maintenance and repair 机械

维修间隔　07.353

time limit of shipment arrival　运达期限　06.094

time mean speed　时间平均速度　03.070

time occupancy ratio　时间占有率　03.101

time table of runs　班次时刻表　06.130

timing controller　定时信号控制机　03.303

timing light　正时灯　05.341

T-intersection　丁字形交叉　02.203

tire changer　轮胎拆装机　05.338

TNI　交通噪声指数　10.048

toe of slope　坡脚　02.258

toll administration　收费管理　01.358

toll booth　收费亭　03.334

toll bridge　收费桥　04.166

toll canopy　收费棚　03.337

toll center　收费中心　03.343

toll display　收费显示器　03.324

toll elasticity　收费弹性　03.342

toll gate　收费卡门　03.333

toll highway　收费公路　02.038

toll house　收费亭　03.334

toll island　收费岛　03.336

toll lane detector　收费车道检测器　03.315

toll lane signal lamp　收费车道信号灯　03.329

toll mode　收费制式　03.344

toll pass ticket　收费通行券　03.326

toll plaza　收费广场　03.338

toll station　收费站　03.335

tongue and groove joint　企口缝　02.463

tonnage of cargo handled　货物操作量　06.317

tools attachment　随机工具　07.335

top and bottom pilot tunneling method　上下导洞法　04.875

top of slope　坡顶　02.257

topographic feature　地貌　01.218

topographic map　地形图　01.214

topographic map of bridge site　桥位地形图　01.304

topographic survey　地形测量　01.213

topography　地形　01.220

topologic theory　拓扑学理论　03.007

top retaining wall　上挡墙　02.276

topsoil　表土　01.294

total highway transport production output value　公路

运输总产值　06.380

total length of bridge　桥梁全长　04.013

total quality control　全面质量管理　01.095

total suspended particles　总悬浮微粒　10.023

total vehicle-days　总车日　06.348

total vehicle-seat-days　总车客位日　06.357

total vehicle-ton-days　总车吨位日　06.356

total width of bridge　桥梁总宽度　04.020

tourist bus ticket price　旅游客车票价　06.478

towed asphalt paver　拖式沥青混合料摊铺机　07.161

towed grader　拖式平地机　07.015

towed scraper　拖式铲运机　07.009

towed sheep-foot roller　拖式羊足碾　07.064

towed vibratory roller　拖式振动压路机　07.070

tower crane　塔式起重机　07.287

towing resistance measurement　拖动阻力测定　07.428

towing test　拖动试验　07.430

township highway　乡公路，＊乡道　01.344

TP　转点　01.196

TQC　全面质量管理　01.095

traction coefficient measurement　附着系数测定　07.437

tractor　牵引车　05.020

tractor road　机耕道　02.012

tractor-trailer combination　汽车列车　05.049

tractor-trailer transport　拖挂运输　06.031

tractor-truck　半挂牵引车　05.021

traffic accessibility　交通可达性　03.136

traffic accident　交通事故　03.214

traffic accident crime　交通肇事罪　03.216

traffic accident prediction　交通事故预测　03.235

traffic accident rate　交通事故率　03.215

traffic-actuated controller　感应信号控制机　03.302

traffic administration cost　交通管理费　01.067

traffic aisle　交通通道　03.262

traffic analysis　交通分析　03.037

traffic behavior　交通行为　03.020

traffic bottleneck　交通瓶颈　03.038

traffic bound　行车碾压　02.448

traffic characteristics　交通特性　01.084

traffic classification　交通分类　03.049

traffic composition　交通组成　01.082

traffic cone　锥形交通路标　03.483

traffic conflict　交通冲突　03.229

traffic congestion　交通拥挤　03.034

traffic converging　交通汇合　03.035

traffic court　交通法庭　03.231

traffic demand　交通需求　03.032

traffic demand forecast　交通需要预测　03.119

traffic density　交通密度　03.084

traffic direction block　交通指路牌　03.482

traffic direction stud　交通指向钮　03.481

traffic dispersion　交通疏导　01.354

traffic diverging　交通分流　03.036

traffic divided blocks　交通隔离墩　03.451

traffic engineering　交通工程学　01.004

traffic facilities　交通设施　03.351

traffic factor　交通因素　03.052

traffic flow　交通流　03.003

traffic flow rate　交通流率　03.053

traffic flow theory　交通流理论　03.004

traffic guardrail　交通护栏　03.431

traffic guidance　交通指挥　03.193

traffic information system　交通信息系统　03.280

traffic intensity　交通强度　01.083

traffic interference　交通干扰　03.063

traffic island　交通岛　02.216

traffic management devices　交通管理设施　03.350

traffic modal choice model　交通方式选择模型
　03.142

[traffic] modal split　交通方式划分　03.110

traffic network　交通网络　03.033

traffic noise　交通噪声　10.047

traffic noise evaluation　交通噪声评价　10.049

traffic noise index　交通噪声指数　10.048

traffic patrolling　交通巡逻　03.194

traffic physiology　交通生理学　01.020

traffic prohibited　禁止通行　03.196

traffic prohibited sign　禁止通行标志　03.396

traffic psychology　交通心理学　01.019

traffic regulations　交通规则　03.208

traffic responsive control　交通感应式控制　03.264

traffic rules　交通规则　03.208

traffic safety　交通安全　03.212

traffic safety facilities　交通安全设施　01.361

traffic sign　交通标志　03.353

traffic signal lamp　交通信号灯　03.307

traffic signal off-line control　交通信号脱机控制
　03.270

traffic simulation　交通仿真　03.017

traffic stability　交通稳定性　03.022

traffic statistics　交通统计学　01.021

traffic surveillance and control center　交通监控中心
　03.278

traffic surveillance and control system　交通监控系统
　03.279

traffic survey　交通调查　01.159

traffic survey chart　交通调查表　03.138

traffic system　交通系统　03.039

traffic vibration　交通振动　10.066

traffic violation　交通违章　03.209

traffic violator　交通违章者　03.210

traffic volume　交通量　03.050

traffic volume adjustment factor　交通量修正系数
　03.267

traffic [volume] assignment　交通量分配　03.116

traffic [volume] distribution　交通量分布　03.117

traffic volume estimating　交通量估计　03.051

traffic volume forecast　交通量预测　03.118

traffic volume measurement　交通量计量　03.062

traffic volume survey　交通量调查　01.160

trailer　挂车　05.022

trailer pick-up transport　甩挂运输　06.032

trailer ton-km ratio　拖运率　06.374

trailer/tractor ratio　挂车配比　06.375

trailer-typed mobile laboratory vehicle　拖车式移动试
　验车　09.197

trailing load　拖挂重量　06.236

transitional grade　缓和坡段　02.114

transition curve　缓和曲线　02.092

transition slab　渐变板　02.474

transition zone of cross section　断面渐变段　02.134

transition zone of curve widening　加宽缓和段
　02.133

transit transport　过境运输　06.016

transmission test　传动试验　07.455

transport administration fee　运输管理费，＊运管费

06.386

transportation economics　运输经济学　01.015

transportation engineering　运输工程学　01.005

transportation network　交通运输网　01.025

transportation system management　交通系统管理
01.192 03.192

transport capacity　运输能力　06.008

transport coefficient　运输系数　06.383

transport elasticity coefficient　运输弹性系数
06.382

transport in circulation　流通过程运输　06.022

transport in production process　生产过程运输
06.021

transport liability period　运输责任期　06.189

transport mark　运输标志　06.162

transport market　运输市场　06.399

transport market control　运输市场管理　06.400

transport service industry　运输服务业　06.406

transport structure　运输结构　06.006

transport system　运输体系　06.007

transport volume forecast　运量预测　06.337

transshipment service package charge　中转劳务包干
费　06.504

transversal distribution measurement of girder deflection　挠度横向分布测量　09.098

transverse beam　横梁　04.275

transverse crack　横向裂缝　02.606

transverse diaphragm　横隔板　04.276

transverse distribution of deflection　挠度横向分布
04.574

transverse distribution of load　荷载横向分布
04.575

transverse drainage　横向排水　02.515

transverse gallery　横洞　04.841

transverse joint　横缝　02.459

transverse ventilation　横向通风　04.910

traveler　旅客，*乘客　06.035

travel expense of handling equipment　装卸机械走行
费　06.495

travelling-beam testing device for deck surface irregularity　桥面平整度测定仪　09.210

travel performance test　行驶性能试验　07.419

travel speed　行程车速　03.076

travel time　行程时间　03.082

travel-time ratio　行程时间比　03.099

traverse　导线　01.178

tremie concreting　[直升]导管浇筑混凝土法
04.771

trencher　挖沟机　07.050

trench test　槽探　01.250

trend of change of vehicle technical conditions　汽车
技术状况变化规律　05.170

trestle　高架桥，*栈桥，*旱桥　04.154

trestle road along cliff　栈道　02.015

trial steel work fixing　钢梁结构试装配　04.737

triaxial shear equipment　三轴剪切仪　09.163

triaxial test　三轴试验　09.017

trimmed tunnel　衬砌隧道　04.857

trip　出行　03.127

trip-distribution　出行分布　03.131

trip end model　出行端点模型　03.135

trip generation　交通生成，*交通发生　03.115

trip interchange model　出行交换模型　03.134

triple acting jack　三作用千斤顶　04.732

trip length　出行长度　03.129

trip rate　出行率　03.128

trip time　出行时间　03.130

truck crane　汽车式起重机　07.286

truck handling safety interval distance　货车装卸安全
间距　06.296

truck loading capacity utilization rate　载重量利用率
06.371

truck load rate　整车运价　06.465

truck-load transport　整车货物运输　06.165

truck mixer　混凝土搅拌输送车　07.191

truck turntable　汽车调头转盘　07.313

true joint　真缝　02.465

trumpet interchange　喇叭形立交　02.233

trunk highway　干线公路　02.035

truss bridge　桁架桥　04.214

trussed arch bridge　桁架拱桥　04.227

TS　直缓点　02.098

T[-shaped] beam　T形梁　04.269

T-shaped rigid frame bridge　T形刚构桥　04.231

T-shaped truss rigid frame bridge　桁式T形刚构桥
04.232

TSM 交通系统管理 03.192

tube-sinking cast-in-situ pile 沉管灌注桩 04.427

tune-up [发动机]检修 05.283

tunnel 隧道 04.830

tunnel alignment by laser 隧道激光导向 04.884

tunnel anti-disaster facilities 隧道防灾设施 04.926

tunnel boring machine 隧道掘进机 07.312

tunnel broadcasting system 隧道广播系统 04.939

tunnel class 隧道级别 04.831

tunnel construction dust controlling 隧道施工防尘 04.902

tunnel construction ventilation 隧道施工通风 04.903

tunnel engineering 隧道工程学 01.012

tunnel entrance brilliance 隧道入口区亮度 03.464

tunnel excavator 隧道挖掘机 07.310

tunnel fire hazard 隧道火灾 04.928

tunnel fire monitoring 隧道火灾监测 04.927

tunnel fire protection system 隧道消防系统 04.924

tunnel floor 隧道底板 04.897

tunnel in earth 土质隧道 04.834

tunneling 隧道工程 04.829

tunneling method 掘进方法 04.864

tunnel in rock 岩石隧道 04.835

tunnel lighting 隧道照明 03.459

tunnel lining 隧道衬砌 04.885

tunnel loader 隧道装载机 07.311

tunnel maintenance 隧道养护 04.947

tunnel muck hauling 出碴 04.882

tunnel operation control center 隧道监控中心 04.934

tunnel portal 隧道洞门，＊隧道口 04.847

tunnel power supply system 隧道供电系统 04.938

tunnel roof 隧道顶板 04.896

tunnel roof fall 隧道坍顶 04.899

tunnel's discharge pollution 隧道排气污染 10.031

tunnel shield 盾构 04.940

tunnel side wall 隧道边墙 04.898

tunnel signal 隧道信号 04.933

tunnel support 隧道支撑 04.890

tunnel survey 隧道测量 04.883

tunnel telephone system 隧道电话系统 04.937

tunnel through 隧道贯通 04.900

tunnel through error 隧道贯通误差 04.901

tunnel toll 过隧费 01.034

tunnel traffic monitoring system 隧道交通监测系统 04.936

tunnel trunk 隧道洞身 04.848

tunnel ventilation 隧道通风 04.904

tunnel warning installation 隧道报警装置 04.925

tunnel water proofing 隧道防水 04.932

tunnel water sealing by injection 隧道注浆止水 04.931

turn around loop 回车道 02.183

turn crossing 可调头交叉口 03.453

turning 转向 05.092

turning around 掉头 05.093

turning lane 转弯车道 02.168

turning point 转点 01.196

turning test 转向性能试验 07.441

turnover 周转量 06.342

twin ferry boat 双体渡车船 04.962

twin-side heading method 双侧壁导洞法 04.881

twin tunnel 双[孔]隧道 04.840

two-columned pier 双柱式[桥]墩 04.369

two dimension collision 二维碰撞 03.223

two-dimension prestressing 双向预应力 04.623

two-hinged arch bridge 二铰拱桥 04.220

two shaft pier 双柱式[桥]墩 04.369

two shifts run 双班运行 06.079

two-way curved arch bridge 双曲拱桥 04.224

two way curved arch tile 拱波 04.353

two-way loading transport 双程运输 06.025

two-way ramp 双向匝道 02.241

two-way road 双向路 02.167

two-way slab 双向板 04.285

type A partially prestressed concrete A 类部分预应力混凝土 04.615

type approval test 工业性试验 07.409

type B partially prestressed concrete B 类部分预应力混凝土 04.616

tyre depreciation rate 轮胎摊提率 06.434

tyre mileage rating 轮胎行驶里程定额 05.064

tyre noise 轮胎噪声 10.055

tyre recapping rate 轮胎翻新率 05.069

tyre scuffs [轮胎]压印 05.097

tyre skid [轮胎]拖印 05.098

U

U-abutment U形桥台 04.384

ultimate limit state of bridge carrying capacity 桥梁
承载能力极限状态 04.507

ultimate minimum radius of horizontal curve 平曲线
极限最小半径 02.078

ultimate stress 极限应力 04.552

ultrasonic pulse velocity measurement device 超声脉
冲测量仪 09.213

ultrasonic test of concrete strength 混凝土强度超声
测量 09.121

ultra-sonic vehicle detector 超声波车辆检测器
03.290

unconfined compression strength 无侧限抗压强度
08.213

unconfined compression strength test 无侧限抗压强
度试验 09.020

unconstrained operation 非约束运行 03.182

underground asphalt storage 地下沥青贮仓 07.170

underlying stratum 下卧层 01.331

undermining method 暗挖法 04.866

underpass 下穿式立交 02.225

under-reamed pile 扩底桩 04.430

under-reinforced slab with slightly curved bottom 少
筋微弯板 04.282

underwater bulldozer 水下推土机 07.005

under water concreting 水下浇筑混凝土法 04.770

under water pump 潜水泵 07.267

under water rig 潜水钻机 07.254

undisturbed sample 原状土样 01.335

unfavourable season 不利季节 02.586

unidirectional passenger traffic 单向性客流 06.043

unified toll system 统一收费系统 03.345

uniform 标志服装 06.106

uniform load 均布荷载 04.482

unit current repair 总成小修 05.281

unit diagnosis 单元诊断 05.235

unit exchange repair method 总成互换修理法
05.308

unitized transport 成组运输 06.168

unit major repair 总成大修 05.280

unit operation ton 工序吨 06.318

unit repair 总成修理 05.279

unit repair plant 汽车总成修理厂 05.319

unit vehicle accounting 单车核算 06.388

universal container fork lift 通用型集装箱叉车
06.309

universal photo 全能法测图 01.240

unloading bare rib 裸拱卸架 04.696

unloading bare ring 裸拱卸架 04.696

unscheduled maintenance 计划外维护 05.260

unscheduled repair 非计划修理 05.291

unstable traffic flow 不稳定交通流 03.043

unstuffing 掏箱 06.261

upheaval 拥包 02.613

upheaval leveling 铲除拥包 02.665

urban bus transport 城市公共客运 06.060

urban road 城市道路 02.003

urban tunnel 市区隧道 04.833

used parts reconditioning cost 旧件修复成本
06.433

user test 用户试验 07.408

U[-shaped] beam U形梁 04.270

V

vacuum dewatering technique 真空脱水工艺
02.478

vacuum sweeper 真空吸尘清扫车 07.222

vacuum water sucker 真空吸水机械 07.293

valley line 沿溪线 02.063

valuable goods 贵重货物 06.147

valued ticket 定额客票 06.097

valve refacer 磨气门机 05.355

valve seat grinder 磨气门座机 05.354

vane-shear apparatus 十字板剪力仪 09.229

vane shear test 十字板剪切试验 09.021

van trailer 厢式挂车 05.031

van truck 厢式货车 05.007

vapor lock 气阻 05.201

variable load 可变荷载 04.478

variable sign 可变标志 03.363

variable speed-limit control 可变限速控制 03.281

variable speed-limit sign 可变限速标志 03.364

vegetation 植被 01.171

vehicle accident loss 车辆事故损失 06.387

vehicle actuated signal 车辆感应信号 03.261

vehicle allotment 车辆调拨 05.070

vehicle avail ability rate 车辆完好率 05.074

vehicle availability rate 完好车率 06.349

vehicle average technical speed 汽车平均技术速度 05.131

vehicle condition monitoring 汽车状况监控 05.227

vehicle conversion factor 车辆换算系数 01.085

vehicle current repair 汽车小修 05.276

vehicle depreciation 车辆折旧 06.384

vehicle depreciation mileage 车辆折旧里程 05.073

vehicle depreciation rate 车辆折旧率 05.078

vehicle detector 车辆检测器 03.287

vehicle diagnosis 汽车诊断 05.225

vehicle diagnostic station 汽车诊断站 05.240

vehicle durability 汽车耐久性 05.138

vehicle dynamic quality 汽车动力性 05.139

vehicle economical speed 汽车经济车速 05.132

vehicle economic service life 车辆经济使用寿命 05.084

vehicle economy 汽车经济性 05.140

vehicle emission 汽车排放物 10.010

vehicle empty container mileage 汽车空箱行程 06.271

vehicle energy balance 汽车能量平衡 05.157

vehicle energy saving 汽车节能 05.152

vehicle energy utilization factor 汽车能量利用率 05.156

vehicle exhaust analyser 汽车排放分析仪 10.029

vehicle failure 汽车故障 05.187

vehicle fleet accounting 车队核算 06.389

vehicle fuel consumption per hundred kilometers 汽车百车公里油耗 05.158

vehicle fuel consumption per hundred ton-kilometers 汽车百吨公里油耗 05.159

vehicle fuel consumption rating 汽车燃料消耗定额 05.063

vehicle fuel saving technique 汽车节油技术 05.153

vehicle fuel utilization factor 汽车燃料利用率 05.155

vehicle general inspection and diagnosis line 汽车综合诊断线 05.324

vehicle handling stability 汽车操纵稳定性 05.149

vehicle height detector 车辆高度检测器 03.314

vehicle inspection and test 汽车检测 05.226

vehicle inspection and test station 汽车检测站 05.241

vehicle in use 在用汽车 05.054

vehicle lay-off 车辆停驶 05.071

vehicle leasing 车辆租赁 05.076

vehicle line haul hour 出车时间 06.360

vehicle load carrying capacity 汽车容载量 05.135

vehicle loaded container mileage 汽车重箱行程 06.272

vehicle maintainability 汽车维修性 05.151

vehicle maintenance 汽车维护, *汽车保养 05.248

vehicle maintenance and repair 汽车维修 05.242

vehicle maintenance and repair enterprise 汽车维修企业 05.246

vehicle maintenance and repair indices 汽车维修指标 05.245

vehicle maintenance and repair network 汽车维修网点 05.247

vehicle maintenance and repair system 汽车维修制度 05.243

vehicle maintenance classification 汽车维护类别 05.250

vehicle maintenance depot 汽车维护场 05.270

vehicle maintenance engineering 汽车维护工程 05.249

vehicle maintenance grade 汽车维护级别 05.251

vehicle maintenance method 汽车维护方法 05.265

vehicle maintenance norms 汽车维护规范 05.264

vehicle maintenance on universal post 汽车维护定位作业法 05.267

vehicle maintenance station 汽车维护站 05.271

vehicle maintenance technological process 汽车维护工艺过程 05.263

vehicle maintenance technology 汽车维护工艺 05.261

vehicle major repair 汽车大修 05.274

vehicle management system 汽车管理制度 05.059

vehicle medium repair 汽车中修 05.275

vehicle mobility 汽车机动性 05.148

vehicle noise 车辆噪声 10.050

vehicle non-working day 停驶车日 06.353

vehicle non-working rate 车辆停驶率 06.367

vehicle operation 汽车使用 05.117

vehicle operational consumption materials 汽车运行材料 05.061

vehicle operational mode 汽车运行工况 05.119

vehicle operational reliability 汽车使用可靠性 05.137

vehicle operation condition 汽车使用条件 05.118

vehicle operation efficiency 汽车使用效率 05.129

vehicle operation intensity 汽车使用强度 05.128

vehicle operation performance 汽车使用性能 05.134

vehicle operation plan 车辆运行作业计划 06.336

vehicle outfitting 车辆装备 05.080

vehicle overhaul cost 汽车大修理费用 06.432

vehicle passability 汽车通过性 05.150

vehicle payload-container mileage 汽车载箱行程 06.270

vehicle periodic maintenance 汽车定期维护 05.254

vehicle purchase additional fee 车辆购置附加费 01.037

vehicle rational service life 车辆合理使用寿命 05.083

vehicle rational towing load 汽车合理拖载 05.133

vehicle rational utilization 汽车合理使用 05.130

vehicle registration 车辆注册登记 06.405

vehicle remoulding 车辆改造 05.081

vehicle repair 汽车修理 05.272

vehicle repair classification 汽车修理类别 05.273

vehicle repair method 汽车修理方法 05.305

vehicle repair on universal post 汽车修理定位作业法 05.307

vehicle repair plant 汽车修理厂 05.318

vehicle repair technical standard 汽车修理技术标准 05.304

vehicle repair technological process 汽车修理工艺过程 05.293

vehicle repair technology 汽车修理工艺 05.292

vehicle replacement 车辆更新 05.079

vehicle ride comfort 汽车乘坐舒适性 05.146

vehicle running smoothness 汽车行驶平顺性 05.145

vehicle safety 汽车安全性 05.141

vehicle safety inspection and test line 汽车安全检测线 05.323

vehicle scrapping 车辆报废 05.077

vehicle service life 车辆使用寿命 05.082

vehicle storing up 车辆封存 05.072

vehicle technical file 车辆技术档案 05.060

vehicle technical service life 车辆技术使用寿命 05.085

vehicle terminal charge 车辆站务费 06.506

vehicle type detector 车型检测器 03.308

vehicle utilization conveniency 汽车使用方便性 05.144

vehicle wear-out 汽车损耗 05.171

vehicle weight efficiency 汽车重量利用系数 05.136

vehicle working rate 车辆工作率 06.366

vehicle working shift 工作车班 06.352

vehicular bump-integrator 车载式颠簸累积仪 09.181

vehicular gap 车间净距 03.090

ventilation by axial flow fan 轴流式通风 04.915

ventilation by exhaust fan 排风机通风 04.913

ventilation by force draft 射流纵向通风 04.914

ventilation design 通风设计 04.906

ventilation efficiency 通风效率 04.919

ventilation installation 通风装置 04.922

ventilation monitoring 通风监测 04.921

ventilation tower 通风塔 04.918

ventilation type 通风方式 04.905

ventilation volume 通风量 04.920

verge 路肩 02.142

vertical alignment 纵面线形 02.110

vertical clearance 净高 02.126

vertical curb 立缘石 02.153

vertical curve 竖曲线 02.120

vertical drainage 竖向排水 02.516

very heavy traffic 特重交通 02.383

viaduct 高架桥,＊栈桥,＊旱桥 04.154

viameter 路面平整度测定仪 09.160

vibrating compaction 振动压实 02.302

vibrating feeder 振动给料机 07.094

vibrating screen 振动筛 07.085

vibration 振动 10.064

vibration drum exciting force measurement 振动轮激振力测定 07.461

vibration excited by cutting off holding rope 放松拉索激振 09.105

vibration excited by dropping weight 落物激振 09.106

vibration excited by explosive action 爆破激振 09.107

vibration excited by moving truck 行车激振 09.103

vibration excited by truck jumping from threshold on deck 跳车激振 09.104

vibration excitor 激振器 09.201

vibration isolate ditch 隔振沟 10.070

vibration measurement of driver cab instrument board 驾驶室仪表板振动测定 07.448

vibration measurement of driver seat 驾驶员座位振动测定 07.447

vibration pile driver extractor 振动沉拔桩机 07.247

vibration reducing measure 减振措施 10.069

vibratory plate compactor 振动平板夯 07.071

vibratory roller 振动压路机 07.061

vibratory tamper 振动冲击夯 07.072

vibrostand 结构振动试验台 09.195

Vicat apparatus 维卡稠度仪 09.157

violation 违章行为 06.411

viscoelasticity theory 粘弹性理论 02.376

viscosity 粘滞度 08.225

viscosity test 粘滞度试验 09.064

visibility 能见度 03.506

vision with both driver and object moving 全动视力 03.509

vision with driver moving 人动视力 03.507

vision with object moving 物动视力 03.508

vista 公路景观 02.692

visual acuity 视觉敏锐度 03.529

visual inspection 检视 05.295

void ratio 孔隙比 08.177

volatile matter content 挥发分含量 08.241

volatile matter content test 挥发分含量试验 09.077

volume 运量 06.341

vortex shedding effect 涡流效应 04.572

voussoir 拱砌块 04.701

VSL anchorage VSL锚具 04.643

V type snow plough 犁式除雪机 07.218

W

wall heel 墙踵 02.267

wall in advance of arched roof tunneling method 先墙后拱法 04.876

wall toe 墙趾 02.266

warehouse ［货物］仓库 06.254

warning post 示警柱 03.454

warning sign 警告标志 03.354

warp 翘曲 04.568

washout 水毁 02.645

waste 弃土 02.291

waste bank 弃土堆 02.293

waste bank refarming 弃土反耕 10.085

waste bank treatment 废方处理 10.075

water absorptivity test 吸水率试验 09.031

water and soil conservation 水土保持 10.071

water bound macadam 水结碎石路面,＊马克当路面 02.430

water cement ratio 水灰比 08.259

water content 含水量 08.193

water content test 含水量试验 09.001

water curing [保护层]水养法 02.668

water drop 跌水 02.536

water erosion 水侵蚀 02.641

water gate bridge 闸门桥 04.247

water head 水头 04.091

water insulation course 隔水层 02.357

water level 水位 04.090

water permeability 透水性 02.504

water pollution 水质污染 10.035

water proof asphalt-felt 油毡防水层 04.296

water proof asphalt mastic 沥青胶砂防水层，＊沥青玛琋脂防水层 04.297

water proof asphalt membrane 沥青薄膜防水层 04.294

water proof coating 防水涂层 04.293

water proof layer 防水层 04.292

water proof polymer fabrics 聚合织物防水层 04.295

water quality 水质 10.034

water reducing agent 减水剂 08.145

water stability 水稳性 08.235

water stratum 含水层 01.259

water surface gradient 水面比降 04.060

water tank for loading 加载水箱 09.209

wave height 浪高 04.109

waybill 行车路单 06.409

wear allowance 磨损限值，＊容许磨耗 04.602

wear hardness testing apparatus 耐磨硬度试验仪 09.169

wearing 磨损 02.622

wearing course 磨耗层 02.344

wear limit 极限磨损 05.175

wear of vehicle parts 汽车零件磨损 05.172

wear process 磨损过程 05.173

wear rate 磨损率 05.177

wear test 磨损试验 07.390

weather detector 气象检测器 03.310

weathering steel 耐候钢 08.119

weaving area 交织区 03.181

weaving section capacity 交织区通行能力 03.171

weaving traffic flow 交织交通流 03.045

web 梁腹，＊腹板 04.535

weep-hole 泄水孔 02.541

weighing system 称重系统 07.141

weight-balanced abutment 衡重式桥台 04.393

weighted platform 加载平台 09.200

weight limit sign 限重标志 03.376

welded joint strength test [钢筋]焊接接头强度试验 09.090

welded steel bridge 焊接钢桥 04.190

wet dust collector 湿式除尘器 07.133

wetted perimeter 湿周 04.062

wetting and drying test 干湿试验 09.045

wet traffic accident 雨天事故 03.232

wharf 码头 04.952

wheel alignment meter [汽车]车轮定位仪 05.337

wheel bulldozer 轮胎式推土机 07.003

wheel crane 轮胎式起重机 07.285

wheel dynamic balancer 车轮动平衡仪 05.339

wheel load 轮载 01.091

wheel loader 轮胎式装载机 07.054

wheel lock up 车轮抱死 05.218

3-wheel roller 三轮压路机 07.059

wheel rutting test 车辙试验 09.081

wheel spinning 车轮滑转 05.102

wheel-type asphalt paver 轮胎式沥青混凝土摊铺机 07.158

wheel type single bucket excavator 轮胎式单斗挖掘机 07.043

wheel-type soil stabilizer 轮胎式稳定土搅拌机 07.114

wheel-type stabilized soil road-mixer 轮胎式稳定土搅拌机 07.114

widening 拓宽 02.575

width-span ratio 宽跨比 04.021

wild animals protection sign 保护野生动物标志 10.084

wild animals refuge area 野生动物保护区 10.083

willow fascine 柴排 04.141

wind accelerating dike 风力加速堤 02.678

3-wing bit 三翼钻头 07.258

wind bracing 抗风支撑 04.354

wind-cable 风缆 04.706

wind erosion 风蚀 02.652

wind fairing 风嘴 04.544

wind induced oscillation 风致摆动 04.573

wind load 风荷载 04.487

wind pressure distribution graph 风压分布图 04.586

wind resisting stability 抗风稳定性 04.584

wing wall 翼墙，*耳墙 04.383

wire cold-drawn test 钢丝冷拔试验 09.088

wire grouping 钢丝束制作 04.724

wire guy 钢丝牵索 08.135

wire mesh laying machine 钢丝网铺设机 07.206

wooden bridge 木桥 04.199

wood-truss centering 桁式木拱架 04.691

wood wedge for centering unloading 卸架木楔 04.699

workability 和易性，*工作度 08.261

working ability of vehicle 汽车工作能力 05.165

working-day cost 台班使用费 07.331

working vehicle-days 工作车日 06.351

W-type guardrail W型护栏 03.435

X

X-typed anchorage XM锚具 04.645

Y

yield ahead sign 前方让路标志 03.380

yielding sign 让路标志 03.368

Y-intersection Y形交叉 02.204

Y-shaped pier Y形[桥]墩 04.371

Y-typed anchorage YM锚具 04.644

Z

zebra crossing 斑马线 03.417

zero air void ratio 饱和孔隙比 08.178

zero section 零断面 02.131

zoning rate 区域运价 06.470

инженерия зксплоатация автомобилъного транспорта （俄） 汽车运用工程 01.014

汉 英 索 引

A

安全措施　safety measure　03.213
安全岛　refuge island　02.219
安全事故　safety misadventure　07.341
安全系数　safety factor　04.529
安装拆卸费　mounting and dismounting cost　07.338

按质赔偿　compensation for loss　06.118
暗涵　buried culvert　04.803
暗挖法　undermining method　04.866
凹式平板挂车　cranked platform trailer　05.028
凹形竖曲线　concave vertical curve　02.123

B

八字形桥台　flare wing-walled abutment　04.385
白垩土　chalky clay　01.292
百米桩　100m stake　01.190
摆式仪　portable pendulum tester　09.178
摆式支座　rocker bearing　04.453
斑马线　zebra crossing　03.417
班车客运　scheduled bus transport　06.057
班车线路　regular service route　06.056
班次　number of runs　06.064
班次密度　density of runs　06.067
班次时刻表　time table of runs　06.130
搬移费　moving charge　06.498
板　slab　04.279
板端错台　faulting of slab ends　02.627
板肋拱桥　slab-rib arch bridge　04.226
板桥　slab bridge　04.207
板式橡胶支座　laminated rubber bearing　04.448
板体断裂　slab rupture　02.629
板体翘曲　slab warping　02.628
板体温度翘曲应力　slab stress due to thermal warping　02.410
板桩　sheet pile　04.428
板桩围堰　sheet pile cofferdam　04.765
拌和　mixing　02.443
拌和功率测定　mixing power measurement　07.470
拌和宽度测定　mixing width measurement　07.467
拌和深度测定　mixing depth measurement　07.468

伴随过程参数　accompanying parameters of operation process　05.168
半地下沥青贮仓　semi-underground asphalt storage　07.171
半电池　half cell　09.219
半定向式立交　semi-directional interchange　02.232
半盾构　semi-shield　04.942
*半封闭　partial control of access　02.236
半感应式信号控制　semi-actuated signal control　03.266
半刚性基层　semi-rigid base　02.491
半挂车　semi-trailer　05.024
半挂汽车列车　semi-trailer train　05.051
半挂牵引车　tractor-truck　05.021
半横向通风　semi-transverse ventilation　04.911
半价票　half price ticket　06.476
半苜蓿叶形立交　partial clover-leaf interchange　02.229
半山洞　half tunnel　02.253
半填半挖断面　cut and fill section　02.128
半压力式涵洞　inlet submerged culvert, partial pressure culvert　04.809
半永久性桥　semi-permanent bridge　04.179
包车客运　chartered bus transport　06.059
包车运价　charter rate　06.471
包干计费里程　chartered pay mileage　06.451
包装储运标志　package indicative mark　06.163

包装货物　packed goods　06.138

剥层翻修　peeled resurfacing　02.672

剥落试验　stripping test　09.038

薄壳取样器　thin-walled tube sampler　09.233

薄膜加热试验　thin-film heating test　09.071

薄膜养生　membrane curing　02.484

薄膜养生液　membrane curing solution　08.152

保护层　protection course　02.346

[保护层]水养法　water curing　02.668

保护野生动物标志　wild animals protection sign　10.084

保价运输　insured transport　06.170

保留金　retention money　01.137

保温挂车　thermal insulated trailer　05.038

保温货车　thermal insulated vehicle　05.008

饱和度　degree of saturation　08.270

饱和孔隙比　zero air void ratio　08.178

饱和流量　saturation volume　03.160

饱和流率　saturation volume rate　03.161

饱水率试验　saturated water content test　09.039

暴雨径流　rainstorm run-off　04.079

暴雨强度　rainstorm intensity　04.080

爆破激振　vibration excited by explosive action　09.107

爆破漏斗　blasting crater　02.326

爆破振动　blasting vibration　10.068

爆破作业　blasting operation　02.325

爆燃　detonation　05.210

爆炸夯　explosion rammer　07.074

*爆震　detonation　05.210

背景噪声　background noise　10.045

*贝克曼梁　deflectometer, Benkelman beam　09.186

贝雷桥　Bailey bridge　04.260

被动安全性　passive safety　05.143

*本底噪声　background noise　10.045

崩坍　collapse, fall　01.264

泵站排水　pumping drainage　02.526

比表面　specific surface　08.183

比较线　alternative line　02.067

比重　specific gravity　08.199

比重试验　specific gravity test　09.004

闭式联合碎石机组　close type crushing plant

07.098

边沟　side ditch　02.517

边坡　side slope　02.256

边坡冲蚀　slope erosion　02.639

边坡冲刷防治　slope erosion treatment　10.073

边坡平台　plain stage of slope　02.263

边坡稳定　slope stabilization　10.076

边坡修整　slope trimming　02.658

边缘钢筋　edge reinforcement　02.472

边缘线　edge line　03.416

扁壳桥　shell bridge　04.229

扁千斤顶　flat jack　09.202

便道　temporary road　02.026

便桥　detour bridge　04.181

变坡点　grade change point　02.119

变速车道　speed change lane　02.169

*变形钢筋　deformed bar　08.121

变性混凝土　metamorphic concrete　08.110

标称直径　nominal diameter　08.138

标底　engineer's estimate　01.118

*标高　elevation　01.207

标贯装置　standard penetration equipment　09.231

标志　sign　03.352

标志服　safety marked coat　02.591

标志服装　uniform　06.106

标志视认性　sign legibility　03.359

标准车辆荷载　standard truck loading　04.504

标准贯入试验　standard penetration test, SPT　09.026

标准筛　standard sieves　09.164

标准设计　standard design　01.071

标准轴载　standard axle load　02.380

表观流速　apparent velocity　04.056

表面粗糙度　surface roughness　02.365

表面构造　surface texture　02.367

表面构造深度　surface texture depth　02.368

表面宏观构造　surface macro texture　02.369

表面巨观构造　surface mega texture　02.371

表面摩擦系数　surface friction coefficient　02.366

表面平整度　surface evenness　02.364

表面微观构造　surface micro texture　02.370

*表面纹理　surface texture　02.367

表上作业法　table dispatching method　06.213

表土　topsoil　01.294

冰冻地区　frost region　01.168

冰冻深度　frost depth　01.169

冰丘　ice hummock　01.273

冰雪路面驾驶　driving on snowy and icy road　05.114

冰压力　ice pressure　04.494

冰障　ice jam　02.651

冰锥　ice cone　01.274

玻璃钢护栏　glass fiber reinforced plastic fence　03.442

玻璃钢桥　glass fiber reinforced plastic bridge　04.202

玻璃微珠　micro glass bead　08.172

玻璃纤维　glass fiber　08.153

拔桩　pile extracting　04.763

波纹钢桥面　corrugated steel deck　08.136

波纹管涵　corrugated-metal pipe culvert　04.811

波形梁护栏　corrugated beam barrier　03.438

博什烟度　Bosh smoke unit　10.021

＊泊车　parking　03.105

泊松比　Poisson's ratio　08.255

补充客票　additional ticket　06.098

补坑　patching　02.664

补强层　strengthening course　02.352

不可预见费　contingencies　01.135

不利季节　unfavourable season　02.586

不排水下沉沉井　sinking open caisson by undrained dredging　04.744

不稳定交通流　unstable traffic flow　03.043

布袋除尘器　bag dust collector　07.132

布氏硬度试验　Brinell hardness test　09.049

部分负荷工况　partial load mode　05.127

部分控制进入　partial control of access　02.236

部分预应力混凝土　partially prestressed concrete　04.613

C

擦伤　scratching　05.178

材料拌和均匀度测定　material mixing uniformity measurement　07.472

材料试验法　materials test methods　01.073

材料输送机械　material conveyer　07.099

材料性能　characteristic of materials　08.175

参数诊断　parameter diagnosis　05.230

残留响应区　residual response zone　09.115

仓储理货费　storage charge　06.503

仓库面积使用率　storage area utilization ratio　06.260

操作力测定　operating force measurement　07.417

操作行程测定　operating distance measurement　07.418

糙率　coefficient of roughness　04.064

槽探　trench test　01.250

侧壁导洞法　side heading method　04.880

侧滑　side skidding　05.220

侧靠式渡船　side-mooring ferry boat　04.964

侧面集装箱叉车　side container fork lift　06.308

侧限抗压强度试验　confined compression strength test　09.019

侧向碰撞　side collision　03.221

侧向视野　field of lateral vision　03.503

侧向余宽　lateral clearance　02.157

侧向最小安全间距　minimum safe lateral clearance　05.101

测点　observation point　01.180

测力柱　load measurement column　09.205

测量　survey　01.174

测斜仪　inclinometer　09.225

测站　instrument station　01.179

层理　stratification　01.322

层铺法　layer spread method　02.435

层状土　stratified soil　01.332

岔道标志　cross-buck sign　03.378

差别运价　differential rate　06.461

差热分析　differential thermal analysis　09.022

柴捆　fagot　04.142

柴捆坝　fagot dam　04.143

柴排　mattress, willow fascine　04.141

柴油机排烟　diesel smoke　10.019

柴油机喷油泵试验台　diesel fuel injection pump tester　05.345

柴油机喷油定时测定器　diesel injection timing tester　05.342

柴油机烟度计　diesel smoke meter　05.336

柴油桩锤　diesel pile hammer　07.239

掺加料　admixture　08.141

铲除拥包　upheaval leveling　02.665

铲斗式清岩机　rock shovel　07.035

铲运机　scraper　07.007

颤振　flutter　04.569

场地稳定性评价　site stability evaluation　01.337

常规试验　normal test　07.391

常水位　mean water level　04.096

长大笨重货物　heavy and bulky goods　06.143

长货挂车　pole trailer　05.040

长隧道　long tunnel　04.854

长途客车　intercity bus　05.017

长柱试验机　long column testing machine　09.193

厂拌法　plant-mixing method　02.437

厂矿道路　factory and mine road　02.005

超车　overtaking　05.099

超车车道　overtaking lane, passing lane　02.181

超车视距　overtaking sight distance　02.108

超高　superelevation　02.138

超高横坡度　superelevation slope　02.139

超高缓和段　superelevation run-off　02.141

超高作业　over-height stacking operation　06.300

超静定桥梁结构　statically indeterminate bridge structure　04.526

超深稳定土搅拌机　super-deep soil stabilizer　07.116

超声波车辆检测器　ultra-sonic vehicle detector　03.290

超声脉冲测量仪　ultrasonic pulse velocity measurement device　09.213

超限货物　oversize or overweight goods　06.144

超远作业　over-distance stacking operation　06.299

超载预压　surcharge preloading　02.320

潮湿类型　dampness type　01.166

潮水位　tidal level　04.101

车次　serial number of bus run　06.065

车道　lane　02.159

车道分布　lane distribution　03.055

车道分界线　lane line　03.415

车道荷载　lane load　04.485

车道偏移　lane shift　03.188

车道平衡　lane balance　03.186

车道收费机　lane toll machine　03.325

车道通行能力　lane capacity　03.169

车道系数　coefficient of lanes　02.404

车道占有率　lane occupancy ratio　03.100

车队　platoon　03.001

车队核算　vehicle fleet accounting　06.389

车吨产量　annual output per vehicle tonnage capacity　06.377

车间净距　vehicular gap　03.090

车辆报废　vehicle scrapping　05.077

车辆调拨　vehicle allotment　05.070

车辆封存　vehicle storing up　05.072

车辆改造　vehicle remoulding　05.081

车辆感应信号　vehicle actuated signal　03.261

车辆高度检测器　vehicle height detector　03.314

车辆更新　vehicle replacement　05.079

车辆工作率　vehicle working rate　06.366

车辆购置附加费　vehicle purchase additional fee　01.037

车辆合理使用寿命　vehicle rational service life　05.083

车辆换算系数　vehicle conversion factor　01.085

车辆活塞作用　piston action of moving vehicles　04.923

车辆技术档案　vehicle technical file　05.060

车辆技术使用寿命　vehicle technical service life　05.085

车辆检测器　vehicle detector　03.287

车辆经济使用寿命　vehicle economic service life　05.084

车辆平均技术等级　mean grade of vehicle technical condition　05.075

车辆使用寿命　vehicle service life　05.082

车辆事故损失　vehicle accident loss　06.387

车辆停放指数　parking index　03.201

车辆停驶　vehicle lay-off　05.071

车辆停驶率　vehicle non-working rate　06.367

车辆完好率　vehicle avail ability rate　05.074

车辆允许噪声标准　allowable vehicle noise standard　10.051

车辆运行作业计划 vehicle operation plan 06.336
车辆噪声 vehicle noise 10.050
车辆站务费 vehicle terminal charge 06.506
车辆折旧 vehicle depreciation 06.384
车辆折旧里程 vehicle depreciation mileage 05.073
车辆折旧率 vehicle depreciation rate 05.078
车辆注册登记 vehicle registration 06.405
车辆装备 vehicle outfitting 05.080
车辆租赁 vehicle leasing 05.076
车流返回波 backward wave in traffic flow 03.018
车流起动波 starting wave in traffic flow 03.021
车流停驶波 stopping wave in traffic flow 03.019
车轮抱死 wheel lock up 05.218
车轮动平衡仪 wheel dynamic balancer 05.339
车轮滑转 wheel spinning 05.102
车身矫正机 body and frame straightener 05.368
车速 speed 03.067
车速限制 speed limit 03.187
车头间距 space headway 03.088
车头时距 [time] headway 03.089
车位 loading/unloading lot 06.258
车厢容积 carriage box volume capacity 06.237
车型检测器 vehicle type detector 03.308
车行道 carriageway 02.160
车载式颠簸累积仪 vehicular bump-integrator 09.181
车辙 rut 02.616
车辙试验 wheel rutting test 09.081
沉淀池 sedimentation basin 02.525
沉管灌注桩 tube-sinking cast-in-situ pile 04.427
沉积节理 sedimentation joint 01.321
沉降差 differential settlement 04.659
沉井基础 open caisson foundation 04.403
沉井刃脚 caisson cutting edge 04.404
沉埋法 immersed tunneling method 04.869
衬砌隧道 trimmed tunnel 04.857
撑杆 boat pole 04.969
称重系统 weighing system 07.141
城间道路 interurban road 02.004
城市出入口公路 city approach highway 02.031
城市道路 urban road 02.003
城市公共客运 urban bus transport 06.060
城市客车 city bus 05.016

成组运输 unitized transport 06.168
乘车安全岛 loading island 02.188
*乘客 passenger, traveler 06.035
承包人 contractor 01.102
承台 bearing platform, pile cap 04.440
承运 acceptance of consignment 06.178
承运货物代理费 freight forwarder agency charge 06.505
承运人 carrier 06.173
承载板 bearing plate 09.158
承载板试验 plate-bearing test 09.023
承重层 supporting course 02.353
持力层 bearing stratum 01.330
持续性客流 continuous passenger traffic 06.039
驰道 royal road [Qin Dynasty], ancient drive way 02.017
驰振 galloping 04.570
冲击夯 shocking rammer 07.073
冲击力 impact force 04.490
冲击韧度试验 impact toughness test 09.047
冲击韧度试验仪 impact toughness apparatus 09.168
冲击式凿岩机 shocking jack hammer 07.025
冲击系数 impact factor 04.530
冲击系数测定 impact factor evaluation 09.118
冲击旋转式凿岩机 shocking rotary jack hammer 07.027
冲积土 alluvial soil 01.297
冲孔成桩法 pile by percussion drill method 04.754
冲刷 scouring, erosion 02.643
冲刷深度 scouring depth 04.118
冲刷系数 coefficient of scouring 04.119
冲抓法 percussion and grabbing method 04.755
重复运输 reflux of freight 06.029
抽风机 exhauster 07.150
抽样试验 sample test 07.407
稠度界限 consistency limit 08.197
稠度试验 consistency test 09.054
初步设计 preliminary design 01.042
初测 preliminary survey 01.176
初期养护 initial maintenance 02.567
出碴 tunnel muck hauling 04.882
出厂证明试验 certification test 07.406

出车时间 vehicle line haul hour 06.360
出口标志 exit sign 03.404
出口预告标志 advance exit sign 03.405
出口匝道控制 exit ramp control 03.273
*出入境运输 international transport 06.015
出水口 outlet 02.539
出行 trip 03.127
出行长度 trip length 03.129
出行端点模型 trip end model 03.135
出行分布 trip-distribution 03.131
出行交换模型 trip interchange model 03.134
出行率 trip rate 03.128
出行时间 trip time 03.130
出租汽车 taxi 05.019
出租汽车客运 taxi transport 06.061
除尘装置 dust collector 07.130
除雪 snow removing 02.681
除雪机 snow remover 07.216
储备通行能力 reserve capacity 03.173
储存稳定度试验 storage stability test 09.080
触变性 thixotropy 08.198
触探 sounding, cone penetration test 01.254
穿束机 strand pulling machine 07.296
传动试验 transmission test 07.455
传力杆 dowel bar 02.468
船舶撞击力 ship collision force 04.492
船渡费 ferry toll 01.035
锤击沉桩法 pile driving method by hammer 04.760
锤击式碎石机 hammer crusher 07.081
锤式冲抓斗 hammer grab 07.256

磁感式沉降标 magnetic tell tale 09.227
磁感应式车辆检测器 magnetic vehicle detector 03.289
磁卡收费机 magnetic card toll machine 03.327
磁卡通行券 magnetic card toll pass ticket 03.328
磁性测裂计 magnetic crack detector 09.214
次高级路面 sub-high class pavement 02.335
粗级配 coarse gradation 08.188
粗粒土 coarse-grained soil 08.060
粗砂 coarse sand 08.078
促凝剂 coagulator 08.149
促凝压蒸试验 accelerated setting autoclave test 09.057
窜油 oil pumping 05.209
脆点 brittle point 08.229
脆点试验 brittle point test 09.072
*存车 parking 03.105
存车搭乘 park and drive 03.107
存车换乘 park and ride 03.106
存在型车辆检测器 presence vehicle detector 03.292
搓板 corrugation 02.618
错车道 passing bay 02.173
错车洞 passing bay in tunnel 04.842
错乘 taking wrong bus 06.108
错台 slab staggering 02.626
错位交叉 staggered intersection, offset intersection 02.205
错运 misshipment 06.222
错装 misloading 06.220

D

搭接钢板接缝 lapped steel plate joint 04.308
打入桩 driven pile 04.439
打桩 pile driving 04.759
打桩分析仪 pile driving analyzer, PDA 09.224
打桩机 pile driver 07.237
*大车道 cart road 02.011
大跨径桥 long span bridge 04.169
大气环境 air environment 10.008
大桥 great bridge 04.168

大头螺栓 stud bolt 08.133
大修工程费 major maintenance cost 01.066
大循环运行 full circulation operation 06.075
大中修周期 maintenance period 02.573
大宗货物 mass goods 06.153
戴马格接缝 Demag joint 04.313
带式输送机 belt conveyer 07.100
带状地形图 belt topographic map 01.215
代办站 agency 06.127

代理人 agent 01.104

贷款偿还能力 loan repay ability 01.051

袋装砂井 sand bag well 02.317

怠速工况 idling mode 05.120

怠速排放污染物 pollutants of idle speed emission 10.013

单车产量 annual output per vehicle 06.376

单车道 single lane 02.163

单车核算 unit vehicle accounting 06.388

单程运输 one-way loading transport 06.026

单斗挖掘机 power shovel, single bucket excavator 07.038

单铰拱桥 single-hinged arch bridge 04.221

单路通道排队 single-channel queue 03.030

单室箱梁 single cell box girder 04.265

单索面斜拉桥 single plane cable stayed bridge 04.240

单向板 one-way slab 04.284

单向推力墩 single direction thrusted pier 04.374

单向性客流 unidirectional passenger traffic 06.043

单向匝道 one-way ramp 02.240

单行路 one-way road 02.166

单行线标志 one-way sign 03.397

单叶平转桥 single-leaf swing bridge 04.252

单叶竖旋桥 single-leaf bascule bridge 04.255

单元诊断 unit diagnosis 05.235

单柱式[桥]墩 single-columned pier, single shaft pier 04.368

单桩 individual pile, single pile 04.413

单桩承载力 bearing capacity of pile 04.596

单作用气动桩锤 single acting pneumatic hammer 07.240

单作用蒸气桩锤 single acting steam hammer 07.242

当量故障率 equivalent failure rate 07.454

当量厚度 equivalent thickness 02.395

*当量交通量 equivalent traffic volume 03.061

当量圆直径 diameter of equivalent circle 02.379

当量轴次 equivalent axles 02.381

挡土墙 retaining wall 02.265

倒车 backing 05.094

倒垛 stack transfer 06.305

倒虹吸涵 inverted siphon culvert 04.813

倒Y形索塔 inverted Y tower 04.325

导洞 heading, pilot 04.846

导洞法 pilot-tunneling method 04.873

导洞与层阶法 heading and bench tunneling method 04.879

导梁 launching nose 04.707

导流岛 channelization island 02.217

导流堤 diversion dike 04.127

导热油沥青熔化装置 asphalt hot oil melter 07.176

导线 traverse 01.178

到达交付 delivery on arrival 06.188

道班房 maintenance gang house 02.588

道班养护 gang maintenance 02.587

道路 road 02.001

道路标识 road markings 03.413

道路标线漆 road mark paint 08.171

道路催眠状态 road hypnosis 03.518

道路反光镜 road reflecting mirror 03.473

道路工程学 road engineering 01.002

道路几何数据收集系统 road geometry data acquisition system 09.173

道路建筑限界 boundary line of road construction 01.347

道路经济分析 road economical analysis 01.055

道路沥青 road bitumen 08.027

道路收费系统 road toll system 03.332

道路水泥 road cement 08.045

道路引导系统 route guidance system 03.284

道路运输 road transport 06.001

灯光信号 light signal 03.241

灯光照明标志 illuminated sign 03.409

等代均布荷载 equivalent uniform distributed load 04.483

等高距 contour interval 01.217

等高线 contour line 01.216

等级公路 standard highway, classified highway 02.024

等速工况 cruising mode 05.123

等外公路 substandard highway 02.025

堤岸冲刷 bank erosion 04.112

低货位 low freight lot 06.256

低级路面 low class pavement 02.337

低净空标志 low clearance sign 03.399

低平板挂车 low-bed trailer 05.027

低水位桥 low water level bridge 04.174

低温模拟试验 simulated cold climate test 07.384

低液限粉土 low liquid limit silt 08.085

低液限粘土 low liquid limit clay 08.086

*迪维达格锚固系统 thread bar anchorage, Dywidag anchorage system 04.649

狄法尔磨耗试验机 Deval abrasion testing machine 09.170

底基层 subbase 02.355

底盘测功机 chassis dynamometer 05.326

底土 subsoil 01.295

底卸式挂车 bottom-dump trailer 05.042

地表水 surface water 02.500

地层 stratum 01.329

地方公路 local highway 02.037

地基承载力 bearing capacity of subsoil 04.595

地理信息系统 geographic information system 01.245

地沥青 asphalt 08.004

地貌 topographic feature 01.218

地面高程 ground elevation 01.211

地面控制点测量 ground control point survey 01.234

地面立体摄影测量 ground stereo-photogrametry 01.233

地名标志 place name sign 03.360

地上沥青贮仓 ground asphalt storage 07.169

地图板显示 map display 03.300

地物 culture 01.219

地下沥青贮仓 underground asphalt storage 07.170

地下连续墙 continuous slurry wall, diaphragm wall 04.443

地下排水 subsoil drainage 02.510

地下水 ground water 01.257

地下水位 ground water level 01.258

地形 topography 01.220

地形测量 topographic survey 01.213

地形图 topographic map 01.214

地震荷载 earthquake load, seismic force 04.488

地震基本烈度 basic earthquake intensity 04.517

地震加速度 seismic acceleration 04.520

*地震力 earthquake load, seismic force 04.488

地震区 seismic region 01.300

地震危险性分析 seismic risk analysis 04.514

地震危险性评定 seismic risk evaluation 09.147

地震震级 earthquake magnitude 04.519

地质测井 geologic logging 01.309

地质调查 geologic survey 01.306

地质剖面图 geologic profile 01.307

地质图例 geologic legend 01.310

地质钻机 exploration drill 07.235

递近递增运价 increasing rate with decreasing distance 06.458

递远递减运价 decreasing rate with increasing distance 06.459

颠簸累积式平整仪 bump-integrator roughometer trailer 09.180

点焊机 spot welder 07.305

点控制 spot control, isolated signal control 03.297

点蚀 pitting 05.180

点速度 spot speed 03.069

垫层 bed course 02.356

垫块支座 pad bearing 04.456

电动冲击夯 electric shocking rammer 07.075

电动卷扬机 electric hoister 07.281

电动凿岩机 electric jack hammer 07.020

电热[法]张拉 electric-heat prestressing 04.720

电热式沥青加热装置 bitumen electrical heating device 07.186

电探 electrical resistivity survey 01.255

电位测量装置 potential test device 09.216

电位轮 potential test wheel 09.217

电液自动找平装置 electronic-hydraulic automatic leveling device 07.368

电子收费系统 electronic toll system 03.349

电子水平仪 electrolevel 09.226

电子自动售票系统 automatic electronic ticket system 06.099

电阻率测量装置 resistivity test device 09.218

电阻应变仪 electric resistance strain gauge 09.240

掉头 turning around 05.093

吊杆 suspender 04.332

吊货工夹具 hoisting fixture 06.284

*吊桥 suspension bridge 04.244

* OD 调查 origin-destination study, OD study 01.161

调查表调查 inquiry form survey 06.132

调车费 dispatch charge 06.488

调度方法 dispatching method 06.210

调度命令 dispatching order 06.211

调度日志 dispatching log 06.209

跌水 hydraulic drop, water drop 02.536

叠桁式拱架 double-layer-truss centering 04.692

丁坝 spur dike 04.129

丁苯橡胶 buna-S, styrene-butadiene rubber, SBR 08.160

丁钠橡胶 buna 08.162

丁字形交叉 T-intersection 02.203

钉板梁桥 pin-connected plywood girder bridge 04.201

顶推法 incremental launching method 04.669

顶推设备 incremental launching device 07.292

定班运行 scheduled run 06.069

定测 location survey 01.177

定额 norm, quota, rating 01.076

定额客票 valued ticket 06.097

定额赔偿 rated compensation 06.223

定期检查 periodical inspection 02.558

定期养护 periodical maintenance 02.569

定时信号控制 fixed-time control 03.272

定时信号控制机 pretimed controller, timing controller 03.303

定线 line location 02.069

定线运行 fixed line operation 06.077

定向爆破 directional blasting 02.331

定向设计小时交通量 directional design hourly traffic volume, DDHV 03.066

定向式立交 directional interchange 02.231

定向行车道标记 directional carriageway marking 03.430

定站客票 fixed ticket 06.096

定站停靠率 rate of stops at scheduled bus stops 06.121

定置试验 static-state test 07.410

动力触探试验 dynamic sounding 09.028

动力放大过程线 dynamic amplification duration curve 09.111

动力固结法 dynamic consolidation method 02.315

动力式弯沉仪 dynaflect 09.189

* 动力系数 impact factor 04.530

动力粘滞度 kinematic viscosity 08.226

动视觉敏锐度 dynamic visual acuity 03.530

动态回弹弯沉 dynamic rebound deflection 02.391

动态三轴试验 dynamic triaxial test 09.018

动态诊断 dynamic diagnosis 05.236

动态轴重检测器 dynamic axle weight detector 03.312

动弯沉 dynamic deflection 02.390

动物通道 animal corridor 03.494

动员费预付贷款 advance mobilization loan 01.134

动载 dynamic load 04.474

冻害 frost damage 02.647

冻融 frost thawing 02.649

冻融试验 freezing and thawing test 09.040

冻土 frozen soil 01.271

冻胀 frost heaving 02.648

抖振 buffeting 04.571

斗轮式挖沟机 bucket wheel trencher 07.052

斗轮式挖掘机 bucket wheel excavator 07.048

斗式混凝土布料机 hopper type concrete spreader 07.196

斗式提升机 bucket elevator 07.102

陡坡涵洞 culvert on steep grade 04.814

独塔斜拉桥 single pylon cable stayed bridge 04.241

读卡机 card reader 03.331

渡槽 aqueduct 04.824

渡口 ferry, ferry crossing 04.948

渡口安全规则 ferry safety regulation 04.971

渡口管理 ferry management 04.970

渡口管理所 ferry house 04.972

渡轮 ferry boat 04.958

渡运码头 ferry wharf 04.953

端靠式渡船 front-mooring ferry boat 04.963

短隧道 short tunnel 04.856

断背曲线 broken back curve 02.086

断层 fault 01.324

断层走向 fault strike 01.325

断火 interruption of firing 05.214

断级配 gap gradation 08.187

断链 broken chain 01.193

断面渐变段 transition zone of cross section 02.134

断头路 dead end highway 02.043

堆垛 stacking up 06.303

堆垛机械 stowing machinery 06.283

堆货场 stock yard 06.257

＊堆积土 slope wash soil 01.298

堆料台 roadside material terrace 02.593

对接式伸缩装置 butt type expansion installation 04.300

对流运输 counter flow transport 06.028

镦头锚 bulb-end anchorage, BBRV anchorage 04.648

墩帽 pier cap, pier coping 04.361

墩前壅水 back-water at pier 04.107

墩身 pier body 04.360

墩式基础 pier-foundation 04.408

吨位 rated tonnage 06.354

盾构 tunnel shield 04.940

盾构法 shield method 04.870

盾构切口环 cutting ring of shield 04.943

盾构支承环 supporting ring of shield 04.944

多层立交 multi-level interchange 02.227

多车道 multi-lane 02.165

多乘员车辆车道 high occupancy vehicle lane 03.238

多次碰撞 multiple collision 03.226

多点同步顶推 incremental launching construction by multipoint jacking 04.670

多斗挖掘机 multi-bucket excavator 07.047

多功能桩架 multi-purpose pile frame 07.245

多级离心泵 multistage centrifugal pump 07.269

多角度交通事故预测法 multiplex traffic accident prediction method 03.236

多路停车控制 multiway stop control 03.269

多路通道排队 multiple-channel queue 03.031

多面临空爆破 open face blasting 02.332

多姆克蜂房接缝 Domke honeycomb joint 04.314

多年冻土 permafrost 01.272

多室箱梁 multi cell box girder 04.266

多塔斜拉桥 multi pylon cable stayed bridge 04.243

多通道数字记录系统 multi channel digital record system 09.244

多用浮箱 multi-function pontoon 04.966

多用养护车 multi-use maintenance truck 07.215

E

颚式碎石机 jaw crusher 07.079

＊耳墙 wing wall 04.383

二次碰撞 secondary collision 03.225

二铰拱桥 two-hinged arch bridge 04.220

二维碰撞 two dimension collision 03.223

F

发车正点率 rate of on-time departure 06.122

发动机大修 engine rebuilding, engine remanufacture 05.282

[发动机]检修 tune-up 05.283

发动机清洗机 engine cleaner 05.366

发动机噪声 motor noise 10.054

发动机诊断仪 engine analyzer 05.340

发光标志 luminous sign 03.406

发货人 shipper 06.174

发卡机 card sender 03.330

发裂 hair-like crack 02.604

发运 pre-departure operation 06.179

发针形曲线 hairpin curve 02.088

乏力 lack of power 05.205

GM法 Guyon-Massonet method 04.583

＊法兰 flange 04.536

翻浆 frost boiling 02.614

反铲挖掘机 backhoe excavator 07.040

反光标识 reflecting marking 03.474

反光标志 reflecting sign 03.407

反光材料 reflective materials 08.173

反光路钮 reflecting button 03.476

反力式制动试验台 reaction type brake tester 05.328

反滤层 inverted filter 02.530

反射裂缝 reflection crack 02.601

* 反向分析 retrogressive analysis for bridge erection 04.779

反向曲线 reverse curve 02.085

反循环钻机 reverse circulation drill 07.252

反循环钻孔法 reverse circulation boring method 04.757

反压护道 loading berm 02.261

泛滥区 flooded area 04.049

泛油 bleeding 02.611

方位角 azimuth 01.198

方向分布 direction distribution 03.054

方向角 direction angle 01.199

防波堤 break-water 04.133

防尘设施 dustproof facilities 10.033

防冻剂 antifreezing agent 08.147

防护林带 shelter belt 02.697

防护栅 guard fence, safety fence 03.490

防滑层 skid resistant course 02.349

防滑处理 anti-skid treatment 02.669

防沙坝 sand protection dike 02.679

防沙林 sand protection green 02.677

防沙设施 sand protection facilities 02.675

防牲畜护栏 cattle fence 03.493

防水层 water proof layer 04.292

防水涂层 water proof coating 04.293

防眩板 anti-glare panel 03.478

防眩屏 anti-glare screen 03.479

防雪设施 snow protection facilities 02.682

防雪栅 snow-fence 02.684

防噪设施 acoustic treatment facilities 10.058

防震挡块 anti-knock block, restrain block 04.379

防撞垫 bumper 04.967

防撞墩 crash bearer 03.446

放射性货物 radioactive goods 06.141

放松拉索激振 vibration excited by cutting off holding rope 09.105

放样 setting out 01.141

非常荷载 abnormal load 04.480

非定班运行 non-scheduled run 06.070

* 非互通式立交 grade separation without ramps 02.223

非机动车辆 non-power-driven vehicle 05.003

非机动车运输 non-power-driven vehicle transport 06.004

非集合模型 disaggregation model 03.014

非计划修理 unscheduled repair 05.291

非交织交通流 nonweaving traffic flow 03.046

非离子乳化沥青 non-ionic emulsified bitumen 08.015

非排土桩 non-displacement pile 04.434

非通航桥孔 non-navigable span 04.012

非营业性运输 own account transport 06.020

非雨天事故 dry traffic accident 03.233

非预应力钢筋 non-prestressed reinforcement 04.625

非约束运行 unconstrained operation 03.182

废车堆积场 auto-wrecking yard 10.089

废方处理 waste bank treatment 10.075

废气滤清器 exhaust scrubber 10.028

废气排放标准 exhaust emission standard 10.012

费用效益分析 cost/benefit analysis 01.056

费油 excessive fuel consumption 05.207

酚含量 phenol content 08.243

酚含量试验 phenol content test 09.078

分包 subcontracting 01.127

分包人 subcontractor 01.103

分道角区 gore area 02.211

分段分环结合砌拱 laying arch by sections and rings 04.679

分段砌拱 laying arch by sections 04.677

分隔带 separator 02.145

分隔式车行道 divided carriageway 02.161

分隔行驶公路标志 divided highway sign 03.384

分环砌拱 laying arch by rings 04.678

分拣作业 sorting operation 06.312

分界标志 boundary sign 03.361

分离离合器滑行 coasting with clutch disengaged 05.108

分离式立交 grade separation without ramps 02.223

分离式路基 separated subgrade 02.254

分流交通量 diverging traffic volume 03.184

218

分流运输　diversion transport　06.030
分期修建　stage construction　01.030
分散性客流　dispersive passenger traffic　06.045
分水岭　dividing ridge　04.076
*FFT 分析仪　fast Fourier transform analyzer, FFT analyzer　09.246
分运　partite transport　06.180
粉料仓　filler bin　07.145
粉料秤　filler weigher　07.146
粉料计量给料器　filler metering feeder　07.147
粉料撒布机　filler spreader　07.120
粉煤灰　fly ash　08.047
粉煤灰水泥混凝土　fly-ash cement concrete　08.102
粉土质砂　silty sand　08.082
粉状货挂车　bulk tanker trailer　05.043
封闭交通　close to traffic　01.353
封闭式收费系统　closed toll system　03.347
封层　seal coat　02.350
封锚　sealing-off and covering anchorage　04.734
峰值响应　peak response　09.112
风尘环境模拟试验　simulated dust environment test　07.389
风荷载　wind load　04.487
风积土　neolian soil　01.296
风景区道路　park-way　02.007
风缆　wind-cable　04.706
风力加速堤　wind accelerating dike　02.678
风蚀　wind erosion　02.652
风压分布图　wind pressure distribution graph　04.586
风致摆动　wind induced oscillation　04.573
风嘴　wind fairing　04.544
缝隙水　interstitial water　01.260
扶壁式桥台　counterfort abutment, buttressed abutment　04.392
扶臂式挡土墙　counterfort retaining wall　02.271
辐射式公路　radial highway　02.027
服务区　service area　02.190
服务区标志　service area sign　03.402
服务手续费　service fee　06.501

服务水平　level of service, LOS　03.162
服务水平量度参数　effectiveness measures of LOS　03.163
服务隧道　service tunnel, acess adit　04.837
服务巡逻　service patrol　03.283
*浮吊　floating crane　07.291
浮力　buoyancy　04.496
浮漂度　floatability　08.236
浮漂度试验　floatability test　09.070
浮桥　floating bridge　04.249
浮式起重机　floating crane　07.291
浮水设备　floating equipment　07.271
浮箱　floating box　07.272
浮运沉井　floating caisson　04.742
浮运架桥法　bridge erection by floating method　04.674
*弗氏锚　cone anchorage, Freyssinet cone anchorage　04.641
辅道　auxiliary road　02.032
辅助标志　auxiliary sign　03.358
*辅助车道　auxiliary lane　02.179
辅助墩　auxiliary pier　04.377
辅助设施费　cost of ancillary facility　07.339
*辅助隧道　service tunnel, acess adit　04.837
辅助装卸费　auxiliary handling charge　06.497
辅助装卸作业　auxiliary handling operation　06.295
腐殖土　muck　01.287
复传力法预应力　prestressing by subsequent bond　04.620
复合式水泥混凝土路面　composite type concrete pavement　02.456
复曲线　compound curve　02.083
复曲线点　point of compound curve, PCC　02.102
*腹板　web　04.535
腹拱　spandrel arch　04.352
负荷模拟试验　simulated load test　07.382
负响应区　negative response zone　09.114
附加车道　auxiliary lane　02.179
附具摊销费　attachment cost　07.336
附着系数测定　traction coefficient measurement　07.437

G

改建桥 reconstructed bridge 04.165

改性沥青 modified asphalt 08.029

改钻机 drills steel sharpener 07.030

钙电极快速测定法 calcium electric rapid determination method 09.060

盖板涵 slab culvert 04.807

盖梁 bent cap 04.363

干密度 dry density 08.201

干容量 dry unit weight 08.203

干湿试验 wetting and drying test 09.045

干线公路 trunk highway, arterial highway 02.035

干硬性混凝土 dry concrete 08.104

干燥－搅拌筒 drying-mixing drum 07.137

干燥类型 dry type 01.164

干燥筒 drier 07.136

*杆件 member, element 04.533

杆件缩短损失 loss due to concrete member shortening 04.637

杆件弯曲损失 loss due to bending of concrete member 04.638

感应信号控制机 traffic-actuated controller 03.302

刚度 rigidity, stiffness 04.554

刚构桥 rigid frame bridge 04.230

刚架拱桥 rigid-framed arch bridge 04.223

刚接梁法 rigid connected beam method 04.580

刚性横梁法 rigid cross beam method 04.578

刚性护栏 stiff safety fence 03.439

刚性路面 rigid pavement 02.450

钢板护栏 steel guardrail 03.437

钢板梁桥 steel plate girder bridge 04.210

钢材涂料 steel coating 08.174

钢材外观检验 steel visual inspection 09.087

钢材验收试验 steel acceptance test 09.086

钢拱支撑 steel-arched support 04.895

钢骨混凝土桥 rolled shape steel reinforced concrete bridge 04.185

钢管混凝土桥 steel pipe-encased concrete bridge 04.187

钢管桩 steel pipe pile 04.423

钢铰线 strand 08.130

钢筋 steel bar 08.120

钢筋保护层测定仪 cover protectometer 09.237

钢筋电位测量 bar potential measurement 09.133

钢筋电阻率测量 bar resistivity measurement 09.132

钢筋对焊机 steel-bar butt welder 07.303

钢筋镦头机 steel-bar header 07.297

[钢筋]焊接接头强度试验 welded joint strength test 09.090

钢筋混凝土护栏 reinforced concrete fence 03.441

钢筋混凝土路面 reinforced concrete pavement 02.452

钢筋混凝土桥 reinforced concrete bridge 04.183

[钢筋]混凝土桩 concrete pile 04.419

钢筋切断机 steel-bar cutter 07.300

钢筋调直机 steel-bar straightener 07.301

钢筋弯曲机 steel-bar bender 07.302

钢筋锈蚀 steel-bar corrosion 04.608

钢筋锈蚀测定计 bar corrosion activity indicator 09.236

钢筋锈蚀活动性评定 bar corrosion activity evaluation 09.130

钢筋锈蚀三因素模型 three-factor model of bar corrosion 09.131

钢筋预应力松弛损失 loss due to tendon relaxation 04.636

钢缆索 cable 08.137

钢梁结构试装配 trial steel work fixing 04.737

钢桥 steel bridge 04.188

钢桥就位 seating steel bridge on supports 04.739

钢桥拼装 assembling of steel bridge 04.738

钢丝缠束机 steel wire strand strapping machine 07.298

钢丝冷拔试验 wire cold-drawn test 09.088

钢丝牵索 wire guy 08.135

钢丝束 bundle of steel wire 08.129

钢丝束制作 wire grouping 04.724

钢丝网铺设机 wire mesh laying machine 07.206

钢索护栏　cable guardrail　03.432

钢围图　steel waling　04.768

钢纤维　steel fiber　08.155

钢斜拉桥　steel deck cable stayed bridge　04.237

钢桩　steel pile　04.422

港口立交桥　estuarial crossing　04.153

港湾式停车处　parking bay　02.184

杠杆原理法　lever principle method　04.576

高程　elevation　01.207

高承台桩基础　high-rise platform pile foundation　04.411

高峰小时系数　peak hour factor, PHF　03.056

高杆照明　high mast lighting　03.457

高级路面　high class pavement　02.334

高架道路　elevated road　02.014

高架桥　viaduct, trestle　04.154

高跨比　depth-span ratio　04.029

高岭土　kaolin　01.288

高强度钢丝　high tensile steel wire　08.128

高强混凝土　high strength concrete　08.107

高强螺栓　high strength bolt　08.132

高水位桥　high water level bridge　04.172

高速公路　freeway　02.020

高速公路监控　freeway surveillance and control　03.276

高速公路主线控制　freeway mainline control　03.277

高速构造深度仪　high speed texture meter　09.185

高温模拟试验　simulated hot climate test　07.385

高压油泵　high-pressure oilpump　07.294

高液限粘土　high liquid limit clay　08.089

高原气候模拟试验　simulated highland climate test　07.386

高站台　high platform　06.255

戈壁　gobi　01.269

格构比拟　grillage analogy　04.610

格栅压路机　grid roller　07.063

隔离网　separation net　03.492

隔离栅　separation fence　03.491

隔离障碍　sovereignty nuisance　10.088

隔声道　sound isolate road　10.060

隔声绿化带　sound isolate greenbelt　10.063

隔水层　water insulation course　02.357

隔温层　thermal insulation course　02.358

隔音墙　acoustic barrier　10.059

隔振沟　vibration isolate ditch　10.070

给料机　feeder　07.089

跟车理论　car-following theory　03.006

*GBM工程　highway standardized and beautified project　02.691

工程材料学　engineering material　01.013

工程测量学　engineering surveying　01.010

工程地质略图　engineering geologic sketch　01.308

工程地质条件分析　engineering geologic condition analysis　09.144

工程地质学　engineering geology　01.009

工程地质遥感测量　remote sensing of engineering geology　01.247

工程废水　construction waste water　10.036

工程间接费　indirect expense of project　01.063

工程决算　final account of project　01.148

工程量清单　bill of quantities, BOQ　01.132

工程塑料　engineering plastic　08.163

工程预备费　reserve fund of project　01.064

工程直接费　direct expense of project　01.062

工程质量管理　engineering quality control　01.094

*工具锚具　clamper　04.640

工期　construction time limit　01.149

工效学　ergonomics　01.018

工形梁　I[-shaped] beam　04.268

工序吨　unit operation ton　06.318

工业废渣　industrial solid waste　08.046

工业性试验　type approval test　07.409

工作车班　vehicle working shift　06.352

工作车日　working vehicle-days　06.351

*工作度　workability　08.261

工作过程参数　parameters of operation process　05.167

*工作架　scaffold, falsework　04.704

*工作锚具　anchorage　04.639

工作照度试验　operating luminance test　07.451

功率自动分配　automatic power apportioning　07.363

公共客车票价　bus ticket price　06.479

公共汽车站　bus station　02.186

公共汽车终点站　bus terminal　02.187

公路 highway 02.002

公路标准化美化工程 highway standardized and beautified project 02.691

公路病害 highway distress 02.596

公路残值 highway residual value 01.068

公路大修 highway major maintenance 02.572

公路渡口 highway ferry 04.949

公路服务设施 highway service facilities 01.362

公路附属设施 highway appurtenance 01.360

公路改善 highway improvement 02.574

公路改线 highway relocation 02.579

公路工程概算定额 approximate estimate norm of highway project 01.078

公路工程估算指标 estimate index of highway project 01.077

公路工程技术标准 highway technical standard 01.070

公路工程监理 highway engineering supervision 01.098

公路工程学 highway engineering 01.003

公路工程预算定额 budgetary norm of highway project 01.079

公路功能 highway function 01.086

公路功能设计 highway functional design 02.060

公路荷载标准 highway load standard 01.087

公路环境 highway environment 10.001

公路环境保护设计 highway environmental protection desing 10.006

公路环境影响评价 highway environmental impact evaluation 10.007

公路基本建设程序 highway capital construction procedure 01.038

公路集装箱中转站 highway container transfer terminal 06.253

公路几何设计 highway geometric design 02.048

公路技术档案 highway technical file 02.564

公路技术管理 highway technical administration 01.069

公路建设管理 highway construction administration 01.024

公路建设基金 highway construction fund 01.031

公路交叉 highway intersection 02.196

公路交通 highway and transportation 01.001

公路交通方式 highway traffic mode 03.109

公路交通公害 highway traffic nuisance 10.004

公路交通规划 highway and transportation planning 03.108

公路交通监理 highway traffic supervision 01.359

公路景观 highway landscape, vista 02.692

公路景观设计 highway landscape design 02.062

公路勘测规程 highway reconnaissance and survey regulations 01.072

公路客运网 bus transport network 06.062

公路里程 highway mileage 02.547

公路路线编号 code number of highway route 01.339

公路路政管理 highway administration 01.338

公路旅客运输 highway passenger transport 06.034

公路绿化 highway planting, highway greening 02.695

公路美学 highway aesthetics 01.022

公路美学设计 highway esthetic design 02.061

公路排水 highway drainage 02.499

公路桥 highway bridge 04.147

公路施工管理 highway construction management 01.093

公路施工环境影响 environmental effects in highway construction 10.005

公路使用者费用 highway user's cost 01.057

公路使用者效益 highway user's benefit 01.058

公路数据库 highway data bank 02.565

公路隧道 highway tunnel 04.832

公路铁路立交 highway-railway grade separation 02.246

公路通行能力 highway capacity 03.159

公路透视图 highway perspective view 02.055

公路网 highway network 01.026

公路网规划 highway network planning 01.027

公路网密度 density of highway network 01.028

公路限界架 highway boundary frame 01.363

公路小修[保养] highway routine maintenance 02.570

公路养护 highway maintenance 02.544

公路养护质量等级 quality level of highway maintenance 02.689

公路用地　right-of-way　01.346

公路运输　highway transport　06.002

公路运输车辆　highway transport vehicle　05.001

公路运输车辆管理　highway transport vehicle management　05.058

公路运输法规　highway transport statute　06.413

公路运输管理　highway transport management　06.333

公路运输计划　highway transport plan　06.334

公路运输经济运距　break-even distance for highway transport　06.346

公路运输净产值　net highway transport production output value　06.381

公路运输量　highway transport volume　06.340

公路运输行业管理　administration of highway transport industry　06.398

公路运输行政管理　administration of highway transport　06.397

公路运输总产值　total highway transport production output value　06.380

公路灾害　highway disaster　02.584

公路灾害防治　prevention and cure against highway disaster　02.585

公路造价　highway construction cost　01.061

公路中修　highway intermediate maintenance　02.571

公路自然区划　highway natural zoning　01.092

公司平均燃料经济性　corporation average fuel economy, CAFE　05.154

公铁两用桥　combined highway and railway bridge, bi-purposed bridge　04.148

拱　arch　04.278

拱背[线]　extrados　04.345

拱波　two way curved arch tile　04.353

拱顶　arch crown　04.342

拱顶石　arch crown block　04.702

拱腹[线]　intrados, soffit　04.344

拱涵　arch culvert　04.806

*拱环　arch ring　04.340

拱架　arch centering, arch center　04.686

拱架卸落　centering unloading　04.695

拱架预压　preloading centering　04.697

拱脚斜块　arch skew block　04.703

拱肋　arch rib　04.341

拱砌块　voussoir　04.701

拱桥　arch bridge　04.215

拱圈　arch ring　04.340

拱圈封顶　closure of arch ring　04.683

拱圈施工缝　construction joint for arch ring　04.682

拱圈应力调整　arch ring stress adjustment　04.681

拱上侧墙　spandrel wall　04.349

拱上横墙　spandrel cross wall　04.350

拱上结构　spandrel structure　04.348

拱上立柱　spandrel column　04.351

拱式钢拱架　steel arch centering　04.688

拱推力　arch thrust　04.498

拱形桥台　arched abutment　04.389

拱胀　blow up　02.630

拱座　arch support　04.343

共振筛　resonance screen　07.087

勾头丁坝　bend-ended spur dike　04.132

构件　member, element　04.533

构件更换　component replacement　04.800

构件支撑　member support　04.891

构造节理　structure joint　01.320

构造深度仪　texture meter　09.183

鼓风机　blower　07.149

[古]丝绸之路　the silk road　02.019

*骨料　aggregate　08.048

故障分布　failure distribution　05.223

故障机理　failure mechanism　05.196

故障率　failure rate　05.222

故障模式　failure mode　05.197

故障树　failure tree, failure branch chart　05.224

固定式装卸机械　fixed handling machinery　06.278

固定信息标志　fixed-message sign　03.366

固定支座　fixed bearing　04.445

固结　consolidation　02.314

固结试验　consolidation test　09.016

固结系数　consolidation coefficient　08.211

固结仪　consolidometer　09.156

刮板式混凝土布料机　slat type concrete spreader　07.198

刮板输送机　slat conveyer　07.101

刮伤　scoring　05.179

挂车　trailer　05.022

挂车配比　trailer/tractor ratio　06.375

挂钩牵引功率测定　drawbar power measurement 07.434

挂钩牵引力测定　drawbar pull measurement 07.433

挂钩牵引效率测定　drawbar efficiency measurement 07.435

挂篮　cradle　04.708

挂梁　suspended beam　04.272

挂摘钩　coupling and uncoupling of hooks　06.297

观景台　sightseeing stand　02.192

*PE管　polyethylene pipe　08.157

*PVC管　polyvinyl chloride pipe　08.158

管道灌浆　duct grouting　04.733

管道摩擦损失　loss due to duct friction　04.630

管道排水　pipe drainage　02.522

管涵　pipe culvert　04.805

管式护栏　pipe guardrail　03.433

管涌　piping　02.644

管制标志　regulatory traffic sign　03.373

管柱基础　colonnade foundation　04.406

惯性式制动试验台　inertial type brake tester 05.329

灌/压浆设备　grouting equipment　07.275

灌浆　slurry penetration　02.447

灌浆截水墙　grouted cut-off wall　04.441

灌浆帷幕　grouted curtain　04.442

灌注桩钻孔机　cast-in-place concrete pile rig 07.249

贯入法　penetration method　02.442

贯入仪　penetration test apparatus　09.154

光测弹性试验装置　photoelastic test installation 09.194

光度控制系统　light control system　03.470

光化学烟雾　photochemical smog　10.009

光面钢筋　plain bar　08.123

光弹性试验　photoelastic analysis　09.094

光纤标志　optic fiber sign　03.408

光纤测裂计　optic fiber sensor for crack monitor 09.215

硅尘混凝土　silica fume concrete　08.112

硅酸盐水泥　portland cement　08.041

龟裂　alligator crack　02.603

轨模式混凝土摊铺机　rail form concrete paver 07.193

贵重货物　valuable goods　06.147

贵重行包　precious belongings　06.084

辊筒式碎石机　roll crusher　07.083

滚动筛　drum screen　07.088

滚动阻力测定　rolling resistance measurement 07.427

滚浆法　rolling slurry process　02.667

滚轮静线压力测定　drum static linear pressure measurement　07.460

滚轮式钻头　roller bit　07.259

滚上滚下集装箱叉车　roll on/roll off container fork lift　06.307

滚筒式沥青混合料搅拌设备　drum-mixing asphalt plant　07.129

滚轴支座　roller bearing　04.452

滚装运输　roll on/roll off transport　06.033

*国道　national trunk highway　01.341

国道主干线　national main trunk line　01.340

国防公路　national defence highway　01.345

国际竞争性招标　international competitive bidding, ICB　01.114

国际平整度指数　international roughness index, IRI 02.702

国际运输　international transport　06.015

国家干线公路　national trunk highway　01.341

国内竞争性招标　local competitive bidding, LCB 01.115

国内投标人优惠　preference for domestic bidders 01.123

过顶式装碴机　over-head loader　07.309

*过渡式路面　intermediate class pavement　02.336

过境公路　through highway　02.030

过境交通　through traffic　03.114

过境运输　transit transport　06.016

过路费　road toll　01.032

过桥费　bridge toll　01.033

过热　overheating　05.200

过湿类型　excessive dampness type　01.167

过水断面　discharge flow cross section　04.051

过水孔径　discharge opening of bridge　04.052

过水路面　ford　02.543

过隧费　tunnel toll　01.034

过隧经历时间　duration of tunnel passage　04.935

H

含蜡量　paraffin content　08.239

含蜡量试验　paraffin content test　09.073

含泥量试验　silt content test　09.033

含砂低液限粉土　sandy silt of low liquid limit　08.084

含砂低液限粘土　sandy clay of low liquid limit　08.087

含砂高液限粘土　sandy clay of high liquid limit　08.088

含水层　water stratum　01.259

含水量　water content　08.193

含水量试验　water content test　09.001

含油率　bitumen content　08.249

涵[洞]　culvert　04.801

涵洞八字墙　culvert wing wall　04.818

涵洞出水口　culvert outlet　04.820

涵洞跌水　culvert water drop　04.821

涵洞洞身长度　culvert barrel length　04.827

涵洞沟床　gully bed neighboring to culvert　04.822

涵洞进水口　culvert inlet　04.819

涵洞孔径　culvert aperture　04.828

涵洞口隆起　culvert end lift　04.826

涵洞口铺砌　culvert inlet/outlet apron　04.823

涵洞一字墙　culvert straight end wall　04.817

涵口沉降缝　settlement joint in culvert　04.825

*旱桥　viaduct, trestle　04.154

焊接钢桥　welded steel bridge　04.190

夯实　tamping　02.299

航空摄影测量　aerial photogrametry　01.232

航摄基线　aerophoto base line　01.235

航摄像片判读　aerophoto interpretation　01.238

毫秒爆破　millisecond delay blasting　02.329

好路率　rate of good level highway　02.690

荷载　load　04.465

荷载横向分布　transverse distribution of load　04.575

荷载系数　load factor　04.531

荷载效应　load effect　04.503

荷载组合　loading combinations　04.502

核子动态湿度密度仪　nuclear moisture and density dynamic meter　09.153

核子静态湿度密度仪　nuclear moisture and density static meter　09.152

核子湿度密度仪　nuclear moisture and density meter　09.151

和易性　workability　08.261

合成坡度　resultant gradient　02.118

合成橡胶支座　elastomeric pad bearing　04.447

合乘车辆车道　car pool lane　03.239

*合流端　merging end　02.221

合流交通量　converging traffic volume　03.185

合同　contract　01.126

合同管理　contract management　01.128

合同通用条件　general conditions of contract　01.130

合同运输　contract transport　06.024

合同专用条件　special conditions of contract　01.131

河岸　river bank　04.089

河槽　river channel　04.085

河槽天然冲刷　natural scour of channel　04.111

河槽压缩　channel contraction　04.103

河床　river bed　04.084

河床比降　river bed gradient　04.061

河床铺砌　river bed paving　04.137

河道调查　river survey　04.038

河堤　levee　04.134

河滩　flood land　04.088

黑洞效应　black-hole effect　03.469

黑棉土　black cotton soil　01.282

桁架拱桥　trussed arch bridge　04.227

桁架桥　truss bridge　04.214

桁式 T 形刚构桥　T-shaped truss rigid frame bridge　04.232

桁式木拱架　wood-truss centering　04.691

横洞　transverse gallery　04.841

横断面　cross section　02.124

横断面测量　cross section survey　01.204

横缝　transverse joint　02.459

横隔板　transverse diaphragm　04.276

横梁　transverse beam, cross beam　04.275

横伸挂车　adjustable track trailer　05.046

横系杆　lateral bracing　04.543

横向承载桩地基系数法　subsoil reaction modulus method for laterally loaded pile　04.652

横向间断光源　laterally intermittent illuminant 03.467

横向力系数测试仪　sideway force coefficient routine investigation machine, SCRIM　09.179

横向裂缝　transverse crack　02.606

横向排水　transverse drainage　02.515

横向通风　transverse ventilation　04.910

横向悬砌拱法　laying arch by transverse overhanging method　04.680

衡重式挡土墙　balance weight retaining wall 02.269

衡重式桥台　weight-balanced abutment　04.393

恒载　dead load　04.476

洪峰流量　peak flow discharge　04.068

洪积土　diluvial soil　01.299

洪水　flood　02.501

洪水调查　flood survey　01.172

洪水径流　flood run-off　04.081

洪水流量　flood discharge　04.067

洪水频率　flood frequency　04.071

洪水位过程线　flood level duration curve　04.074

洪水重现期　return period of flood　04.073

宏观交通模型　macro-traffic model　03.009

红灯时间　red time　03.246

红粘土　red clay　01.284

红外线车辆检测器　infra-red vehicle detector 03.291

红外线烘干装置　infrared oven stand　05.370

后评价　post-assessment　01.046

后张法　post-tensioning method　04.723

后张自锚法预应力　self-anchored post tensioning prestressing　04.622

呼吸性货物　respiratory goods　06.149

湖沥青　lake asphalt　08.007

弧焊机　arc welder　07.304

护岸　bank protection, bank revetment　04.136

护拱　back haunch fillet of arch　04.347

护坡　slope protection　02.259

护坡道　berm　02.260

护墙　guard wall　02.262

护坦　apron　04.140

护柱　guard post　03.450

护桩　reference stake　01.191

互通式立交　interchange　02.226

互通式立交收费站　interchange toll station　03.340

滑溜测量仪　skidometer　09.177

滑模浇筑混凝土　slip-form concreting　04.718

滑模式混凝土路缘铺筑机　slipform concrete curb paver　07.204

滑模式混凝土摊铺机　slipform concrete paver 07.194

滑坡　landslide　01.263

滑行　coasting　05.105

滑行工况　coasting mode　05.125

滑行距离测定　coasting distance measurement 07.432

滑行试验　coasting test　07.429

滑行阻力测定　coasting resistance measurement 07.431

划线机　road marking machine　07.228

环境保护　environmental protection　01.017

环境荷载　environmental load　04.481

环境模拟试验　simulated environment test　07.383

环境损害　environmental hazard　10.003

环境污染　environmental pollution　10.002

环境噪声标准　environmental noise standard 10.046

环境振动标准　environmental vibration standard 10.065

环形道路　ring road　02.028

环形交叉　roundabout, rotary intersection　02.206

环形立交　rotary interchange　02.234

环形线圈式车辆检测器　loop vehicle detector 03.288

环形匝道　loop ramp　02.244

环氧树脂　epoxy resin　08.031

*缓冲垫　bumper　04.967

缓冲护栏　cushion guardail　03.436

缓和坡段　transitional grade　02.114

缓和曲线　transition curve　02.092

缓和照明　adaptation lighting　03.456

缓凝剂　retarder　08.150

缓坡缘石　lip curb　02.155

缓圆点　point of spiral to curve, SC　02.099

缓直点　point of spiral to tangent, ST　02.101

换档　gear shifting　05.089

换算交通量　equivalent traffic volume　03.061

换算重量　converted weight　06.448

换算周转量　converted turnover　06.343

换土　replacement of earth　02.295

荒漠土　desert soil　01.289

黄土　loess　01.280

灰分含量　ash content　08.242

灰分含量试验　ash content test　09.079

灰色控制系统　grey control system　03.016

灰色系统理论　grey system theory　03.015

挥发分含量　volatile matter content　08.241

挥发分含量试验　volatile matter content test　09.077

回车道　turn around loop　02.183

回程系数　return factor　06.195

回火　back fire　05.212

回砂　sand sweeping　02.660

回砂机　sand recovering machine　07.226

回送空箱　return empty container　06.265

回弹模量　modulus of resilience　02.399

回弹弯沉　rebound deflection　02.388

回弹仪　rebound tester, Schmidt hammer　09.211

回填土　back fill　02.294

回头曲线　switch back curve, reverse loop　02.087

回旋线　clothoid　02.093

毁灭性破坏　catastrophic collapse　04.607

会车　meeting　05.100

汇合端　merging end　02.221

汇流标志　converging sign　03.382

汇水面积　catchment area　04.077

浑水冲刷　scour with sediment motion　04.117

混合交通　mixed traffic　03.002

混合料　mixture　08.090

混合料斗式升送机　mixture hopper elevator　07.156

混合料刮板升送机　mixture slat elevator　07.155

混合式收费系统　mixed toll system　03.348

混凝土泵送法　concreting by pumping　04.711

混凝土表面泌水　surface weeping of concrete　02.477

混凝土布料机　concrete spreader　07.195

混凝土弹性压缩损失　loss due to elastic compression of concrete　04.634

混凝土电热养生　electric curing of concrete　04.714

混凝土封底　bottom sealing by concreting　04.772

混凝土管桩　concrete pipe pile　04.424

混凝土回弹试验　rebound test of concrete　09.122

混凝土浇筑　concreting　04.710

混凝土搅拌机　concrete mixer　07.190

混凝土搅拌设备　concrete mixing plant　07.187

混凝土搅拌输送车　truck mixer　07.191

混凝土梁滑模浇筑设备　concrete beam slipform device　07.277

混凝土流动性试验　concrete fluidity test　09.085

混凝土路面拉毛机　concrete pavement texturing machine　07.200

混凝土路面切缝机　concrete pavement joint cutting machine　07.201

混凝土路面清缝机　concrete pavement joint cleaning machine　07.202

混凝土路面填缝机　concrete pavement joint sealing machine　07.203

混凝土路面整面机　concrete pavement finisher　07.199

混凝土氯化物含量测量　concrete chloride content measurement　09.126

混凝土强度超声测量　ultrasonic test of concrete strength　09.121

混凝土收缩损失　loss due to concrete shrinkage　04.633

混凝土输送泵车　concrete pump truck　07.205

混凝土摊铺列车　concrete paving train　07.192

混凝土碳化试验　concrete carbonation test　09.128

混凝土斜拉桥　concrete deck cable stayed bridge　04.236

混凝土徐变损失　creep loss of concrete　04.635

混凝土养生　concrete curing　04.712

混凝土蒸汽养生　steam curing of concrete　04.713

混凝土钻孔内窥镜检查　concrete coring hole

inspecting by endoscope 09.124

混装 mixed loading 06.156

混装修理法 depersonalized repair method 05.310

*活动桥 movable bridge 04.250

活动伸缩缝支座 sliding expansion bearing 04.457

活动支座 expansion bearing, movable bearing 04.446

活塞式水泵 piston pump 07.266

活载 live load 04.475

火管式沥青熔化装置 asphalt firepipemelter 07.174

货差 freight shortage 06.218

货差率 shortage rate of goods 06.232

*货车 [motor] truck 05.005

货车动态 operation status of truck 06.215

货车装卸安全间距 truck handling safety interval distance 06.296

货流 freight flow 06.199

货流图 freight traffic diagram 06.200

货票 freight bill 06.183

货损 freight damage 06.219

货损率 damage rate of goods 06.231

货位 freight lot 06.259

货物包装 goods packing 06.158

货物标志 goods mark 06.160

[货物]仓库 warehouse 06.254

货物操作量 tonnage of cargo handled 06.317

[货物]堆存吨天数 storage ton-days 06.331

货物分等运价 classified rates for goods 06.466

货物流量 freight traffic volume 06.201

货物流时 freight traffic time 06.203

货物流向 freight traffic direction 06.202

[货物]平均堆存天数 average storage days 06.332

货物平均运距 freight average haul distance 06.344

货物清单 freight list 06.182

[货物]托盘 pallet 06.244

货物运单 bill of lading 06.181

货物运输 freight transport 06.017

货物运输杂费 miscellaneous charges for freight transport 06.491

货物站务费 freight terminal charge 06.502

货物装卸 goods handling 06.273

货物装卸费 goods handling charge 06.496

货物装卸量 loading/unloading volume 06.316

货源 freight source 06.192

货源调查 freight source survey 06.193

货源信息 information of freight source 06.198

货运调度 dispatching of freight transport 06.204

货运及时率 rate of timely freight 06.234

货运拒赔 reject of freight claims 06.227

货运理赔 actions for freight claims 06.226

货运联运站 intermodal freight terminal 06.252

货运生产平衡 freight production balance 06.194

货运枢纽站 freight hub terminal 06.250

货运索赔 freight claims 06.225

货运索赔时限 period of limitation for freight claims 06.228

货运信息中心 freight information center 06.251

货运业务 freight transport business 06.171

货运站 freight terminal 06.248

货运质量 freight transport quality 06.216

货运质量事故频率 frequency of freight quality accidents 06.230

货运质量指标 freight quality indicator 06.229

J

击实试验 compaction test 09.012

击实仪 compaction test apparatus 09.150

*击振 buffeting 04.571

基本通行能力 basic capacity 03.164

基本运价 basic rate 06.455

基本折旧率 depreciation rate 07.332

基层 base course 02.354

基础沉降 foundation settlement 04.658

基础工程学 foundation engineering 01.011

基础件修理 basic part repair 05.286

基础隆胀力 heave force of foundation 04.499

基础埋置深度 embedment depth of foundation 04.035

基底应力扩散 stress dispersal beneath footing

04.657

基坑井点排水法 foundation pit well point drainage method 04.773

基线 base line 01.181

基岩 bedrock 01.314

基桩病害检测系统 foundation pile diagnosis system 09.223

机动车辆 power-driven vehicle 05.002

机动车运输 power-driven vehicle transport 06.003

机动渡车船 powered ferry boat 04.959

机耕道 tractor road 02.012

*机械操作 machine operation 07.318

机械操作证 operation certificate of machinery 07.344

机械出租 renting of machine 07.347

机械故障 failure of machine 07.323

机械故障率 failure rate of machinery 07.330

机械管理责任制 responsibility system of machinery management 07.343

机械化程度 level of mechanization 07.348

机械技术经济定额 technical economic quota of machinery 07.346

机械技术状况 technical condition of machine 07.319

机械利用率 operation rate of machinery 07.329

机械生产定额 production quota of machinery 07.345

机械事故 machinery accident 07.340

机械事故处理 accident handling of machinery 07.342

机械停修台日 suspensive daily shift of machine 07.355

机械通风 mechanical ventilation 04.908

机械挖掘机 cable operated excavator 07.045

机械完好率 availability rate of machinery 07.328

机械维护 maintenance of machine 07.320

机械维修间隔 time interval of machine maintenance and repair 07.353

机械维修作业项目 items of machine maintenance and repair 07.354

机械系列 series of machine 07.351

机械型谱 model spectrum of machine 07.350

机械修理 repair of machine 07.324

机械运转 machine operation 07.318

机械运转记录 operating record of machine 07.352

机械装卸 mechanical handling 06.276

机械装卸作业比重 share of machine handling operation 06.320

机械装卸作业量 machine handling operation volume 06.319

*畸变 distortion 04.567

积雪标杆 snow[deposit] markerpost 03.480

激光准直挠度测量 laser alignment deflection measurement 09.100

激光自动找平装置 laser automatic leveling device 07.369

激振器 vibration excitor 09.201

极限间隙 clearance limit 05.302

极限磨损 wear limit 05.175

极限应力 ultimate stress 04.552

集合模型 aggregation model 03.013

集结性客流 aggregate passenger traffic 06.044

集料 aggregate 08.048

集料剥落 stripping of aggregate 02.609

集散道路 collector-distributor road 02.033

集水槽 gully 02.534

集中荷载 concentrated load 04.484

集中排水 concentrated drainage 02.533

集装袋 contain bag 06.246

集装箱包箱运价 charter rate by number of containers 06.481

*集装箱比容 container specific volume capacity coefficient 06.240

集装箱叉车 container fork lift 06.306

集装箱吊具 container spreader 06.311

集装箱挂车 container trailer 05.032

集装箱化率 containerization ratio 06.347

集装箱货车 container carrier 05.012

集装箱空箱运价 rate of empty container 06.483

集装箱栏板挂车 container dropside trailer 05.035

集装箱内容积 container volume capacity 06.238

集装箱平板挂车 container platform trailer 05.034

[集装]箱容利用率 container volume capacity utilizing ratio 06.241

[集装]箱容系数 container specific volume capacity coefficient 06.240

集装箱维修费 container maintenance charge 06.500

[集装]箱位 container lot 06.243

集装箱箱次费 per container charge 06.484

集装箱运输 container transport 06.169

[集装]箱载重利用率 container load capacity utilizing ratio 06.242

集装箱中转费 container transit charge 06.485

集装箱重箱运价 rate of loaded container 06.482

集装箱专用挂车 container flatframe trailer 05.033

集装箱装卸机械 container handling machinery 06.281

集装箱自重系数 coefficient of container dead weight 06.239

集装箱自装卸挂车 self-loading container trailer 05.047

急流槽 chute 02.535

级配 gradation 08.185

级配集料基层 graded aggregate base 02.490

级配路面 graded aggregate pavement 02.433

级配曲线 grading curve 08.191

技术服务 technical service 05.252

技术检验 technical checking 05.294

技术设计 technical design 01.043

技术性试验 technical test 07.401

季节性货源 seasonal freight source 06.197

季节性客流 seasonal passenger traffic 06.041

季节性维护 seasonal maintenance 05.255

计程包车运价 charter rate by distance 06.473

计费里程 charged mileage 06.449

计费重量 charged weight 06.447

计划调度 planned dispatching 06.205

计划任务书 planning assignment 01.041

计划外维护 unscheduled maintenance 05.260

计划修理 scheduled repair 05.288

计划预防维修制度 scheduled preventive maintenance and repair system 05.244

计划运输 planned transport 06.023

计价标准 charge standard 06.446

计件行包 piece luggage 06.082

计量与支付 measurement and payment 01.136

计日工 daywork labor 01.142

计时包车运价 charter rate by time 06.472

计算跨径 calculated span 04.015

计算矢高 calculated rise of arch 04.028

计算行车速度 design speed 01.081

计箱装卸费 handling charge by container number 06.493

忌装 avoidance of mixed loading 06.157

夹板拱架 centering by clamped planks in arch shape 04.693

夹具 clamper 04.640

夹片式锚具 strand tapered anchorage 04.642

加固地基 consolidated subsoil 02.312

加减速延误 acceleration-deceleration delay 03.098

加筋土挡土墙 reinforced earth retaining wall 02.275

加劲杆 stiffener 04.537

* 加劲肋 stiffener 04.537

加劲梁 stiffening girder 04.538

加劲腋 haunch 04.539

加宽缓和段 transition zone of curve widening 02.133

加宽转角式交叉口 intersection with widened corners 02.209

加气剂 air entraining agent 08.146

加强氯丁橡胶板缝 strengthened neoprene plate joint 04.311

加热损失 heating loss 08.234

加热损失试验 heating loss test 09.067

加热型塑料 hot laid plastic 08.169

加速 speeding up 05.090

加速车道 acceleration lane 02.170

加速工况 accelerating mode 05.122

加速滑行 accelerating-coasting 05.109

加速试验 accelerated test 07.392

加速性能试验 acceleration test 07.423

加速噪声 acceleration noise 03.025

加载反力架 reaction frame for loading 09.196

加载平台 weighted platform 09.200

加载水箱 water tank for loading 09.209

加州承载比 California bearing ratio, CBR 02.413

加州承载比试验 California bearing ratio test, CBR test 09.024

加桩 additional stake 01.192

假缝 dummy joint 02.466

价格调整 price adjustment 01.138

价目 tariff classification 06.454

架桥设备 bridging equipment 07.276

架式举升机 crossing lift 05.360

驾驶操纵能力 driving operation ability 03.521

驾驶技能 driving skill 05.111

驾驶技术 driving technique 05.112

驾驶模拟器 driving simulator 05.116

驾驶能力 driving ability 05.110

驾驶疲劳 driving fatigue 03.520

驾驶适应性 driving adaptability 03.526

驾驶室视野试验 test for field of vision of driver cab 07.450

驾驶室仪表板振动测定 vibration measurement of driver cab instrument board 07.448

驾驶习惯 driving habits 03.522

驾驶信息 driving information 03.500

驾驶行为 driving behavior 03.523

驾驶员心理和生理反应 driver psychological and physiological reaction 03.525

驾驶员行为特性 driver behavior pattern 03.524

驾驶员座位振动测定 vibration measurement of driver seat 07.447

监理工程师 supervising engineer 01.105

监理工程师代表 supervising engineer's representative 01.106

坚井 shaft 04.843

坚琴形拉索 harp-type stay cable 04.330

尖拱法封顶 closure by wedging-in crown 04.684

间接视野 indirect field of vision 03.505

间接效益 indirect benefit 01.054

间歇式交通量观测站 intermittent traffic count station 03.122

间歇式沥青混合料搅拌设备 batch asphalt mixing plant 07.127

检测性试验 detective test 07.405

检查井 manhole 02.540

检视 visual inspection 05.295

碱性集料反应 alkali-aggregate reaction 02.635

简支梁桥 simple-supported beam bridge 04.204

剪草机 mower 07.233

剪力 shear 04.558

剪力分布 shear distribution 04.562

剪力键 shear key 04.542

剪力滞后 shear lag 04.563

*剪滞 shear lag 04.563

减少交通事故效益 benefit from accidents reducing 01.060

减水剂 water reducing agent 08.145

减速 slowing down 05.091

减速车道 deceleration lane 02.171

减速工况 decelerating mode 05.124

减速过桥 passing bridge with reduced speed 01.356

减振措施 vibration reducing measure 10.069

鉴定试验 evaluation test 07.404

件货 piece goods 06.139

渐变板 transition slab 02.474

渐变段 tapered section 02.210

渐变故障 gradual failure 05.193

渐近稳定性 asymptotic stability 03.024

焦油沥青 tar 08.022

胶合木桥 glued timber bridge 04.200

交叉角 intersection angle 02.198

交叉口负荷系数 load factor of intersection 03.253

交叉口视距 sight distance of intersection 02.212

交叉口通行能力 intersection capacity 03.168

交叉口延误 intersection delay 03.095

交点 intersection point, IP 01.194

交互式系统 alternate system 03.260

交通安全 traffic safety 03.212

交通安全设施 traffic safety facilities 01.361

交通标志 traffic sign 03.353

交通冲突 traffic conflict 03.229

交通岛 traffic island 02.216

交通调查 traffic survey 01.159

交通调查表 traffic survey chart 03.138

*交通发生 trip generation 03.115

交通法庭 traffic court 03.231

交通方式划分 [traffic] modal split 03.110

交通方式选择模型 traffic modal choice model 03.142

交通仿真 traffic simulation 03.017

交通分隔 segregation of traffic 03.204

交通分类 traffic classification 03.049

交通分流　traffic diverging　03.036
交通分析　traffic analysis　03.037
交通干扰　traffic interference　03.063
交通感应式控制　traffic responsive control　03.264
交通隔离墩　traffic divided blocks　03.451
交通工程学　traffic engineering　01.004
交通管理费　traffic administration cost　01.067
交通管理设施　traffic management devices　03.350
交通规则　traffic rules, traffic regulations　03.208
交通护栏　traffic guardrail　03.431
交通汇合　traffic converging　03.035
交通监控系统　traffic surveillance and control system　03.279
交通监控中心　traffic surveillance and control center　03.278
交通可达性　traffic accessibility　03.136
交通量　traffic volume　03.050
交通量调查　traffic volume survey　01.160
交通量分布　traffic [volume] distribution　03.117
交通量分配　traffic [volume] assignment　03.116
交通量估计　traffic volume estimating　03.051
交通量计量　traffic volume measurement　03.062
交通量修正系数　traffic volume adjustment factor　03.267
交通量预测　traffic volume forecast　03.118
交通流　traffic flow　03.003
交通流理论　traffic flow theory　03.004
交通流率　traffic flow rate　03.053
交通密度　traffic density　03.084
交通瓶颈　traffic bottleneck　03.038
交通强度　traffic intensity　01.083
交通设施　traffic facilities　03.351
交通生成　trip generation　03.115
交通生理学　traffic physiology　01.020
交通事故　traffic accident　03.214
交通事故率　traffic accident rate　03.215
交通事故隐患　hidden peril of accident　03.234
交通事故预测　traffic accident prediction　03.235
交通疏导　traffic dispersion　01.354
交通特性　traffic characteristics　01.084
交通通道　traffic aisle　03.262
交通统计学　traffic statistics　01.021
交通网络　traffic network　03.033

交通违章　traffic violation　03.209
交通违章者　traffic violator　03.210
交通稳定性　traffic stability　03.022
交通系统　traffic system　03.039
交通系统管理　transportation system management, TSM　03.192
交通线网优化　optimization of traffic line and network　03.040
交通心理学　traffic psychology　01.019
交通信号灯　traffic signal lamp　03.307
交通信号脱机控制　traffic signal off-line control　03.270
交通信息系统　traffic information system　03.280
交通行为　traffic behavior　03.020
交通需求　traffic demand　03.032
交通需要预测　traffic demand forecast　03.119
交通巡逻　traffic patrolling　03.194
交通因素　traffic factor　03.052
交通拥挤　traffic congestion　03.034
交通运输网　transportation network　01.025
交通噪声　traffic noise　10.047
交通噪声评价　traffic noise evaluation　10.049
交通噪声指数　traffic noise index, TNI　10.048
交通肇事罪　traffic accident crime　03.216
交通振动　traffic vibration　10.066
交通指挥　traffic guidance　03.193
交通指路牌　traffic direction block　03.482
交通指向钮　traffic direction stud　03.481
交通组成　traffic composition　01.082
交织交通流　weaving traffic flow　03.045
交织区　weaving area　03.181
交织区通行能力　weaving section capacity　03.171
搅拌器　mixer　07.142
铰接梁法　hinge connected beam method　04.579
铰接式压路机　articulated roller　07.069
铰接式自卸车　articulated truck　07.012
铰式支座　knuckle bearing　04.460
脚手架　scaffold, falsework　04.704
角钢镶边接缝　angle steel edged joint　04.306
角砾　angular gravel　08.075
角裂　corner break　02.625
角隅钢筋　corner steel　02.473
[轿车]车身测量整形机　body and frame measure

and correct system 05.367

轿车运载[半挂]车 car carrier semitrailer 05.048

接缝 joint 02.457

接缝破损 joint failure 02.623

*接近端 approach nose 02.220

接力运输 relay transport 06.012

阶梯式涵洞 stepped culvert 04.816

截水沟 intercepting ditch 02.519

*节点板 gusset plate 04.541

节段预制拼装法 segmental precast erection method
04.671

节间 panel 04.534

节油率 fuel saving ratio 06.378

结点板 gusset plate 04.541

结构钢 structural steel 08.118

结构体系转换 structure system transform 04.527

结构完整性 structural integrity 04.528

结构延展性 structure ductility 04.604

结构振动试验台 shaking table for structure test,
vibrostand 09.195

结合料 binder 08.001

界面滑动接触 sliding contact of interface 02.402

界面连续接触 continuous contact of interface
02.401

界面排水 interface drainage 02.529

借土 borrow earth 02.290

金属[材料] metal [material] 08.116

金属滚轴带氯丁橡胶板接缝 metal rollers with
neoprene plate joint 04.312

金属网 metal mesh 08.134

金属纤维 metal fiber 08.154

紧急灭火机 emergency fire extinguisher 04.945

紧急制动 emergency braking 05.096

进水口 inlet 02.537

禁令标志 prohibitory sign 03.355

禁运货物 contraband goods 06.154

禁止超车标志 overtaking prohibited sign 03.392

禁止调头标志 no U-turn sign 03.389

禁止驶入标志 no entry sign 03.394

禁止停车 parking prohibited 03.197

禁止停放标志 parking ban sign 03.398

禁止通行 traffic prohibited 03.196

禁止通行标志 traffic prohibited sign 03.396

禁止右转弯标志 no right turn sign 03.390

禁止左转弯 no left turn 03.195

禁止左转弯标志 no leftturn sign 03.391

浸渍混凝土 impregnated concrete 08.113

劲性骨架混凝土桥 skeleton reinforced concrete
bridge 04.186

精轧螺纹锚 thread bar anchorage, Dywidag anchorage system 04.649

经常维修费 routine repair cost 07.333

经济车速 economic speed 03.079

经济调查 economy survey 01.158

经济净现值 economic net present value, ENPV
01.048

经济跨径 economic span 04.017

经济内部收益率 economic internal rate of return,
EIRR 01.049

经济通行能力 economical capacity 03.172

经济效益费用比 economic benefit cost ratio, EBCR
01.047

经营许可证 business certificate 06.407

井探 shaft test 01.251

警察巡逻 police patrol 03.282

警告标志 warning sign, caution sign 03.354

景观栽植 landscape planting 02.694

静定桥梁结构 statically determinate bridge structure
04.525

静力触探试验 static cone penetration test 09.027

静力压拔桩机 static pile press extractor 07.248

静视觉敏锐度 static visual acuity 03.531

静水护壁钻孔法 static water head hole-boring
method 04.749

静态轴重检测器 static axle weight detector
03.313

静压沉桩法 pile jacking-in method 04.758

静载 static load 04.473

境内出行 local trip 03.133

境内交通 local traffic 03.112

径流 run-off 04.078

径流面积 run-off area 04.082

径流系数 run-off coefficient 04.083

径向固结系数 radial consolidation coefficient
08.212

净高 vertical clearance 02.126

净空 clearance 02.125

净空高度 clearance height 04.025

净跨径 clear span 04.016

净宽 horizontal clearance 02.127

旧件修复成本 used parts reconditioning cost 06.433

旧路技术改造 technical reformation of existing road 02.578

旧桥 existing bridge 04.163

旧线清除机 marking cleaning machine 07.232

就车修理法 personalized repair method 05.311

就地灌注桩 cast-in-situ pile 04.426

*就地浇筑法 cast on scaffolding method, cast-in-place method 04.668

局部故障 partial failure 05.189

局部稳定性 local stability 03.023

矩形梁 rectangular beam 04.271

举升泊车机 parking lift 05.361

聚丙烯筋带 polypropylene belt 08.168

聚合物改性沥青 polymer modified asphalt 08.030

聚合织物防水层 water proof polymer fabrics 04.295

聚氯乙烯管 polyvinyl chloride pipe 08.158

聚四氟乙烯板 polyfluortetraethylene plate 08.159

聚四氟乙烯支座 polytetrafluoroethylene bearing, PTFE bearing 04.450

聚乙烯管 polyethylene pipe 08.157

聚乙烯沥青装置 polyethylene asphalt unit 07.185

巨粒土 over coarse-grained soil 08.059

锯齿形码头 saw-tooth type wharf 04.954

锯缝 sawn joint 02.469

卷管机 tendon duct winding machine 07.299

掘进方法 tunneling method 04.864

绝对基面 absolute datum 01.208

均布荷载 uniform load, distributed load 04.482

军用桥 military bridge 04.152

峻岭 steep mountain 01.225

竣工 completion of construction 01.145

竣工测量 final survey 01.183

K

喀斯特 karst 01.261

开标 bid opening 01.120

开槽机 grooving machine 07.057

开放交通 open to traffic 01.352

开放式收费系统 open toll system 03.346

开工 starting of construction 01.140

开级配 open gradation 08.190

开裂荷载 cracking load 04.468

开启查验 luggage opening for inspection 06.092

开启桥 movable bridge 04.250

开式联合碎石机组 open type crushing plant 07.097

开停业管理 entry/exit regulation 06.401

开业报批程序 entry application and approval procedure 06.403

开业条件 entry conditions 06.402

勘探 exploration 01.249

抗剥落剂 anti-stripping agent 08.143

抗风稳定性 wind resisting stability 04.584

抗风支撑 wind bracing 04.354

抗滑桩 anti-slide pile 04.432

抗老化剂 anti-ager 08.144

抗裂性 crack resistance 04.597

抗折强度 rupture strength 02.405

抗撞墩 anti-collision pier 04.375

颗粒分析试验 grain size test 09.005

颗粒组成 grain composition 08.180

可搬式沥青混合料搅拌设备 movable asphalt mixing plant 07.125

可搬式稳定土厂拌设备 movable stabilized soil mixing plant 07.109

可编程序数据采集器 programmable data logger 09.243

可变标志 variable sign 03.363

可变荷载 variable load 04.478

可变限速标志 variable speed-limit sign 03.364

可变限速控制 variable speed-limit control 03.281

可变向车道 reversible lane 02.180

可变向中心车道线 reversible center lane line 03.421

可变信息标志　changeable message sign　03.365

可插车空档　acceptance gap　03.026

可调头交叉口　turn crossing　03.453

可靠度基准期　reliability datum period　04.513

可靠性试验　reliability test　07.397

可能通行能力　possible capacity　03.165

可行性研究　feasibility study　01.040

刻槽　carving groove　02.481

客车　bus　05.015

客车调度　bus dispatching　06.073

客车定员　rated passenger number　06.063

客车运行周期　bus run cycle　06.074

客车正班率　rate of on-schedule bus runs　06.120

客流　passenger traffic, passenger flow　06.036

客流图　passenger traffic diagram　06.049

客位　rated seat　06.355

客源　passenger source　06.050

客运班期　bus schedule　06.068

客运包车票　charter bill　06.101

客运单证　document of passenger transport　06.095

客运动态记录单　bus running record　06.102

客运拒赔　reject of passenger claims　06.116

客运理赔　actions for passenger claims　06.115

客运索赔　passenger claims　06.114

客运索赔时限　period of limitation for passenger claims　06.117

客运站站务工作量　passenger terminal work load　06.051

客运质量管理　quality management of passenger transport　06.105

啃边　edge failure　02.619

坑槽　pot holes　02.617

坑探　pit test　01.252

*空车里程　deadhead mileage　06.359

空车行程　deadhead mileage　06.359

空腹拱桥　open spandrel arch bridge　04.216

空间平均速度　space mean speed　03.071

空气动力稳定性　aerodynamic stability　04.585

空气幕法下沉沉井　sinking open caisson by injected

air curtain　04.746

空气压缩机　air compressor　07.306

空箱里程　empty container mileage　06.453

空心板桥　hollow slab bridge　04.208

空心[桥]墩　hollow pier　04.366

孔隙比　void ratio　08.177

孔隙率　porosity　08.176

孔隙水压力计　pore pressure cell　09.228

控制室　control house　07.152

枯水位　low water level　04.097

库场作业　storage area operation　06.302

跨河桥　river-crossing bridge　04.150

跨径　span　04.009

跨径长　span length　04.014

跨区客流　inter-regional passenger traffic　06.037

跨线桥　flyover　04.151

块料路面　block pavement　02.424

块石　angular boulder, block stone　08.066

块石拱桥　block stone arch bridge　04.198

块石基层　Telford base　02.488

*快车道　fast lane　02.176

快裂乳化沥青　quick breaking emulsified bitumen　08.009

快凝液体沥青　rapid-curing liquid asphalt　08.017

快速傅里叶变换分析仪　fast Fourier transform analyzer, FFT analyzer　09.246

快速公共交通系统　rapid transit system　03.141

快速试验　prompt test　07.394

宽跨比　width-span ratio　04.021

矿粉　mineral powder　08.053

矿料　mineral aggregate　08.050

矿山法　mine tunneling method　04.868

矿渣基层　slag base　02.497

矿渣水泥　slag cement　08.042

亏油率　fuel deficit ratio　06.379

扩大初步设计　enlarged preliminary design　01.044

扩大基础　spread foundation　04.402

扩底桩　under-reamed pile　04.430

L

拉铲挖掘机　dragline excavator　07.042

拉杆　tie bar　02.467

拉缸　cylinder scoring　05.215

拉力试验　tension test　09.046

拉毛　broom-finish　02.479

拉压支座　tension-compression bearing　04.459

喇叭形立交　trumpet interchange, 3-leg interchange　02.233

栏板挂车　dropside trailer　05.025

栏式缘石　barrier curb　02.154

缆索吊装法　erection with cableway　04.661

缆索起重机　cable crane　07.289

缆索散开室　[steel] cable splay chamber　04.339

浪高　wave height　04.109

劳动工资管理　labour wage management　06.394

老化　ageing　08.269

雷达测速器　radar speedometer　03.294

累计当量轴次　accumulative equivalent axles　02.382

肋板桥　ribbed slab bridge　04.209

肋拱桥　ribbed arch bridge　04.225

肋腋板　slab with haunched ribs　04.283

A 类部分预应力混凝土　type A partially prestressed concrete　04.615

B 类部分预应力混凝土　type B partially prestressed concrete　04.616

冷拔钢丝　cold-drawn wire　08.131

冷拌法　cold mixing method　02.439

冷藏挂车　refrigerated trailer　05.039

冷藏货车　refrigerated vehicle　05.009

冷拉钢筋　cold-stretched steel bar　08.124

冷拉率　cold-drawn rate　09.089

冷拉率控制张拉　prestressing by cold-drawn rate　04.727

冷料输送机　cold aggregate conveyer　07.135

冷磨合　cold breaking-in　05.299

冷铺法　cold laid method　02.441

冷漆划线机　cold paint road marking machine　07.229

冷弯试验　cold bent test　09.050

梨形丁坝　pear-ended spur dike　04.131

犁式除雪机　V type snow plough　07.218

犁式稳定土搅拌机　blade plough soil stabilizer, blade plough stabilized soil road-mixer　07.117

* 犁扬机　elevation grader　07.016

离地间隙测定　ground clearance measurement　07.416

离合器滑转　clutch slippage　05.216

离析　segregation　08.266

离心力　centrifugal force　04.489

离心式斗式提升机　centrifugal bucket elevator　07.103

离心式水泵　centrifugal pump　07.264

理货　tallying of cargoes　06.185

理论弯沉系数　theoretical deflection coefficient　02.392

理论运价　theoretical rate　06.460

里程碑　mile stone, kilometer stone　02.194

里程标　mileage post　02.193

* 里程利用率　operation mileage rate　06.370

砾　gravel　08.074

砾类土　gravelly soil　08.061

历史洪水位　historic flood level　01.173

例行维护　routine maintenance　07.321

* 立交　grade separation　02.222

立面标记　object marking　03.427

立体交叉　grade separation　02.222

立缘石　vertical curb　02.153

立柱式交通标志　post traffic sign　03.372

粒径　grain size　08.179

粒料　granular material　08.049

粒料改善土路　aggregate treated earth road　02.434

粒料稳定土基层　aggregate stabilized soil base　02.498

沥青　bitumen, asphalt　08.003

沥青薄膜防水层　water proof asphalt membrane　04.294

沥青保温罐　asphalt tank with heating device

07.144

沥青泵 asphalt pump 07.164

沥青表面处治 bituminous surface treatment 02.420

沥青秤 asphalt weigher 07.143

沥青抽提仪 bitumen extractor 09.165

沥青稠度 bitumen consistency 08.222

沥青改性剂 asphalt modifier 08.142

沥青罐车 asphalt truck 07.167

沥青贯入式路面 bituminous penetration pavement 02.418

沥青混合料 bituminous mixture 08.091

沥青混合料搅拌设备 asphalt mixing plant 07.124

沥青混合料再生拌和设备 asphalt mixture recycling mixing plant 07.212

沥青混凝土混合料 bituminous concrete mixture 08.092

沥青混凝土路面 bituminous concrete pavement 02.416

沥青混凝土摊铺机 asphalt paver 07.157

沥青混凝土摊铺机试验 asphalt paver test 07.476

沥青胶砂 asphaltmastic 08.095

沥青胶砂防水层 water proof asphalt mastic 04.297

沥青流量计 asphalt flow meter 07.151

沥青路面 bituminous pavement 02.415

沥青路面红外线加热器 asphalt pavement infrared heater 07.210

沥青路面火焰加热器 asphalt pavement flame heater 07.209

沥青路面加热机 asphalt pavement heater 07.208

沥青路面就地再生机 asphalt pavement *in-situ* recycling machine 07.211

* 沥青玛琋脂 asphaltmastic 08.095

* 沥青玛琋脂防水层 water proof asphalt mastic 04.297

沥青喷洒机 asphalt sprayer 07.162

沥青熔化加热装置 asphalt melting and heating device 07.172

沥青乳化机 bitumen emulsifying machine 07.178

沥青乳化设备 bitumen emulsifying plant 07.182

* 沥青乳液 emulsified bitumen, asphalt emulsion 08.008

沥青洒布机试验 asphalt distributor test 07.473

沥青洒布宽度测定 distribution width measurement 07.474

沥青洒布量测定 asphalt distribution measurement 07.475

沥青砂 sand asphalt 08.094

沥青碎石混合料 bituminous macadam mixture 08.093

沥青碎石路面 bituminous macadam pavement 02.417

沥青稀浆 asphalt slurry 08.096

沥青罩面 asphalt overlay 02.670

沥青贮仓 asphalt storage 07.168

力矩 moment, bending moment 04.557

力矩包络线 moment envelope 04.561

力矩分配法 moment distribution method 04.560

联动交通信号 coordinated traffic signals 03.258

联动控制 coordinated control 03.299

联合碎石机组 crushing plant 07.095

联合投标 joint ventures bidder 01.116

联合运输 intermodal transport 06.011

联机控制 on-line control 03.271

联结层 binder course 02.340

联络线 linking-up line 02.044

* 联运 intermodal transport 06.011

连杆校正器 connecting rod aligner 05.351

连杆轴承镗床 connecting rod bearing boring machine 05.352

连拱作用 continuous arch effect 04.591

连接器 coupler 04.650

连续刚构桥 continuous rigid frame bridge 04.234

连续拱桥 continuous arch bridge 04.222

连续梁桥 continuous beam bridge 04.205

连续配筋混凝土路面 continuous reinforced concrete pavement 02.453

连续砌拱 laying arch continuously 04.676

连续桥面 continuous deck 04.289

连续式交通量观测站 continuous traffic count station 03.121

连续式沥青混合料搅拌设备 continuous asphalt mixing plant 07.128

连续性交通流 continuous traffic flow 03.047

连续性模型 continuity model 03.010

链板给料机　slat feeder　07.092

链斗式挖沟机　chain bucket trencher　07.051

*链滑轮　chain hoist, chain block　07.279

链式起重机　chain hoist, chain block　07.279

梁　beam, girder　04.262

梁腹　web　04.535

梁高　beam depth, girder depth　04.026

梁内导管空隙真空加压试验　duct void examination by vacuum pressure test　09.125

梁桥　beam bridge, girder bridge　04.203

梁式钢拱架　steel beam centering　04.687

梁式起重机　beam crane　07.290

*梁腋　haunch　04.539

料场调查　materials field investigation　01.162

料石　ashlar, dressed stone　08.070

料石拱桥　ashlar arch bridge　04.197

[列车]折叠　jack knifing　05.104

裂缝　crack　02.597

裂缝分布图　crack pattern　04.600

裂缝灌注　crack grouting　02.662

裂缝控制　crack control　04.598

裂缝宽度限值　limit of crack opening　04.599

裂缝声发射检测　crack detecting by acoustic emission　09.123

裂隙粘土　fissured clay　01.290

林区道路　forest road　02.006

临界 V/C 比　critical V/C ratio　03.091

临界车速　critical speed　03.068

临界风速　critical wind speed　04.587

临界荷载　critical load　02.409

临界空挡　critical gap　03.027

临界密度　critical density　03.087

临界压曲荷载　critical buckling load　04.500

临时工程　temporary works　01.144

临时性桥　temporary bridge　04.180

菱形立交　diamond interchange　02.230

零担[货物]运输　less-than-truck-load transport　06.166

零担货运站　less-than-truck-load terminal　06.249

零担运价　less-than-truck-load rate　06.464

零断面　zero section　02.131

零件检验分类　inspection and classification of parts　05.296

零件清洗机　parts washer　05.365

零件修理　parts repair　05.285

硫侵蚀　sulphur corrosion　02.636

流冰　drift ice　04.124

流量　flow discharge　04.066

流量过程线　flow duration curve　04.070

流砂　quick sand　01.266

流水压力　flowing water pressure　04.495

流速　current velocity　04.054

流体动力学理论　hydrodynamics theory　03.005

流通过程运输　transport in circulation　06.022

流向　current direction　04.065

流域　river basin　04.075

流值　flow value　08.247

漏乘　missing a bus　06.107

漏浆　leakage　02.485

漏装　neglected loading　06.221

露骨　bare surface　02.610

路拌法　road-mixing method　02.436

路拌机械试验　road-mixing machine test　07.466

路拌式稳定土搅拌机　soil stabilizer, stabilized road-mixer　07.113

路边停车　side parking, curb parking　03.153

路槽　road trough　02.360

路侧广播　roadside radio broadcasting　03.296

路产　highway property　01.348

路床　road bed　02.359

路岛端　approach nose　02.220

路堤　embankment　02.249

路段通行能力　road section capacity　03.167

路幅　roadway　02.158

路拱　crown　02.135

路拱横坡　crown slope, cross fall　02.137

路拱曲线　crown curve　02.136

路滑标志　skid sign　03.387

路基　subgrade　02.247

[路基]沉降　subsidence　02.633

路基冲刷防治　subgrade erosion protection　10.074

路基防护　subgrade protection　02.285

路基工程　subgrade engineering　02.248

路基含水量　subgrade moisture content　02.283

路基加固　subgrade strengthening　02.286

路基精整机　subgrade fine trimmer　07.123

路基临界高度　critical height of subgrade　02.284

路基排水　subgrade drainage　02.511

路基强度　subgrade strength　02.279

路基设计高程　design elevation of subgrade　02.281

路基稳定性　subgrade stability　02.278

路肩　shoulder, verge　02.142

路肩缺口　shoulder gap　02.640

＊路口视距　sight distance of intersection　02.212

路况　highway condition　02.552

路况登记　highway condition registration　02.554

路况调查　highway condition investigation　02.553

路况检查　highway condition inspection　02.555

路况巡视　highway condition patrol　02.556

路面　pavement　02.333

路面板唧泥　slab pumping　02.634

路面病害摄影组合仪　photographic road survey group　09.175

路面补强　pavement strengthening　02.581

路面补强设计　pavement strengthening design　02.411

路面沉陷　pavement depression　02.632

[路面]单向排水　single-slope drainage　02.527

路面调查　pavement investigation　02.561

路面定期调查　pavement periodical investigation　02.563

路面翻修　pavement recapping　02.580

路面工程学　pavement engineering　01.006

路面管理系统　pavement management system, PMS　02.698

路面滑溜　surface skidding　02.595

路面积水　surface gathered water　02.594

路面激光测试仪　laser road surface tester　09.174

路面检测评价系统　pavement monitor and evaluation system　02.699

路面结构　pavement structure　02.338

路面结构组合设计　pavement structural composition design　02.412

[路面]抗滑试验　anti-skid test　09.084

路面抗滑性能　pavement skid resistance condition, PSRC　02.703

路面快速加载试验机　accelerated loading facility, ALF　09.190

[路面]裂缝度　cracking ratio　02.599

[路面]裂缝率　cracking rate　02.598

路面排水　pavement drainage　02.512

路面疲劳寿命　fatigue life of pavement　02.373

路面平整度测定仪　viameter　09.160

路面评价模型　pavement evaluation model　02.701

路面破损率　pavement damage ratio　02.688

路面铺装率　paved ratio　02.551

路面强度　pavement strength　02.387

路面清扫机　pavement sweeping machine　07.221

路面曲率仪　surface-curvature apparatus　09.159

路面设计　pavement design　02.362

路面使用质量指标　pavement operating quality index　02.687

[路面]双向排水　double-slope drainage　02.528

路面随机调查　pavement random investigation　02.562

路面特性　pavement characteristics　02.363

路面透水度测定仪　surface permeameter　09.161

路面文字标记　pavement lettering marking　03.428

路面铣削机　pavemill　07.207

路面再生　pavement recycling　02.673

路面状况指数　pavement condition index　02.700

路钮　button　03.475

路旁绿化带　roadside green belt　02.696

路堑　cutting　02.251

路容　highway appearance　02.560

路上停车点　road parking lot　03.200

路外停车　off-road parking　03.154

路线　route　02.045

路线测量　route survey　01.184

路线计算机辅助设计　route computer aided design　02.054

路线控制点　control point of route　02.051

路线连续性　route continuity　02.052

路线平面图　route plan　01.185

路线三维空间设计　route three-dimensional space design　02.057

路线设计　route design　02.046

路线线形　route alignment　02.047

路线优化设计　route optimum design　02.058

路缘带　marginal strip　02.150

[路]缘石　curb　02.151

路障　roadblock　02.592

旅客　passenger, traveler　06.035

旅客波动系数　passenger number fluctuation coefficient　06.054

旅客流量　passenger traffic volume　06.046

旅客流时　passenger traffic time　06.047

旅客流向　passenger traffic flow direction　06.048

旅客票价　ticket price　06.474

旅客平均运距　passenger average haul distance　06.345

旅客日发送量　daily dispatched passenger number　06.052

旅客一条龙服务　bus streamlined service　06.104

旅客意外伤害　passenger unexpected injury　06.110

旅客意外伤害保险　insurance of passenger unexpected injury　06.111

旅客运输　passenger transport　06.018

旅客运输杂费　miscellaneous charges for passenger transport　06.492

旅客运输质量　bus service quality　06.103

旅客正运率　passengers transport regularity rate　06.123

旅客最高聚集人数　maximum gathered passenger number　06.053

旅游客车票价　tourist bus ticket price　06.478

履带式单斗挖掘机　crawler single bucket excavator　07.044

履带式沥青混凝土摊铺机　crawler-type asphalt paver　07.159

履带式起重机　crawler crane　07.284

履带式推土机　crawler bulldozer　07.002

履带式稳定土搅拌机　crawler soil stabilizer, crawler stabilized soil road-mixer　07.115

履带式液压凿岩台车　crawler hydraulic bench drill　07.029

履带式装载机　crawler loader　07.055

履约担保　performance bond　01.124

氯丁橡胶　polymeric chloroprene rubber　08.161

氯化物含量沿深度分布测量　chloride content depth profile measurement　09.127

绿波带　green wave band　03.249

绿灯时差型式　offset pattern　03.255

绿灯时间　green time　03.245

绿化带　green belt　02.156

绿化里程　planting mileage, greening mileage　02.550

绿信比　split, green ratio　03.248

卵石　cobble　08.067

轮渡站　highway ferry station　04.951

轮迹横向分布系数　coefficient of wheel tracking transverse distribution　02.403

轮廓标线　delineation line　03.420

轮胎拆装机　tire changer　05.338

轮胎翻新率　tyre recapping rate　05.069

轮胎式单斗挖掘机　wheel type single bucket excavator　07.043

轮胎式沥青混凝土摊铺机　wheel-type asphalt paver　07.158

轮胎式起重机　wheel crane　07.285

轮胎式推土机　wheel bulldozer　07.003

轮胎式稳定土搅拌机　wheel-type soil stabilizer, wheel-type stabilized soil road-mixer　07.114

轮胎式装载机　wheel loader　07.054

轮胎摊提率　tyre depreciation rate　06.434

[轮胎]拖印　tyre skid　05.098

轮胎行驶里程定额　tyre mileage rating　05.064

轮胎压路机　pneumatic tyred roller　07.060

[轮胎]压印　tyre scuffs　05.097

轮胎噪声　tyre noise　10.055

轮载　wheel load　01.091

螺杆式空气压缩机　screw air compressor　07.308

螺栓扭角终拧法　bolt final twisting by twist angle method　04.741

螺栓扭矩终拧法　bolt final twisting by torque method　04.740

螺纹钢筋　deformed bar　08.121

螺旋式除雪机　screw snow remover　07.220

螺旋式混凝土布料机　screw type concrete spreader　07.197

螺旋输送机　screw conveyer　07.106

螺旋线　spiral curve　02.094

逻辑诊断　logistic diagnosis　05.232

裸拱卸架　unloading bare rib, unloading bare ring　04.696

裸装货物　naked goods　06.137

落锤　drop hammer　07.238

落锤夯　drop weight rammer　07.077

落锤式弯沉仪 falling weight deflectometer, FWD 09.188

落物激振 vibration excited by dropping weight 09.106

洛杉矶磨耗试验机 Los Angeles abrasion testing machine 09.171

洛氏硬度试验 Rockwell hardness test 09.048

M

麻面 surface pockmark 02.607

码头 wharf 04.952

码头引道 approach to ferry 04.956

码头引桥 bridge approach to ferry 04.957

*马克当路面 water bound macadam 02.430

马歇尔劲度 Marshall stiffness 08.246

马歇尔稳定度 Marshall stability 08.245

马歇尔稳定度试验 Marshall stability test 09.082

埋式接缝 buried joint 04.304

埋置式桥台 buried abutment 04.391

麦康奈尔螺旋线 McConnell's curve 02.095

慢裂乳化沥青 slow breaking emulsified bitumen 08.011

慢凝液体沥青 slow-curing liquid asphalt 08.019

慢速车混入率 slow vehicle mixed rate 03.179

漫水桥 submersible bridge 04.175

盲沟 blind drain, blind ditch 02.532

*猫道 catwalk 04.705

猫眼 cat eyes 03.477

*BBRV 锚 bulb-end anchorage, BBRV anchorage 04.648

锚锭板式挡土墙 anchored bulkhead retaining wall 02.274

锚墩 anchor pier 04.376

锚杆式挡土墙 tie rod anchored retaining wall 02.273

锚杆支护 anchor bolt support 04.893

锚固区 anchorage zone 04.651

锚具 anchorage 04.639

OVM 锚具 oriental cone anchorage 04.646

VSL 锚具 VSL anchorage 04.643

XM 锚具 X-typed anchorage 04.645

YM 锚具 Y-typed anchorage, post-tensioning strand group anchorage 04.644

锚具变形损失 loss due to anchorage deformation 04.631

锚具温差损失 loss due to anchorage temperature difference 04.632

锚索 anchor cable 04.329

锚桩 anchor pile 04.416

锚座 anchor socket 04.326

锚碇 anchorage 04.334

锚碇板 anchor plate 04.336

锚碇板式桥台 anchored bulkhead abutment 04.394

锚[碇]跨 anchor span 04.333

锚碇体 anchor body, anchor block 04.335

毛尔伸缩装置 Mauer expansion installation 04.315

毛接法 coarse joining method 02.666

毛细水 capillary water 02.502

毛细水上升高度 capillary water height 08.209

铆接 riveting 04.735

铆接钢桥 riveted steel bridge 04.189

*帽梁 bent cap 04.363

梅氏曲线 May-curve 10.027

煤沥青 coal tar 08.023

每车平均吨位 average tonnage per vehicle 06.364

每车平均客位 average seats per vehicle 06.365

每升燃料作业量测定 determination of production quantity per litre fuel consumption 07.446

门到门运输 door-to-door transport 06.013

门口效应 gate effect 03.511

门桥 portal frame 04.965

门式交通标志 overhead traffic sign 03.370

门式起重机 gantry crane 07.288

门形索塔 portal framed tower 04.323

*蒙脱石 montmorillonite 01.291

蒙脱土 montmorillonite 01.291

密度试验 density test 09.002

密封条三维伸缩装置 three-dimension expansion installation with sealing strip 04.317

密级配　dense gradation　08.186

免费行包　free-of-charge luggage　06.487

面层　surface course　02.339

苗圃　nursery　02.590

民用车辆　civilian vehicle　05.053

敏感性分析　susceptivity analysis　01.052

明洞　open cut tunnel　04.845

明涵　open culvert　04.802

明渠排水　gutter drainage　02.521

明挖法　open cut method　04.865

明挖基础　open cut foundation　04.401

蘑菇形开挖法　mushroom-type tunneling method
　04.878

模板　formwork　04.716

模量比　modulus ratio　02.400

模拟试验　simulated test　07.379

模型试验　model test　07.376

膜式水泵　diaphragm pump　07.265

磨光　polishing　02.621

磨耗层　wearing course　02.344

磨耗度　abrasiveness　08.251

磨耗试验　abrasion test　09.036

磨合　breaking-in　05.298

磨气门机　valve refacer　05.355

磨气门座机　valve seat grinder　05.354

磨损　wearing　02.622

磨损过程　wear process　05.173

磨损率　wear rate　05.177

磨损试验　wear test　07.390

磨损限值　wear allowance　04.602

磨钻机　bits grinders　07.031

摩擦系数测定仪　friction tester　09.176

摩擦桩　friction pile　04.435

抹光　finishing　02.476

木桥　timber bridge, wooden bridge　04.199

木支撑　timbering support　04.894

木桩　timber pile　04.421

苜蓿叶形立交　clover-leaf interchange　02.228

N

耐候钢　weathering steel　08.119

耐久性试验　durability test　07.398

耐磨硬度试验仪　wear hardness testing apparatus
　09.169

挠度　deflection　04.556

挠度横向分布　transverse distribution of deflection
　04.574

挠度横向分布测量　transversal distribution measure-
　ment of girder deflection　09.098

内侧车道　fast lane　02.176

内窥镜　endoscope　09.222

内陆还箱站　inland container depot　06.266

内摩擦角　internal friction angle　08.221

内摩擦系数　internal friction coefficient　08.219

内啮合齿轮沥青泵　inside gear asphalt pump
　07.166

内燃凿岩机　motor jack hammer　07.019

内粘聚力　internal cohesion　08.220

内纵梁　interior beam, interior stringer　04.273

能见度　visibility　03.506

泥灰结碎石路面　clay-lime bound macadam
　02.432

泥浆泵　mud pump　07.260

泥浆护壁钻孔法　slurry hole-boring method
　04.748

泥浆套法下沉沉井　sinking open caisson by slurry
　coating　04.745

泥结碎石路面　clay bound macadam　02.431

泥沙起动流速　sediment-moving incipient velocity
　04.058

泥石流　mud avalanche, debris flow　01.270

泥炭　peat　01.278

逆向拆除分析　retrogressive analysis for bridge erec-
　tion　04.779

年第三十位最大小时交通量　annual thirtieth high-
　est hourly traffic volume, 30HV　03.060

年均温差　annual mean temperature difference
　04.590

年平均日交通量　annual average daily traffic
　volume, AADT volume　03.058

年最大小时交通量　annual maximum hourly traffic
　volume　03.057

粘层　tack coat　02.342

粘稠沥青　asphalt cement　08.021

粘附　adhesion　05.181

粘附力　adhesion　08.238

粘结的预应力筋　bonded tendon　04.626

粘结力　cohesion　08.237

粘结力试验　cohesion test　09.075

粘弹性理论　viscoelasticity theory　02.376

粘土质砂　clayey sand　08.083

粘滞度　viscosity　08.225

粘滞度试验　viscosity test　09.064

碾压　rolling　02.301

碾压混凝土路面　rolling compacted concrete pavement, RCCP　02.455

碾压速度　rolling speed　02.305

*牛腿　bracket　04.540

扭曲　distortion　04.567

农用运输汽车　agricultural truck　05.014

农作物蒙尘　crops covering by dust　10.032

O

偶然荷载　accidental load　04.479

P

爬坡车道　climbing lane　02.172

爬坡车道标志　climbing lane sign　03.393

爬坡性能试验　climbing ability test　07.440

排队规则　queue rule　03.029

排队理论　queue theory　03.028

排风机通风　ventilation by exhaust fan　04.913

排风竖井　blowing-out shaft　04.917

排架式满布木拱架　full span wooden bent centering　04.689

排架桩墩　pile-bent pier　04.370

排架桩基础　pile-bent foundation　04.410

排涝桥　flood relief bridge　04.177

排气净化器　exhaust purifier　10.030

排气消声器　exhaust muffler　10.053

排气噪声　exhaust noise　10.052

排水板法　sheet drainage　02.318

排水工程　drainage works　02.509

排水沟　drainage ditch　02.520

排水能力　drainage ability　02.505

排水砂垫层　drainage sand mat　02.319

排水设计　drainage design　02.508

排水设施　drainage facility　02.507

排水系统　drainage system　02.506

排土桩　displacement pile　04.433

抛锚　breakdown on the way　05.103

抛石防护　riprap protection　04.145

抛坍爆破　collapse blasting　02.330

抛掷爆破　throwing blasting　02.327

抛掷清扫车　abandon sweeper　07.224

跑偏　pulling to one side　05.219

泡沫沥青装置　foam asphalt unit　07.184

赔偿率　rate of freight compensation　06.233

配料给料装置　aggregate feeder units　07.134

喷浆机　concrete injector, cement gun　07.314

喷锚支护　shotcrete and rock bolt support　04.892

喷漆房　spray booth　05.369

喷气式除雪机　air-jet snow remover　07.219

喷燃器　burner　07.148

喷洒沥青　asphalt distribution　02.446

喷射混凝土　shotcrete　08.115

喷射混凝土衬砌　shotcrete lining　04.889

喷嘴式沥青乳化机　nozzle bitumen emulsifying machine　07.181

喷嘴试验器　nozzle tester　05.343

盆地　basin　01.227

盆式橡胶支座　pot rubber bearing　04.449

膨胀力　swelling force　08.217

膨胀量　swelling amount　08.216

膨胀试验　swelling test　09.010

膨胀水泥　expansive cement　08.043

膨胀土　expansive soil, swelling soil　01.283

膨胀系数　coefficient of expansion　04.532

疲劳荷载　fatigue load　04.469

疲劳设计　fatigue design　04.521

疲劳试验　fatigue test　09.052

疲劳限度　fatigue limit　04.601

皮带称重装置　conveyer belt scale　07.153

皮带给料机　belt feeder　07.090

偏角　deflection angle　01.197

偏角法　deflection angle method　01.201

偏心受压法　eccentric compression method　04.577

片理　schistosity　01.323

片石　rubble　08.069

漂浮物　drifter　04.121

漂石　boulder　08.065

拼箱　mixed stuffing　06.263

拼箱货　less than container load, LCL　06.269

拼装式衬砌　precast lining　04.887

频爆式推土机　blasting bulldozer　07.004

频率仪　frequency recorder　09.212

贫混凝土　lean concrete　08.105

贫混凝土基层　lean concrete base　02.487

平板挂车　platform trailer, flat trailer　05.026

平板支座　plate bearing　04.454

平地机　grader　07.013

平地升送机　elevation grader　07.016

平缝　flat joint　02.464

平衡悬浇法　cast-in-place balancing cantilever method　04.665

平衡悬拼法　precast balancing cantilever method　04.663

平均车日行程　average vehicle daily travel　06.368

平均车数　average number of vehicles　06.361

平均车速　average speed　03.074

平均故障间隔时间　mean time interval between failures　07.452

平均日交通量　average daily traffic volume, ADT　03.064

平均时间有效率　mean time effective rate　07.453

平均总吨位　average total tonnage　06.362

平均总客位　average total seats　06.363

平均纵坡　average gradient　02.115

平均最低水位　mean lowest water level　04.095

平均最高水位　mean highest water level　04.094

平面测量　plan survey　01.186

平面交叉　at-grade intersection　02.199

[平面]交叉口　intersection, crossing, junction　02.197

平面示意图　plan sketch　02.059

平面线形　horizontal alignment　02.074

平曲线　horizontal curve　02.076

平曲线横净距　lateral clear distance of horizontal curve　02.081

平曲线极限最小半径　ultimate minimum radius of horizontal curve　02.078

平曲线加宽　curve widening　02.132

平曲线最小半径　minimum radius of horizontal curve　02.077

平曲线最小长度　minimum length of horizontal curve　02.080

平滩水位　flood land line water level　04.102

*平旋桥　swing bridge　04.251

平原区　plain terrain　01.221

平缘石　flush curb　02.152

平转桥　swing bridge　04.251

评标　evaluation of bids　01.119

坡长限制　grade length limit　02.117

坡道标志　slope sign　03.385

*坡道桥　ramp bridge　04.162

坡顶　top of slope　02.257

坡度损失　loss in grade　02.116

坡积土　slope wash soil　01.298

坡脚　toe of slope　02.258

坡面景观　slope landscape　02.693

坡桥　bridge on slope　04.161

破冰体　ice apron　04.378

破坏荷载　failure load　04.471

破碎带　fracture zone　01.328

铺草皮　sodding　02.656

铺砌　pitching　02.449

铺砂　sanding　02.659

铺砂法试验　sand patch test　09.083

*普氏击实试验　Proctor compaction test　09.013

普通货物　general goods　06.134

普通客票价　ordinary ticket price　06.477

Q

期望车速 desired speed 03.080

起爆器 firing device 07.316

起步 starting 05.088

起动延误 starting delay 03.094

起拱点 springing 04.346

起迄点调查 origin-destination study, OD study 01.161

起重量 hoisting capacity 06.291

起重设备 hoisting equipment 07.278

企口缝 tongue and groove joint, rabbet 02.463

气动岩石破碎机 pneumatic rock breaker 07.032

气动凿岩机 pneumatic rock drill 07.021

*气缸压力表 compression gauge 05.344

气力输送机 air conveyer 07.107

气腿式凿岩机 air-rider jack hammer 07.022

气象检测器 weather detector 03.310

气象站 meteorologic station 04.046

气象资料 meteorologic data 04.041

气压沉箱基础 pneumatic caisson foundation 04.405

气硬性 air hardening 08.265

气阻 vapor lock 05.201

弃土 waste 02.291

弃土堆 waste bank 02.293

弃土反耕 waste bank refarming 10.085

汽车 motor vehicle, automobile 05.004

汽车安全检测线 vehicle safety inspection and test line 05.323

汽车安全性 vehicle safety 05.141

汽车百车公里油耗 vehicle fuel consumption per hundred kilometers 05.158

汽车百吨公里油耗 vehicle fuel consumption per hundred ton-kilometers 05.159

*汽车保养 vehicle maintenance 05.248

汽车不良技术状况 bad technical condition of vehicle 05.164

汽车操纵轻便性 easiness of vehicle control 05.147

汽车操纵稳定性 vehicle handling stability 05.149

[汽车]侧滑试验台 side slip tester 05.332

[汽车]车轮定位仪 wheel alignment meter 05.337

汽车乘坐舒适性 vehicle ride comfort 05.146

汽车大修 vehicle major repair 05.274

汽车大修返修率 return rate of vehicle major repair 05.317

汽车大修费用定额 cost quota of vehicle major repair 05.066

汽车大修间里程 average mileage between vehicle major repairs 05.067

汽车大修理费用 vehicle overhaul cost 06.432

汽车大修平均在厂车日 average days in plant during major repair of vehicle 05.313

汽车大修平均在修车日 average days during major repair of vehicle 05.314

汽车代用燃料 alternative motor fuels 05.160

汽车底部自动清洗机 automatic underbody washer 05.364

汽车电磁波干扰 automobile electro-magnetic interfere 10.087

汽车调头转盘 truck turntable 07.313

汽车定期维护 vehicle periodic maintenance 05.254

汽车动力性 vehicle dynamic quality 05.139

汽车工作能力 working ability of vehicle 05.165

汽车故障 vehicle failure 05.187

汽车管理制度 vehicle management system 05.059

汽车合理使用 vehicle rational utilization 05.130

汽车合理拖载 vehicle rational towing load 05.133

汽车货运装卸站 motor transport handling station 06.292

汽车机动性 vehicle mobility 05.148

汽车极限技术状况 limiting technical condition of vehicle 05.169

汽车技术经济定额 techno-economic rating for vehicle operation 05.062

汽车技术状况 technical condition of vehicle 05.162

汽车技术状况变化规律 trend of change of vehicle technical conditions 05.170

汽车技术状况参数 parameters of vehicle technical condition 05.166

汽车驾驶 automobile driving 05.086

汽车检测 vehicle inspection and test 05.226

汽车检测站 vehicle inspection and test station 05.241

汽车节能 vehicle energy saving 05.152

汽车节能装置 fuel saving device 05.161

汽车节油技术 vehicle fuel saving technique 05.153

汽车经济车速 vehicle economical speed 05.132

汽车经济性 vehicle economy 05.140

汽车举升机 car lift 05.358

汽车空箱行程 vehicle empty container mileage 06.271

汽车列车 tractor-trailer combination 05.049

汽车零件磨损 wear of vehicle parts 05.172

汽车轮渡 car ferrying 04.950

汽车门式自动清洗机 gate type automatic car washer 05.363

汽车耐久性 vehicle durability 05.138

汽车能量利用率 vehicle energy utilization factor 05.156

汽车能量平衡 vehicle energy balance 05.157

汽车排放分析仪 vehicle exhaust analyser 10.029

汽车排放物 vehicle emission 10.010

[汽车]排气分析仪 exhaust gas analyzer, emission analyzer 05.335

汽车平均技术速度 vehicle average technical speed 05.131

[汽车]前大灯试验仪 head light tester 05.334

汽车清洗机 car washer 05.362

汽车燃料利用率 vehicle fuel utilization factor 05.155

汽车燃料消耗定额 vehicle fuel consumption rating 05.063

汽车容载量 vehicle load carrying capacity 05.135

汽车使用 vehicle operation 05.117

汽车使用方便性 vehicle utilization conveniency 05.144

汽车使用可靠性 vehicle operational reliability 05.137

汽车使用强度 vehicle operation intensity 05.128

汽车使用条件 vehicle operation condition 05.118

汽车使用效率 vehicle operation efficiency 05.129

汽车使用性能 vehicle operation performance 05.134

汽车式起重机 truck crane 07.286

[汽车]速度[表]试验台 speedometer tester 05.333

汽车损耗 vehicle wear-out 05.171

汽车通过性 vehicle passability 05.150

汽车完好技术状况 good technical condition of vehicle 05.163

汽车维护 vehicle maintenance 05.248

汽车维护场 vehicle maintenance depot 05.270

汽车维护定位作业法 vehicle maintenance on universal post 05.267

汽车维护方法 vehicle maintenance method 05.265

汽车维护工程 vehicle maintenance engineering 05.249

汽车维护工艺 vehicle maintenance technology 05.261

汽车维护工艺过程 vehicle maintenance technological process 05.263

汽车维护规范 vehicle maintenance norms 05.264

汽车维护级别 vehicle maintenance grade 05.251

汽车维护类别 vehicle maintenance classification 05.250

汽车维护流水作业法 flow method for vehicle maintenance 05.266

汽车维护生产纲领 production program of vehicle maintenance 05.268

汽车维护站 vehicle maintenance station 05.271

汽车维护周期 interval between vehicle maintenances 05.269

汽车维护作业 operation of vehicle maintenance 05.262

汽车维修 vehicle maintenance and repair 05.242

汽车维修费用定额 cost quota of vehicle maintenance and repair 05.065

汽车维修工具 instrument of vehicle maintenance and repair 05.322

汽车维修工艺设备 technological equipment of vehicle maintenance and repair 05.321

汽车维修平均费用 average costs of vehicle maintenance and repair 05.316

汽车维修平均工时 average man-hours of vehicle maintenance and repair 05.315

汽车维修企业 vehicle maintenance and repair enterprise 05.246

汽车维修网点 vehicle maintenance and repair network 05.247

汽车维修性 vehicle maintainability 05.151

汽车维修指标 vehicle maintenance and repair indices 05.245

汽车维修制度 vehicle maintenance and repair system 05.243

汽车小修 vehicle current repair 05.276

汽车小修频率 frequency of vehicle current repair 05.068

汽车行驶平顺性 vehicle running smoothness 05.145

汽车修理 vehicle repair 05.272

汽车修理厂 vehicle repair plant 05.318

汽车修理定位作业法 vehicle repair on universal post 05.307

汽车修理方法 vehicle repair method 05.305

汽车修理工艺 vehicle repair technology 05.292

汽车修理工艺过程 vehicle repair technological process 05.293

汽车修理技术标准 vehicle repair technical standard 05.304

汽车修理类别 vehicle repair classification 05.273

汽车修理流水作业法 flow method for vehicle repair 05.306

汽车修理生产纲领 production program of vehicle repair 05.312

汽车运价构成 formation of motor transport tariff 06.444

汽车运价管理体制 motor transport rate management system 06.463

汽车运价率 motor transport tariff rate 06.445

汽车运价体系 motor transport tariff system 06.443

汽车运输 motor vehicle transport 06.005

汽车运输边际成本 motor transport marginal cost 06.439

汽车运输车辆大修折旧 motor transport vehicle overhaul depreciation 06.426

汽车运输车辆费用 motor transport vehicle cost 06.436

汽车运输车型成本 motor transport cost by vehicle type 06.424

汽车运输沉落成本 motor transport sunk cost 06.441

汽车运输成本 motor transport cost 06.414

汽车运输成本范围 motor transport cost scope 06.420

汽车运输成本分析 motor transport cost analysis 06.416

汽车运输成本计算单位 motor transport cost accounting unit 06.419

汽车运输成本计算对象 motor transport cost accounting object 06.418

汽车运输成本降低额 motor transport cost reduction amount 06.431

汽车运输成本降低率 motor transport cost reduction rate 06.430

汽车运输成本结构 motor transport cost structure 06.422

汽车运输成本控制 motor transport cost control 06.438

汽车运输成本项目 motor transport cost item 06.421

汽车运输单车成本 motor transport unit vehicle cost 06.423

汽车运输辅助生产成本 motor transport auxiliary production cost 06.428

汽车运输机会成本 motor transport opportunity cost 06.440

汽车运输计划成本 motor transport planned cost 06.415

汽车运输企业成本 motor transport enterprise cost 06.425

汽车运输预测成本 motor transport forecast cost 06.417

汽车运输运营成本 motor transport operation cost 06.435

汽车运输站场设施 terminal and yard facilities of motor transport 06.247

汽车运输装卸成本 motor transport handling cost 06.427

汽车运输综合成本 motor transport comprehensive cost 06.429

汽车运行材料 vehicle operational consumption materials 05.061

汽车运行工况 vehicle operational mode 05.119

汽车运用工程 инженерия эксплоатация автомобильного транспорта(俄) 01.014

汽车载箱行程 vehicle payload-container mileage 06.270

汽车诊断 vehicle diagnosis 05.225

汽车诊断站 vehicle diagnostic station 05.240

汽车中修 vehicle medium repair 05.275

汽车重量利用系数 vehicle weight efficiency 05.136

汽车重箱行程 vehicle loaded container mileage 06.272

汽车专项修理厂 specialty vehicle repair shop 05.320

汽车专用公路 expressway, motorway 02.022

汽车状况监控 vehicle condition monitoring 05.227

汽车综合诊断线 vehicle general inspection and diagnosis line 05.324

汽车总成修理厂 unit repair plant 05.319

汽车钻机 drill truck 07.236

[汽车]最小转弯半径 minimum turning radius 02.079

汽缸珩磨机 cylinder honing machine 05.353

牵引车 tractor 05.020

牵引功率比油耗测定 specific drawbar power fuel consumption measurement 07.438

牵引试验 drawbar test 07.425

铅粒 particulate lead 10.016

铅橡胶支座 lead-rubber bearing 04.451

千斤顶 jack 07.282

千斤顶法封顶 closure by jacking and sealing-off crown 04.685

前方让路标志 yield ahead sign 03.380

前方视野 field of front vision 03.502

前方停车标志 stop ahead sign 03.379

前方信号标志 signal ahead sign 03.381

前墙 front wall 04.382

前置标志 advance sign 03.369

前轴负荷测定 front axle-load measurement 07.414

潜水泵 under water pump 07.267

潜水设备 diving equipment 07.274

潜水钻机 under water rig 07.254

浅基础 shallow foundation 04.399

浅埋隧道 shallow-buried tunnel 04.839

嵌花式路面 mosaic pavement 02.427

嵌岩桩 socketed pile 04.437

墙趾 wall toe 02.266

墙踵 wall heel 02.267

强化试验 intensified test 07.393

强制怠速工况 forced idling mode 05.121

强制式混凝土搅拌设备 forced concrete mixing plant 07.188

强制式稳定土搅拌设备 forced stabilized soil mixing plant 07.111

强制性交通流 forced traffic flow 03.044

抢修工程 rush-repair work 02.582

敲缸 piston slap 05.211

桥侧人行道 bridge sidewalk 04.318

桥墩 pier 04.359

桥墩分水尖 pier break-water 04.138

桥墩局部冲刷 local scour around pier 04.115

桥涵顶入法 jacking-in method of culvert or subsurface bridge 04.675

[桥基]沉降 settlement 02.631

桥基沉降观测 bridge foundation settlement observation 09.134

桥基冲刷 scouring at bridge foundation 04.113

桥基稳定性评定 bridge foundation stability evaluation 09.146

桥孔压缩 bridge opening contraction 04.104

桥[梁] bridge 04.146

桥梁安装监测 bridge erection monitoring 04.775

桥梁安装容许误差 bridge erection tolerance 04.780

桥梁编号 bridge numbering 04.791

桥梁标准设计 standard design of bridge 04.006

桥梁病害诊断 bridge defect diagnosis 09.095

桥梁病害整治 bridge fault repairing 04.792

桥梁测试车 bridge testing laboratory vehicle 09.198

桥梁承载能力 load carrying capacity of bridge 04.593

桥梁承载能力极限状态 ultimate limit state of bridge carrying capacity 04.507

[桥梁]动力响应试验 bridge response to forced vibration 09.102

桥梁动载试验 bridge dynamic loading test 09.097

桥梁墩台防撞 collision prevention of pier and abutment 04.794

桥梁翻新 bridge retrofitting 04.799

桥梁方案设计 bridge conceptual design 04.462

桥梁概率极限状态设计法 probabilistic limit state design method of bridge 04.510

桥梁工程学 bridge engineering 01.007

桥梁管理系统 bridge management system 04.783

桥梁荷载系数设计法 load factor design method of bridge 04.509

桥梁合拢 closure of bridge structure 04.774

桥梁基础 bridge foundation 04.398

桥梁极限状态设计 limit state design of bridge 04.506

桥梁技术档案 bridge technical file 04.786

桥梁计算机辅助设计 computer aided design for bridge, CAD for bridge 04.609

桥梁加固 bridge strengthening 04.797

桥梁监测系统 bridge monitoring system 04.785

桥梁检查 bridge inspection 04.787

桥梁检查规则 bridge inspection regulation 04.788

桥梁检查类别 bridge inspection category 04.789

桥梁检查周期 bridge inspection cycle 04.790

桥梁建筑高度 construction height of bridge 04.018

桥梁结构安装控制 bridge structure erection control 04.776

桥梁结构可靠度 reliability of bridge structure 04.512

桥梁结构设计 bridge structure design 04.463

桥梁静载试验 bridge static loading test 09.096

桥梁抗震加固 bridge aseismatic strengthening 04.798

桥梁抗震设计 aseismatic design of bridge 04.515

桥梁抗震稳定性 aseismatic stability of bridge 04.516

桥梁空间结构 space structure for bridge 04.524

桥梁栏杆 bridge railing 04.319

桥梁脉动测量 bridge pulsation measurement 09.110

桥梁美学 bridge aesthetics 01.023

桥梁模型风洞试验 bridge model wind tunnel test 09.093

桥梁模型试验 bridge model test 09.092

桥梁耐久性 bridge durability 04.594

桥梁挠度曲线 bridge deflection curve 09.099

桥梁平面结构 plane structure for bridge 04.523

桥梁评价系统 bridge evaluation system 04.784

桥梁破坏 bridge collapse 04.606

桥梁破损 bridge failure 04.605

桥梁浅基防护 bridge shallow foundation protection 04.793

桥梁全长 total length of bridge 04.013

桥梁容许应力设计 allowable stress design of bridge 04.505

桥梁设计 bridge design 04.461

桥梁施工 bridge construction 04.660

桥梁使用能力极限状态 serviceability limit state of bridge 04.508

桥梁试运行荷载 test run loading for bridge 04.472

桥梁数据库 bridge data bank 04.782

桥梁水毁 bridge disaster by flood 04.795

桥梁细部设计 bridge detail design 04.464

桥梁限载 bridge load limit 01.355

桥梁验收荷载试验 bridge acceptance loading test 09.091

桥梁养护 bridge maintenance 04.781

桥梁优化设计 optimum design of bridge 04.511

桥梁振型分析 bridge vibration mode analysis 09.109

桥梁自振频率测量 bridge natural frequency measurement 09.108

桥梁总宽度 total width of bridge 04.020

桥梁总体规划 bridge overall planning 04.001

桥面 deck 04.288

桥面板 deck slab 04.280

桥面标高 deck elevation 04.031

桥面单点加载装置 single point loading device on deck 09.208

桥面横坡 deck transverse slope 04.033

桥面净空 clearance above bridge deck 04.022

桥面宽度 bridge deck width 04.019

桥面排水系统 deck drainage system 04.290

桥面平整度测定仪 travelling-beam testing device for deck surface irregularity 09.210

桥面铺装 deck pavement 04.291

桥面伸缩缝 deck expansion joint 04.298

桥面伸缩装置 deck expansion installation 04.299

桥面系 bridge deck system 04.287

桥面纵坡 deck profile grade 04.032

桥名 bridge name 04.003

桥前壅水 back-water in front of bridge 04.106

桥塔 bridge pylon, bridge tower 04.321

桥台 abutment 04.380

[桥]台后回填 back filling behind abutment 04.397

桥头堡 bridge head 04.320

桥头搭板 bridge end transition slab 04.357

桥头回填设计 bridge end backfilling design 04.522

桥头引道 bridge approach 04.356

[桥头]锥坡 conical slope 04.135

桥位 bridge site 04.004

桥位地形图 topographic map of bridge site 01.304

桥位勘测 bridge site survey 01.302

桥位平面图 plan of bridge site 01.303

桥位选择 bridge site selection 04.005

桥下净空 clearance under span 04.023

桥下一般冲刷 general scour at bridge opening 04.114

桥型 bridge type 04.002

桥址断层活动性评定 fault activity evaluation of bridge site 09.148

桥址水文观测 hydrologic observation at bridge site 04.044

桥址稳定性评定 bridge site stability evaluation 09.145

桥轴线纵剖面 profile of bridge axis 01.305

翘曲 warp 04.568

切线长 tangent length 02.089

切线支距法 tangent offset method 01.202

亲水性集料 hydrophilic aggregate 08.072

轻浮货物 light goods 06.142

轻浮行包 light luggage 06.083

轻交通 light traffic 02.386

轻型击实试验 Proctor compaction test 09.013

*轻型桥台 supported type abutment 04.395

轻制沥青 cutback asphalt 08.020

轻质混凝土 light-weight concrete 08.106

倾翻稳定性试验 overturning stability test 07.442

倾斜断层 dip fault 01.326

倾斜角度试验机 overturn angle tester 05.331

清除坍方 removing landslide 02.674

清理边沟 ditch cleaning out 02.657

清水冲刷 scour without sediment motion 04.116

清尾[车]时间 clearance time 03.256

清洗和熏蒸费 cleaning and steaming charge 06.499

清岩机 ballast loader 07.034

晴通雨阻公路 fine-weather highway 02.042

晴雨通车公路 all-weather highway 02.041

球形支座 spherical bearing 04.455

*区间客流 inter-regional passenger traffic 06.037

区界交通调查 cordon traffic survey, cordon counts 03.137

区内交通 intra-zone traffic 03.113

区内客流 intra-regional passenger traffic 06.038

区域运价 zoning rate 06.470

曲线长 curve length 02.090

p-y曲线法 p-y curve method 04.655

*曲线桥 curved bridge 04.160

曲线中点 mid point of curve, MC 02.103

曲轴箱排放物 crankcase emissions 10.015

屈服破坏荷载 failure load by buckling 04.501

驱动元件滑转率测定 driving element slipping rate measurement 07.436

渠化标线 channelizing marking 03.422

渠化交叉口 channelized intersection 02.207

渠化交通 channelized traffic 03.139

取土坑 borrow pit 02.292

全部控制进入 full control of access 02.235

全动视力 vision with both driver and object moving

03.509

全断面开挖法　full-face tunneling method　04.872

* 全封闭　full control of access　02.235

全负荷工况　full load mode　05.126

全感应式信号控制　fully-actuated signal control
03.265

全挂车　full trailer　05.023

全挂汽车列车　full trailer train　05.050

全红信号　all red signal　03.250

仓厚翻修　full depth resurfacing　02.671

全厚式沥青路面　full depth asphalt pavement
02.421

全价票　full price ticket　06.475

全面质量管理　total quality control, TQC　01.095

全能法测图　universal photo　01.240

全气压盾构　all-round pressurized shield　04.941

全球定位系统　global positioning system, GPS
01.243

全套管钻机　all casing drill　07.255

全挖式断面　full cut section　02.130

全预应力混凝土　fully prestressed concrete　04.612

全员劳动生产率　all-personnel labour productivity
06.392

全自动汽车检测系统　computerized vehicle inspection system　05.325

缺火　misfiring　05.213

缺陷　defect　05.185

缺陷责任期　defects liability period　01.155

缺陷责任终止证书　defects liability release certificate
01.156

确定性模型　determinacy model　03.011

群桩　pile group　04.414

R

燃点　burning point　08.231

燃点试验　burning point test　09.066

燃油流量计　fuel consumption gauge　05.346

让车道标志　passing bay sign　03.395

让路标志　yielding sign　03.368

扰动土样　disturbed sample　01.336

扰流板　spoiler　04.355

绕行标志　detour sign　03.400

绕行公路　bypass　02.029

热拌法　hot mixing method　02.438

热混合料贮仓　hot mix silo　07.154

热料仓　hot aggregate bin　07.140

热料提升机　hot aggregate elevator　07.138

热料振动筛　hot aggregate vibrating screen　07.139

热磨合　hot breaking-in　05.300

热铺法　hot laid method　02.440

热漆划线机　hot paint road marking machine
07.230

热塑划线机　thermoplastic road marking machine
07.231

热塑性塑料　thermoplastic　08.170

热弯试验　hot bent test　09.051

热稳性　thermostability　08.233

热应力　thermal stress　04.548

热轧钢筋　hot rolled bar　08.122

人动视力　vision with driver moving　03.507

人工照明过渡　artificial lighting transition　03.463

人机调节系统　man-machine processing system
03.495

人力渡车船　manually operated ferry boat　04.960

人力装卸　manual handling　06.275

人群荷载　pedestrian load　04.486

人行地道　pedestrian underpass　03.487

人行横道　cross walk　03.486

人行桥　pedestrian bridge　04.149

人行天桥　pedestrian overpass　03.488

人造集料　artificial aggregate　08.054

人造雪崩　artificial snow slide　02.685

人字形排水系统　herringbone drainage system
02.531

韧度试验　rupture test　09.041

日常检查　routine inspection　02.557

溶解度　dissolubility　08.232

溶解度试验　dissolubility test　09.068

溶陷　melt sinking　02.650

容许荷载　allowable load　04.470

容许回弹弯沉值　allowable rebound deflection value
02.393

* 容许磨耗　wear allowance　04.602
容许应力　allowable stress　04.551
柔性墩　flexible pier　04.372
柔性护栏　flexible safety fence　03.440
柔性路面　flexible pavement　02.414
蠕变　creep　08.268
乳化剂　emulsifying agent　08.012
乳化沥青　emulsified bitumen, asphalt emulsion　08.008
入口匝道控制　entrance ramp control　03.274

软化点　softening point　08.227
软化点试验[环球法]　softening point test [ringball method]　09.063
软颗粒含量试验　soft grain content test　09.035
软练砂浆强度试验　plastic mortar strength test　09.056
软弱地基　soft ground　02.313
软土　soft soil, mollisoil　01.279
软岩　softrock　01.316

S

撒布　spreading　02.445
撒砂机　sand spraying machine　07.227
洒水车　sprinkler　07.213
三铰拱桥　three-hinged arch bridge　04.219
三轮压路机　3-wheel roller　07.059
三向预应力　three-dimension prestressing　04.624
三翼钻头　3-wing bit　07.258
三轴剪切仪　triaxial shear equipment　09.163
三轴试验　triaxial test　09.017
三作用千斤顶　triple acting jack　04.732
散装货物　bulk goods　06.136
扫砂机　sand sweeping machine　07.225
砂　sand　08.077
砂浆稠度仪　mortar consistency tester　09.166
砂井　sand well　02.323
砂类土　sandy soil　08.062
砂砾　sand gravel　08.076
砂率　sand ratio　08.260
砂土液化　sand liquefaction　01.301
砂桩　sand pile　02.321
* 刹尖　closure by wedging-in crown　04.684
沙害　sand hazard　02.653
沙漠　desert　01.268
* 沙漠土　desert soil　01.289
沙丘　sand dune　01.267
沙障　sand barrier　02.676
沙洲　shoal　04.087
筛分　sieve analysis　08.184
筛分机　sieving machine　07.084
山洞式锚碇　anchor in rock gallery　04.338

山脊　ridge　01.226
山脊线　ridge line　02.064
山岭区　mountain terrain　01.224
山坡线　hill side line　02.065
闪点　flash point　08.230
闪点试验[开口杯法]　flash point test [open cup method]　09.065
闪光信号　flashing signal　03.254
扇形拉索　fan-type stay cable　04.331
商务纠纷　commercial dispute　06.412
商用车辆运行管理系统　commercial vehicle operation management system, CVOM　03.322
商用汽车　commercial vehicle　05.055
上拌下贯式路面　penetration macadam with coated chips　02.419
上部结构　superstructure　04.261
上承式桥　deck bridge　04.155
上挡墙　top retaining wall　02.276
上跨式立交　overpass　02.224
上下导洞法　top and bottom pilot tunneling method　04.875
烧伤　burning　05.183
少筋微弯板　under-reinforced slab with slightly curved bottom　04.282
摄影测量　photogrametry　01.231
射流纵向通风　ventilation by force draft　04.914
射水沉桩法　pile jetting method　04.762
涉水驾驶　fording drive　05.113
设备报废　equipment scrapping　07.358
设备残值　remanent value of equipment　07.359

设备定期维护 equipment periodic maintenance 07.322

设备封存 equipment storing up 07.356

设备更新 equipment replacement 07.357

设备配套 coordinative composition of equipments 07.349

*设计车速 design speed 01.081

设计高程 design elevation 01.210

设计规范 design specification 01.074

设计荷载 design load 01.088

设计洪水频率 design flood frequency 04.072

设计交通量 design traffic volume 01.080

设计烈度 design intensity of earthquake 04.518

设计流量 design discharge 04.069

设计流速 design current velocity 04.055

设计水位 design water level 04.098

设计通行能力 design capacity 03.166

设计小时交通量 design hourly traffic volume, DHV 03.065

伸缩板 expansion plate 04.545

伸缩铲挖掘机 telescopic boom excavator 07.049

伸缩滚轴 expansion roller 04.458

深基础 deep foundation 04.400

深井水泵 deep well pump 07.268

深埋隧道 deep-buried tunnel 04.838

神经机能测定 nerve function measurement 03.496

渗井 seepage well 02.523

渗透水 seepage water 02.503

渗透系数 permeability coefficient 08.210

声发射测量系统 acoustic emission system 09.221

声级计 sound level meter 05.347

生产过程运输 transport in production process 06.021

生产率测定 productive capacity measurement 07.443

生理机能测定 physiological function measurement 03.497

生理疲劳 physiological fatigue 03.519

生石灰 quick lime 08.038

生态 ecology 10.080

生态环境 ecology environment 10.081

生物固沙 sand consolidation with biologic 10.079

生物节律 biorhythms 03.499

牲畜家禽挂车 livestock trailer 05.044

升降桥 lift bridge 04.257

升运式铲运机 elevating scraper 07.010

*省道 provincial trunk highway 01.342

省干线公路 provincial trunk highway 01.342

失控 out of control 05.204

失调 maladjustment 05.203

施工便桥 temporary bridge for construction 04.182

施工步道 catwalk 04.705

施工测量 construction survey 01.182

施工废弃物 construction waste 10.037

施工缝 construction joint 02.462

施工规范 construction specification 01.075

施工荷载 construction load 04.467

施工合同条款 construction contract terms 01.129

施工机械排放物 construction machine emission 10.011

施工机械噪声 construction machinery noise 10.056

施工机械振动 construction machinery vibration 10.067

施工计划 construction plan 01.097

施工路段交通管理 construction section traffic management 03.206

施工区通行能力 construction section capacity 03.174

施工水位 construction water level 04.100

施工图设计 construction drawing design 01.045

施工现场管理 construction site management 01.096

湿地推土机 swamp bulldozer 07.006

湿法养生 moist curing 02.483

湿式除尘器 wet dust collector 07.133

湿陷性黄土 collapsed loess 01.281

湿周 wetted perimeter 04.062

十字板剪力仪 vane-shear apparatus 09.229

十字板剪切试验 vane shear test 09.021

十字形交叉 cross road 02.202

石材硬度 hardness of stone 08.253

石场弃渣 quarry waste 08.052

石方爆破 rock blasting 02.324

石膏 gypsum 08.039

石拱桥 stone arch bridge 04.196

石灰 lime 08.036

石灰粉煤灰土基层 lime-flyash-soil base 02.495

石灰含量测定法 method for determining the lime content 09.059

石灰砂浆 lime mortar 08.099

石灰土基层 lime-soil base 02.493

石灰桩 lime pile 02.322

石料分级 stone classification 08.252

石料加速磨光仪 accelerated stone polishing tester 09.172

石料磨光值 polished stone value 08.254

石料磨光值试验 polished stone value test 09.037

石料摊铺机 aggregate paver 07.121

石笼 gabion 04.144

石桥 stone bridge 04.195

石屑 chips 08.051

石屑撒布机 chip spreader 07.163

石油沥青 petroleum asphalt, asphalt 08.005

拾震器 geophone 09.239

时间平均速度 time mean speed 03.070

时间占有率 time occupancy ratio 03.101

时效处理 time effect treatment 04.728

实腹拱桥 filled spandrel arch bridge 04.217

实时交通控制 real-time traffic control 03.263

实体[桥]墩 solid pier 04.365

实物分隔带 physical separator 03.447

实验室模拟试验 laboratory simulation test 07.380

实用荷载 service load 04.466

实载率 load factor 06.373

矢高 rise of arch 04.027

矢跨比 rise-span ratio 04.030

使用试验 service test 07.375

使用重量测定 operating weight measurement 07.412

使用状态挠度限值 deflection limit for serviceability 04.603

驶出路外事故 run-off-road accident 03.230

驶出匝道 exit ramp 02.239

驶入匝道 entrance ramp 02.238

示警柱 warning post 03.454

事故地点图 accident spot map 03.218

事故死亡率 death rate of accident 03.217

事故易发地点 [accident] black spot 03.220

事后维护 maintenance after failure 05.259

事后修理 repair after failure 05.278

事前修理 repair before failure 05.277

适箱货 container load freight 06.267

*适应照明 adaptation lighting 03.456

市区隧道 urban tunnel 04.833

视比重试验 apparent specific gravity test 09.030

视距 sight distance 02.106

视距三角形 sight triangle 02.213

视觉敏锐度 visual acuity 03.529

视力适应性 adaptation of vision 03.527

视情检修 inspection and repair only as necessary, IROAN 05.290

视情维护 condition-based maintenance 05.258

视情修理 repair on technical conditions 05.289

视线 sight line 02.104

视线不良弯道 blind curve 02.105

视线障碍物 sight obstruction 02.109

视野 field of vision 03.501

*CBR试验 California bearing ratio test, CBR test 09.024

试验环道 test loop road 09.192

试验路 test road 09.191

试验室试验 laboratory test 07.374

试桩 test pile 09.135

收发货标志 shipping mark 06.161

收费车道检测器 toll lane detector 03.315

收费车道信号灯 toll lane signal lamp 03.329

收费弹性 toll elasticity 03.342

收费岛 toll island 03.336

收费公路 toll highway 02.038

收费管理 toll administration 01.358

收费广场 toll plaza 03.338

收费卡门 toll gate 03.333

收费棚 toll canopy 03.337

收费桥 toll bridge 04.166

收费亭 toll booth, toll house 03.334

收费通行券 toll pass ticket 03.326

收费显示器 toll display 03.324

收费站 toll station 03.335

收费制式 toll mode 03.344

收费中心　toll center　03.343

收回赔偿　reclaim of compensation　06.224

收货人　consignee　06.175

收缩　shrinkage　08.267

收缩裂缝　contraction crack　02.600

收缩试验　shrinkage test　09.009

手持式凿岩机　hand-held rock drill　07.023

手推式构造深度仪　minitexture meter　09.184

手摇卷扬机　manual hoist　07.280

受迫振动互谱分析　forced vibration cross-spectrum analysis　09.119

梳式接缝　steel comb joint　04.309

熟石灰　slaked lime　08.037

竖曲线　vertical curve　02.120

竖曲线最小长度　minimum length of vertical curve 02.121

竖向排水　vertical drainage　02.516

竖旋桥　bascule bridge　04.254

数量指标　quantitative index　06.338

数字地面模型　digital terrain model　01.244

甩挂运输　trailer pick-up transport　06.032

甩客　denial of passenger　06.109

栓焊钢桥　bolted and welded steel bridge　04.192

栓接　bolting　04.736

栓接钢桥　bolted steel bridge　04.191

双班运行　two shifts run　06.079

双壁钢围堰钻孔桩基础　double-walled steel cofferdam and bored pile foundation　04.407

双侧壁导洞法　twin-side heading method　04.881

双层渡车船　double-decked ferry boat　04.961

双层钢围图　double-walled steel waling　04.769

双车道　dual lane　02.164

双程运输　two-way loading transport　06.025

双幅车行道　dual carriageway　02.162

双挂汽车列车　double trailer train　05.052

双黄线　double amber lines　03.414

双[孔]隧道　twin tunnel　04.840

双力矩　bi-moment　04.559

双轮荷载　dual wheel loading　02.378

双轮压路机　tandem roller　07.058

双轮振动压路机　tandem vibratory roller　07.066

双曲拱桥　two-way curved arch bridge　04.224

双梳式接缝　double steel comb joint　04.310

双索面斜拉桥　double plane cable stayed bridge 04.239

双塔斜拉桥　double pylon cable stayed bridge 04.242

双体渡车船　twin ferry boat　04.962

双向板　two-way slab　04.285

双向路　two-way road　02.167

双向预应力　two-dimension prestressing　04.623

双向匝道　two-way ramp　02.241

双叶平转桥　double-leaf swing bridge　04.253

双叶竖旋桥　double-leaf bascule bridge　04.256

双柱式[桥]墩　two-columned pier, two shaft pier 04.369

双作用气动桩锤　double acting pneumatic hammer 07.241

双作用千斤顶　double acting jack　04.731

双作用蒸气桩锤　double acting steam hammer 07.243

水道测量　hydrographic survey　04.037

水底隧道　subaqueous tunnel　04.836

水工构造物　hydraulic structure　04.047

水滑现象　hydroplaning phenomena　02.372

水灰比　water cement ratio　08.259

水毁　flood damage, washout　02.645

水结碎石路面　water bound macadam　02.430

水力半径　hydraulic radius　04.063

水力计算　hydraulic computation　04.043

水流导向　current deflecting　04.125

水面比降　water surface gradient　04.060

水泥　cement　08.040

水泥安定性试验　cement soundness test　09.058

水泥标号　cement mark　08.256

水泥混凝土　cement concrete　08.101

水泥混凝土标号　cement concrete mark　08.257

水泥混凝土混合料　cement concrete mixture 08.100

水泥混凝土路面　cement concrete pavement 02.451

水泥混凝土配合比　proportioning of cement concrete 08.258

水泥砂浆　cement mortar　08.098

水泥土基层　cement-soil base　02.494

水平搬运机械　horizontal handling machinery

06.282
水侵蚀　water erosion　02.641
水头　water head　04.091
水土保持　water and soil conservation　10.071
水土流失　erosion loss　10.072
水网地区　dense waterway net region　04.050
水位　water level　04.090
水文测量　hydrologic survey　04.039
水文地质调查　hydrogeological survey　01.170
水文地质学　hydrogeology　01.008
水文分析　hydrologic analysis　04.042
水文站　hydrologic station　04.045
水文资料　hydrologic data　04.040
水稳性　water stability　08.235
水下浇筑混凝土法　under water concreting　04.770
水下推土机　underwater bulldozer　07.005
水硬性　hydraulicity　08.264
水质　water quality　10.034
水质污染　water pollution　10.035
水准测量　leveling survey　01.205
水准点　bench mark, BM　01.206
顺坝　longitudinal dike　04.128
顺序拼装分析　progressive analysis for bridge erection　04.778
[司机]反应距离　[driver] reaction distance　03.510
[司机]感觉反应距离　[driver] perception-reaction distance　03.515
[司机]感觉反应时间　[driver] perception-reaction time　03.514
司机及助手劳动生产率　productivity of driver and assistant　06.393
[司机]判断时间　[driver] judgement time　03.512
[司机]识别距离　[driver] decipherment distance　03.513
四铰管涵　quadri-hinge-pipe culvert　04.812
四向停车　four-way stop　03.190
松动　loosening　05.202
松动爆破　loosening blasting　02.328
松铺厚度　loose laying depth　02.307
松铺系数　coefficient of loose laying　02.309
松散　surface loosening　02.608
松散保护层　loose protection course　02.347

松土机　ripper　07.017
送风竖井　blowing-in shaft　04.916
塑限　plastic limit, PL　08.205
塑限试验　plastic limit test　09.007
塑性混凝土　plastic concrete　08.108
塑性指数　plasticity index, PI　08.206
酸度计　pH meter　09.220
酸碱度值　pH value　08.214
随车调查　on vehicle survey　06.133
随车观测法　moving observer method　03.102
随车液压起重臂　escort hydraulic arm　06.285
随车装卸机械　escort handling machinery　06.280
随机工具　tools attachment　07.335
随机故障　random failure　05.195
随机性模型　stochastic model　03.012
随挖随盖法　cut-and-cover method　04.867
碎落　dehris, avalanche　01.265
碎落台　stage for heaping debris　02.264
碎石　angular cobble, broken stone, crushed stone　08.068
碎石基层　macadam base　02.489
碎石机　crusher　07.078
隧道　tunnel　04.830
隧道报警装置　tunnel warning installation　04.925
隧道边墙　tunnel side wall　04.898
隧道测量　tunnel survey　04.883
隧道长度　length of tunnel　04.850
隧道衬砌　tunnel lining　04.885
隧道底板　tunnel floor　04.897
隧道电话系统　tunnel telephone system　04.937
隧道顶板　tunnel roof　04.896
隧道洞门　tunnel portal　04.847
隧道洞身　tunnel trunk　04.848
隧道防水　tunnel water proofing　04.932
隧道防灾设施　tunnel anti-disaster facilities　04.926
隧道工程　tunneling　04.829
隧道工程学　tunnel engineering　01.012
隧道供电系统　tunnel power supply system　04.938
隧道贯通　tunnel through　04.900
隧道贯通误差　tunnel through error　04.901
隧道广播系统　tunnel broadcasting system　04.939
隧道火灾　tunnel fire hazard　04.928
隧道火灾监测　tunnel fire monitoring　04.927

隧道激光导向　tunnel alignment by laser　04.884
隧道级别　tunnel class　04.831
隧道监控中心　tunnel operation control center　04.934
隧道交通监测系统　tunnel traffic monitoring system　04.936
隧道净空　clearance of tunnel　04.851
隧道掘进机　tunnel boring machine　07.312
＊隧道口　tunnel portal　04.847
隧道口净空　portal clearance　04.852
隧道埋深　buried depth of tunnel　04.849
隧道排气污染　tunnel's discharge pollution　10.031
隧道入口区亮度　tunnel entrance brilliance　03.464
隧道施工防尘　tunnel construction dust controlling　04.902
隧道施工通风　tunnel construction ventilation　04.903
隧道适宜亮度　suitable brilliance in tunnel　03.465
隧道坍顶　tunnel roof fall　04.899
隧道通风　tunnel ventilation　04.904
隧道挖掘机　tunnel excavator　07.310
隧道瓦斯爆炸　gas explosion in tunnel　04.929
隧道瓦斯泄出　gas emission in tunnel　04.930

隧道消防系统　tunnel fire protection system　04.924
隧道信号　tunnel signal　04.933
隧道养护　tunnel maintenance　04.947
隧道照明　tunnel lighting　03.459
隧道支撑　tunnel support　04.890
隧道注浆止水　tunnel water sealing by injection　04.931
隧道装载机　tunnel loader　07.311
损伤　damage　05.186
缩短里程效益　benefit from distance shortening　01.059
缩缝　contraction joint　02.461
缩限　shrinkage limit　08.207
索鞍　cable saddle　04.327
索夹　cable clamp　08.139
索力测定计　cable tension measurement device　09.207
索力测量　cable force measurement　09.101
索力控制　cable force control　04.777
索赔　claims　01.154
索塔　cable support tower　04.322

T

塔式起重机　tower crane　07.287
踏勘　reconnaissance　01.175
台班使用费　working-day cost　07.331
台地　terrace　01.229
台架模拟试验　bench simulation test　07.381
台架试验　bench test　07.372
台口式断面　benched section　02.129
台口式路基　benched subgrade　02.252
台帽　abutment cap, abutment coping　04.362
台身　abutment body　04.381
太阳能沥青熔化装置　asphalt solar energy melter　07.177
坍方　landslide　02.642
坍落度　slump　08.262
坍落度试验　slump test　09.053
坍落度圆锥筒　slump cone　09.167
摊铺　paving　02.444

摊铺材料压实度试验　paved material compaction test　07.482
摊铺厚度测定　paving depth measurement　07.478
摊铺宽度测定　paving width measurement　07.477
摊铺路面平整度试验　paved surface evenness test　07.481
摊铺作业试验　paving operation test　07.479
弹簧现象　springing　02.615
弹街路面　pitching pavement　02.429
弹性半无限地基　elastic semi-infinite foundation　02.377
弹性层状体系理论　elastic layer system theory　02.375
弹性梁支承法　elastic supported beam method　04.581
弹性模量　modulus of elasticity　02.398
弹性模量试验　elastic modulus test　09.043

碳化深度酚酞试验 carbonation depth by phenolph-thalein test 09.129

碳素钢 carbon steel 08.117

碳纤维 carbon fiber 08.156

镗缸机 cylinder boring machine 05.349

掏箱 unstuffing 06.261

陶粒 sintered aggregate 08.055

套班运行 package run 06.078

套管 sheath 08.126

套管护壁钻孔法 casing hole-boring method 04.747

套箱围堰 precast-boxed cofferdam 04.767

特长隧道 super long tunnel 04.853

特大桥 grand bridge 04.167

特殊土 special soil 08.064

特种车辆货物运输 special truck transport 06.167

特种车辆运价 rate for special purpose vehicle 06.468

特种挂车 special trailer 05.030

特种货物 special goods 06.135

特种货物运价 rate for special goods 06.467

特种载货汽车 special truck 05.006

特重交通 very heavy traffic 02.383

*提单 bill of lading 06.181

提取空箱 pick-up empty container 06.264

体内预应力 prestressing with bond 04.617

体外预应力 external prestressing 04.618

替换设备 spare attachment 07.334

天沟 gutter 02.518

天然稠度试验 natural consistency test 09.008

天然地基 natural subsoil 02.311

天然含水量 natural water content 08.194

天然沥青 natural asphalt 08.006

天然演变冲刷 natural scour 04.110

填方 fill 02.288

填缝 joint filling 02.663

填缝板 joint plate 02.471

填缝板接缝 joint with sealing plate 04.303

填缝材料 joint sealing materials 04.302

填缝料 joint filler 02.470

填缝料破损 joint filler failure 02.624

调[配]漆机 color matching system 05.371

调坡 adjusting gradient 02.576

调整路拱 adjusting crown 02.577

调治构造物 regulating structure 04.126

跳板 gangboard 04.968

跳车激振 vibration excited by truck jumping from threshold on deck 09.104

铁路道口标志 railway crossing sign 03.412

铁路平交道口 railway crossing 02.245

停车 parking 03.105

停车饱和度 degree of parking saturation 03.157

停车标志 stop sign 03.374

停车场 parking area, parking 02.185

停车场标志 parking lot sign 03.367

停车车道 parking lane 02.174

停车车辆总数 parking accumulation 03.149

停车车位 parking set 03.143

停车车位短缺 parking deficiency 03.150

停车车位供应量 parking supply 03.151

停车车位过剩 parking surplus 03.152

停车车位需要 demand for parking spaces 03.147

停车持续时间 parking duration 03.145

停车费 parking fee 03.158

停车管理计划 parking management program 03.156

停车规则 parking regulation 03.211

停车计划 parking plan 03.155

停车计时器 parking meter 03.341

停车率 parking rate 03.144

停车视距 stopping sight distance 02.107

停车线 stop line 03.425

停车线延误 stop line delay 03.096

停车延误 stop delay 03.097

停车周转次数 parking turnover 03.146

停储长度 storage length 03.202

停放车位标线 parking space limit marking 03.424

停靠站 bus stop 06.126

停驶车日 vehicle non-working day 06.353

停业报批程序 exit application and approval proce-dure 06.404

停止 stopping 03.104

*通车 open to traffic 01.352

通车里程 mileage open to traffic 02.548

通风方式 ventilation type 04.905

通风监测 ventilation monitoring 04.921

通风量　ventilation volume　04.920

通风设计　ventilation design　04.906

通风塔　ventilation tower　04.918

通风效率　ventilation efficiency　04.919

通风装置　ventilation installation　04.922

通过型车辆检测器　passage vehicle detector　03.293

通航净空　navigable clearance　04.024

通航桥孔　navigable span　04.011

通航水位　navigable water level, NWL　04.099

通视浑浊度仪　all-through turbidimeter　09.248

通视能见度检验仪　all-through visibility tester　09.247

通行能力指数　capacity index　03.175

通用型集装箱叉车　universal container fork lift　06.309

同步交通事故预测法　simultaneous traffic accident prediction method　03.237

同步式系统　synchronous system　03.259

同向曲线　same direction adjacent curve　02.084

桶装沥青熔化装置　asphalt barreled melter　07.173

统计诊断　statistical diagnosis　05.231

统一收费系统　unified toll system　03.345

投标　bidding　01.100

投标担保　bid bond　01.111

投标截止期　deadline of bid　01.117

投标人须知　instruction to bidder　01.110

投标书　bid, tender　01.109

投标书有效期　bid validity　01.113

投标邀请书　invitation to bid　01.112

投资回收期　repayment period of investment　01.050

头尾相撞　rear-end collision　03.228

透层　prime coat　02.343

透视图检验　perspective examination　02.056

透水度试验　pervious test　09.011

透水路堤　permeable embankment　02.250

透水磨耗层　pervious wearing course　02.345

透水性　water permeability　02.504

凸块压路机　padfoot roller　07.062

凸形竖曲线　convex vertical curve　02.122

突起分隔带　raised separator　03.448

突然故障　sudden failure　05.194

图上作业法　graphic dispatching method　06.212

徒步出行　pedestrian trip　03.132

土　soil　08.081

土的成因分类　soil classification by formation　08.058

土的工程分类　engineering classification of soil　08.057

土的粒组分类　soil classification by grain　08.056

土地利用预测模型　land-use forecast model　03.111

土方调配图　cut-fill transition diagram　02.073

土方[工程]　earth work　02.287

土方累积图　mass diagram　02.072

*土工布　geotextile　08.164

土工格栅　geogrid　08.167

土工膜　geomembrane　08.166

土工织物　geotextile　08.164

土基承载能力　subsoil bearing capacity　02.280

土基干湿类型　subsoil moistness classification　01.163

土-结构共同作用　soil-structure interaction　04.592

土路　earth road　02.040

土路肩　earth shoulder　02.143

土牛拱架　earthen centering　04.694

土围堰　earth cofferdam　04.766

土压力　earth pressure　04.493

土圆锥仪　soil cone penetrator　09.155

土质调查　soil survey　01.248

土质隧道　tunnel in earth　04.834

推荐速度　advisory speed　03.072

推土铲运机　scraper-dozer　07.011

推土机　bulldozer　07.001

推移质　bed material load　04.123

拖车式移动试验车　trailer-typed mobile laboratory vehicle　09.197

拖动试验　towing test　07.430

拖动阻力测定　towing resistance measurement　07.428

拖挂运输　tractor-trailer transport　06.031

拖挂重量　trailing load　06.236

拖式铲运机　towed scraper　07.009

拖式沥青混合料摊铺机　towed asphalt paver

07.161

拖式平地机 towed grader 07.015

拖式羊足碾 towed sheep-foot roller 07.064

拖式振动压路机 towed vibratory roller 07.070

拖运率 trailer ton-km ratio 06.374

拖滞 dragging 05.221

托架 bracket 04.540

托运 consignment 06.177

托运计划 consignment plan 06.176

托运人 consignor 06.172

托运行包 consigned luggage 06.085

脱档 spontaneous out-of-gear 05.217

脱档滑行 coasting in neutral 05.106

脱皮 scaling 02.620

驮道 bridle path 02.013

拓宽 widening 02.575

拓宽路口式交叉口 flared intersection 02.208

拓扑学理论 topologic theory 03.007

W

挖方 cut, excavation 02.289

挖沟机 trencher 07.050

挖掘机 excavator 07.037

挖孔灌桩法 installing pile shaft by excavation 04.753

挖孔桩 cast-*in-situ* pile by excavation 04.438

蛙式夯 electric frog rammer 07.076

外侧车道 near-side lane 02.178

外侧分隔带 outer separator 02.146

外加剂 additive 08.140

外距 external distance 02.091

外啮合齿轮沥青泵 outside gear asphalt pump 07.165

外形尺寸测定 overall dimensions measurement 07.411

外纵梁 exterior beam, exterior stringer 04.274

弯沉盆 deflection basin 02.396

弯沉试验 deflection test 09.029

弯沉系数 deflection coefficient 02.394

弯沉仪 deflectometer, Benkelman beam 09.186

弯沉值 deflection value 02.389

弯道标志 curve sign 03.377

*弯矩 moment, bending moment 04.557

*弯矩分配法 moment distribution method 04.560

弯拉强度 flexural-tensile strength 02.408

弯拉应变 flexural-tensile strain 02.407

弯拉应力 flexural-tensile stress 02.406

弯桥 curved bridge 04.160

完好车率 vehicle availability rate 06.349

完好车日 available vehicle-days 06.350

完全故障 complete failure 05.188

晚点 behind schedule 06.072

网裂 net-shaped crack 02.602

往返性客流 round trip passenger traffic 06.042

往复给料机 reciprocating feeder 07.091

往复式空气压缩机 reciprocating compressor 07.307

微分法测图 differential photo 01.241

微观交通模型 micro-traffic model 03.008

微粒物 particles 10.022

微丘区 rolling terrain 01.222

微弯板组合梁桥 composite shell-slab and I-beam bridge 04.246

微型桩 mini-pile 04.429

危桥 bridge in danger 04.164

危险货物 hazardous goods 06.140

危险货物包装标志 package mark for hazardous goods 06.164

违约 default 01.150

违章行为 violation 06.411

桅杆式起重机 derrick crane 07.283

围岩 surrounding rock 04.859

围岩分类 surrounding rock classification 04.863

围岩稳定 stability of surrounding rock 04.861

围岩压力 surrounding rock pressure 04.860

围岩自承能力 surrounding rock self-supporting capacity 04.862

围堰 cofferdam 04.764

维护费 maintenance cost 07.337

维卡稠度仪 Vicat apparatus 09.157

未衬砌隧道 rough tunnel 04.858

位移　displacement　04.555

卫星遥感测量　satellite remote sensing　01.246

温差　temperature difference　04.589

*温差应力　thermal stress　04.548

温度梯度　thermal gradient　04.588

温度巡检箱　temperature scanning unit　09.242

温度自动控制　automatic temperature control　07.365

文物保护范围　cultural relic protection scope　10.086

*纹理深度　surface texture depth　02.368

稳定保护层　stabilized protection course　02.348

稳定材料基层　stabilized material base　02.492

稳定工况试验　steady-state test　07.378

稳定剂　stabilizer　08.151

稳定交通流　stable traffic flow　03.042

稳定土厂拌设备　stabilized soil mixing plant　07.108

稳定土摊铺机　stabilized soil paver　07.122

稳定性　stability　04.553

稳钩　stabilize hooks　06.298

涡流效应　vortex shedding effect　04.572

卧铺客车　motor coach sleeper　05.018

圬工拱桥　masonry arch bridge　04.194

圬工桥　masonry bridge　04.193

污染超限　over-limit pollution　05.206

无侧限抗压强度　unconfined compression strength　08.213

无侧限抗压强度试验　unconfined compression strength test　09.020

无纺布伸缩装置　non-woven fabrics expansion installation　04.316

无纺土工织物　nonweaven geotextile　08.165

无机结合料　inorganic binder　08.035

无铰拱桥　hingless arch bridge　04.218

无粘结的预应力筋　non-bonded tendon　04.628

无损检测　non-destructive test, NDT　09.120

无信号交叉口　non-signalized crossing　02.215

无压力涵洞　inlet unsubmerged culvert, non-pressure culvert　04.810

无噪声打桩技术　noiseless piling technique　10.061

无噪声路面　noiseless pavement　10.062

*无支架吊装法　erection with cableway　04.661

无钻杆反循环钻机　rodless reverse circulation rig　07.263

雾检测器　fog detector　03.311

物动视力　vision with object moving　03.508

物探　geophysical prospecting　01.256

物资管理　material management　06.396

X

细琢石路面　dressed stone pavement　02.426

峡谷　canyon　01.228

下部结构　substructure　04.358

下承式桥　through bridge　04.157

下穿式立交　underpass　02.225

下挡墙　lower retaining wall　02.277

下拉荷载　downdrag, negative friction of pile　04.497

下卧层　underlying stratum　01.331

先拱后墙法　arched roof in advance of wall tunneling method　04.877

先进的长途运输系统　advanced rural transportation system, ARTS　03.321

先进的城市[公共]运输系统　advanced public transportation system, APTS　03.320

吸湿性货物　humidity absorbing goods　06.150

吸水率试验　water absorptivity test　09.031

吸引交通量　absorbed traffic [volume]　03.126

稀浆封层　slurry seal　02.423

稀浆封层机　asphalt slurry seal machine　07.234

*稀释沥青　cutback asphalt　08.020

熄火滑行　coasting with engine off　05.107

铣槽　milling groove　02.482

系杆拱桥　bowstring arch bridge, tied arch bridge　04.228

细度　fineness　08.181

细度模数　fineness modulus　08.182

细级配　fine gradation　08.189

细粒土　fine-grained soil　08.063

细砂　fine sand　08.080

先进的驾驶员信息系统 advanced driver information system, ADIS 03.319

先进的交通管理系统 advanced traffic management system, ATMS 03.317

先进的汽车控制系统 advanced vehicle control system, AVCS, IVCS 03.318

先墙后拱法 wall in advance of arched roof tunneling method 04.876

先张法 pretensioning method 04.722

先钻后扩法 reaming-after-boring method 04.751

鲜活货物 fresh and living goods 06.145

纤维混凝土 fiber concrete 08.111

纤维混凝土路面 fiber concrete pavement 02.454

涎流冰 salivary flow ice 01.275

险情检查 examination of dangerous situation 02.559

现场拌和试验 field mixing test 07.471

现场调度 on-site dispatching 06.207

现场考察 site inspection 01.133

现场试验 field test 07.373

现场压实试验 field compaction test 07.465

现有服务指数 present service index, PSI 02.374

现有公路 existing highway 01.029

*县道 county highway 01.343

县公路 county highway 01.343

限额赔偿 limited compensation 06.119

限速标志 stated-speed sign 03.386

限运货物 restricted goods 06.155

限重标志 weight limit sign 03.376

线控制 line control, linked control 03.298

线形协调 alignment coordination 02.053

线形要素 alignment elements 02.050

线压力 linear pressure 02.304

相对高程 relative elevation 01.209

相对含水量 relative water content 08.195

相对密度试验 relative density test 09.003

相对密实度 relative density 08.200

相位 phase 03.244

相位差 offset, phase offset 03.251

厢式挂车 van trailer 05.031

厢式货车 van truck 05.007

镶合浇筑 match casting 04.715

*镶辑复照图 photo-index 01.237

箱涵 box culvert 04.804

箱梁 box girder 04.264

箱梁型护栏 box-girder fence 03.434

箱内作业叉车 inside container operation fork lift 06.310

箱[形]梁桥 box-girder bridge 04.212

箱形桥台 box type abutment 04.388

乡村道路 rural road 02.008

*乡道 township highway 01.344

乡公路 township highway 01.344

橡胶改性沥青 rubber modified asphalt 08.028

橡胶接缝 rubber joint 04.307

橡胶沥青装置 rubber asphalt unit 07.183

像片索引图 photo-index 01.237

像片镶嵌图 photo-mosaic 01.242

消冰 deicing 02.680

消冰剂污染 deicing agent pollution 10.038

消除粘结的预应力筋 debonded tendon 04.627

消能槛 baffle sill 04.139

*消石灰 slaked lime 08.037

小道 path 02.009

小桥 small bridge 04.171

小石子混凝土 micro-aggregate concrete 08.114

小型车运价 rate for small vehicle 06.469

小型客车票价 ticket price of mini bus 06.480

*小型桩 mini-pile 04.429

小循环运行 partial circulation operation 06.076

楔形缝 keyed joint 04.305

协议书 agreement 01.125

斜撑式满布木拱架 full span wooden inclined strut centering 04.690

斜交叉 skew intersection 02.201

斜交涵洞 skew culvert 04.815

斜交角 ·skew angle 04.036

斜交桥 skew bridge 04.159

斜井 inclined shaft 04.844

斜拉桥 cable stayed bridge 04.235

斜列式停车 diagonal parking 03.199

斜索 stay cable 04.328

斜腿刚构桥 slant legged rigid frame bridge 04.233

斜桩 battered pile, raked pile 04.415

卸架木楔 wood wedge for centering unloading 04.699

卸架砂筒　sand cylinder for centering unloading　04.700

蟹耙式清岩机　rake-up ballast remover　07.036

泄漏　leakage　05.199

泄水孔　weep-hole　02.541

泄水口　drain opening　02.542

新奥法　New Austrian Tunneling Method, NATM　04.871

新泽西护栏　New Jersey safety barrier　03.449

信号处理机　signal processor　09.245

信号控制　signal control　03.240

信号控制交叉口　signalized crossing　02.214

信号控制器　signal controller　03.301

*[信号]时差　offset, phase offset　03.251

信号周期　signal cycle　03.242

W 型护栏　W-type guardrail　03.435

JM12 型锚具　JM12 anchorage　04.647

形变模量　modulus of deformation　02.397

T 形刚构桥　T-shaped rigid frame bridge　04.231

Y 形交叉　Y-intersection　02.204

T 形梁　T[-shaped] beam　04.269

U 形梁　U[-shaped] beam　04.270

T[形]梁桥　T-beam bridge　04.211

Y 形[桥]墩　Y-shaped pier　04.371

U 形桥台　U-abutment　04.384

T 形桥台　T-abutment　04.387

A 形索塔　A-framed tower　04.324

行包　luggage　06.080

行包保管　luggage storage service　06.090

行包标签　luggage label　06.100

行包承运　acceptance of luggage consignment　06.089

行包丢失　loss of luggage　06.112

行包货位　luggage lot　06.129

行包交付　luggage delivery　06.091

行包赔偿率　luggage compensation rate　06.125

行包受理　luggage acceptance　06.088

行包损坏　damage of luggage　06.113

行包托运　luggage consignment　06.087

行包运价　luggage transport rates　06.486

行包正运率　luggages transport regularity rate　06.124

*行车道　carriageway　02.160

行车激振　vibration excited by moving truck　09.103

行车津贴　operation subsidy　06.395

行车路单　waybill　06.409

行车碾压　traffic bound, free bound　02.448

行程车速　travel speed, journey speed　03.076

行程利用率　operation mileage rate　06.370

行程时间　travel time, journey time　03.082

行程时间比　travel-time ratio　03.099

行道树　roadside trees　02.589

行近流速　approach velocity　04.057

行人安全设施　pedestrian safety devices　03.484

行人按钮信号控制机　pedestrian push-button signal controller　03.305

行人管制　pedestrian control　03.207

行人过街标志　pedestrian crossing sign　03.375

行人过街信号灯　pedestrian crossing signal lamp　03.306

行人横穿设施　pedestrian crossing devices　03.485

行人护栏　pedestrian barrier　03.489

行人检测器　pedestrian detector　03.285

行人信号相位　pedestrian signal phase　03.257

行驶时间　running time　03.081

行驶速度　running speed　03.075

行驶性能试验　travel performance test　**07.419**

行驶延误　operational delay　03.093

行驶阻力测定　moving resistance measurement　07.426

性能试验　performance test　07.396

*胸墙　front wall　04.382

休息区　rest area　02.189

休息区标志　rest area sign　03.401

休息娱乐区　recreation area　02.191

修复　reconditioning　05.287

修复工程　rehabilitation work　02.583

修理尺寸　repair size　05.301

*修正击实试验　modified compaction test, modified Proctor test　09.014

锈蚀试验　corrosion test　07.388

需要停车次数　parking demand　03.148

虚交点　imaginary intersection point　01.195

*许用应力　allowable stress　04.551

畜力车道　cart road　02.011

悬臂板　cantilever slab　04.286

悬臂浇筑法　cast-in-place cantilever method
　04.664

悬臂梁桥　cantilever beam bridge　04.206

悬臂拼装法　cantilever erection method　04.662

悬臂式挡土墙　cantilever retaining wall　02.270

悬臂式交通标志　cantilever traffic sign　03.371

悬带桥　suspended ribbon bridge　04.245

悬索桥　suspension bridge　04.244

悬移质　suspended load　04.122

旋风式除尘器　cyclone dust collector　07.131

旋刷刮板清扫车　rotary broom-slat sweeper
　07.223

旋转刀具圆周速度测定　rotary tools circular velocity
measurement　07.469

旋转式凿岩机　rotary jack hammer　07.026

旋转式钻机　swiveling drill　07.250

选线　route selection　02.049

学童过街标志　children crossing sign　03.410

学校标志　school sign　03.411

穴蚀　cavitation　05.184

雪崩防治　snow slide protection　02.683

雪害　snow hazard　02.646

血中酒精浓度　blood-alcohol concentration　03.498

循环调度法　cyclic dispatching method　06.214

循环作业量测定　determination of operating cycle
production quantity　07.445

巡回养护　patrol maintenance　02.568

Y

压槽　rolling groove　02.480

压电式传感器　piezoelectric transducer　09.238

压拉双作用预应力　prestressing with subsequent
compression and tension　04.621

压力盒　pressure cell　09.206

压力式涵洞　outlet submerged culvert, pressure
culvert　04.808

压曲　buckling　04.566

*压屈　buckling　04.566

压实　compaction　02.296

压实度　degree of compaction　02.297

压实度试验　compactness test　09.015

压实度自动检测装置　automatic compactometer
　07.371

压实厚度　compaction depth　02.308

压实机械试验　compaction machine test　07.457

压实宽度测定　compacting width measurement
　07.458

压实能量　compaction energy　02.306

压实深度测定　compacting depth measurement
　07.459

压实速度测定　rolling speed measurement　07.464

压实系数　compacting factor　02.298

压碎值　crushing value　08.250

压碎值试验　crushing value test　09.032

压缩试验　compression test　09.042

压缩压力计　compression gauge　05.344

押运　escorting　06.187

垭口　pass　01.230

烟度排放标准　smoke emission standard　10.020

烟粒　soot　10.018

烟浓度　smoke concentration　10.017

淹没丁坝　submerged spur dike　04.130

淹水地区　inundated district　04.048

盐侵蚀　salt corrosion　02.637

盐蚀模拟试验　salt fog simulated test　07.387

盐胀　salt heaving　02.638

盐渍土　salty soil　01.285

严重故障　major failure　05.191

研究性试验　research test　07.399

岩层　rock stratum　01.317

岩堆　talus　01.262

岩缝填充物　joint-filling material　01.318

*岩溶　karst　01.261

岩石隧道　tunnel in rock　04.835

岩土特性指标试验　geotechnical index property test
　09.025

岩相　lithofacies　01.319

岩心　core　01.333

岩心采取率　percentage of coring　01.334

岩心取样器　core sampler　09.232

延度　ductility　08.228

延度试验 ductility test 09.062

延期 extension of time 01.153

延期违约偿金 liquidated damages 01.151

延误 delay 03.092

延滞费 demurrage 06.489

颜色适应性 color adaptability 03.528

沿溪线 valley line 02.063

验道 road inspection 06.186

验货 inspection of goods 06.184

验收 acceptance 01.146

验收试验 acceptance test 07.403

验收证书 acceptance certificate 01.147

验算荷载 check load 01.089

验证性试验 proving test 07.400

扬尘性货物 dusty goods 06.148

羊足压路机 sheep-foot roller 07.065

阳离子乳化沥青 cationic emulsified bitumen 08.013

氧化沥青 oxidized asphalt 08.026

养护对策 maintenance counterproposal 01.357

养护费用 maintenance cost 01.065

养护工程 maintenance project 02.546

养护管理 maintenance management 02.545

养护里程 maintenance mileage 02.549

养护质量 maintenance quality 02.686

养路道班 maintenance gang 01.351

养路段 maintenance section 01.350

养路费 highway maintenance fee 01.036

养路总段 maintenance division 01.349

样机试验 prototype test 07.395

摇摆筛 swinging screen 07.086

咬粘 seizure 05.182

野生动物保护区 wild animals refuge area 10.083

野外定线 field location 02.070

页岩沥青 shale tar 08.024

业主 employer, owner 01.101

叶桨搅拌式沥青乳化机 paddle bitumen emulsifying machine 07.180

夜间施工噪声 night construction noise 10.057

液罐挂车 tank trailer 05.037

液罐货车 tank truck 05.010

液塑限联合测定仪 liquid-plastic combine tester 09.149

液体沥青 liquid asphalt 08.016

液限 liquid limit, LL 08.204

液限试验 liquid limit test 09.006

液性指数 liquidity index 08.208

液压比例自控系统 automatic hydraulic proportioning system 07.364

液压打桩机 hydraulic pile driver 07.246

液压缓冲护栏 hydraulic cushion guardrail 03.444

液压静力触探仪 hydraulic static cone penetrometer 09.230

液压挖掘机 hydraulic excavator 07.046

液压稳定器 hydraulic pressure stabilizer 09.203

液压压拔套管机 hydraulic casing extractor 07.262

液压岩石破碎机 hydraulic rock breaker 07.033

液压元件试验 hydraulic element test 07.456

液压凿岩机 hydraulic jack hammer 07.024

液压桩头破碎器 hydraulic pile head splitter 07.273

液压自动找平装置 hydraulic automatic leveling device 07.367

一般公路 ordinary highway, mixed traffic highway 02.023

一般故障 minor failure 05.192

一般行包 normal luggage 06.081

一次拌成稳定土搅拌机 single pass soil stabilizer, single pass stabilized soil road mixer 07.119

一次碰撞 primary collision 03.224

一级公路 first class highway 02.021

一维碰撞 one dimension collision 03.222

一氧化碳监测仪 carbon monoxide detector 10.024

一氧化碳允许浓度 carbon monoxide allowable concentration 10.025

一字形桥台 head wall abutment 04.386

移动式沥青混合料搅拌设备 mobile asphalt mixing plant 07.126

移动式联合碎石机组 mobile crushing plant 07.096

移动式稳定土厂拌设备 mobile stabilized soil mixing plant 07.110

移动式装卸机械 mobile handling machinery 06.279

驿道　post road [Tang Dynasty]　02.018

易腐货物　perishable goods　06.146

易碎货物　fragile goods　06.151

溢洪桥　spillway bridge　04.176

异响　abnormal noise　05.198

翼墙　wing wall　04.383

翼缘　flange　04.536

阴离子乳化沥青　anionic emulsified bitumen
　08.014

引道　approach　02.255

引气混凝土　air-entraining concrete　08.109

引桥　approach bridge, approach span　04.008

隐蔽工程验收　hidden work acceptance　01.143

应变计　strain gage　09.204

应变巡检箱　strain scanning unit　09.241

应急电话　emergency telephone　03.295

应急电话标志　emergency telephone sign　03.403

应急管理系统　emergency management system
　03.323

应急停车带　emergency parking strip　02.175

应力波方程法　stress wave equation method
　04.656

应力腐蚀　stress corrosion　04.547

应力集中　stress concentration　04.546

应力控制张拉　prestressing under stress control
　04.726

应力吸收薄膜　stress absorbing membrane　02.341

＊膺架　scaffold, falsework　04.704

膺架浇筑法　cast on scaffolding method, cast-in-
place method　04.668

膺架式架设法　erection with scaffolding method
　04.667

营业性运输　for-hire transport　06.019

营运标志　operation signs　06.410

营运范围　operation area　06.010

营运方式　operation manner　06.009

营运费用　operation cost　06.385

营运里程　operation mileage　06.066

营运汽车　for-hire vehicle　05.056

营运收入　operation revenue　06.390

营运速度　operation speed　06.369

营运线路　operation route　06.055

营运线路图　operational route map　06.131

营运证　operation certificate　06.408

营运支出　operation expenditure　06.391

迎面碰撞　head-on collision　03.227

影响面积　influence area　04.565.

影响线　influence line　04.564

影像地图　photographic map　01.236

硬表层　crust　01.312

硬化　hardening　08.263

硬沥青　pitch　08.025

硬练砂浆强度试验　early-dry mortar strength test
　09.055

硬路肩　hard shoulder　02.144

硬盘　hardpan　01.313

硬岩　hardrock　01.315

拥包　upheaval　02.613

拥挤度　degree of congestion　03.178

壅水　baek-water　04.105

壅水高度　back-water height　04.108

＊永冻土　permafrost　01.272

永久荷载　permanent load　04.477

永久性桥　permanent bridge　04.178

用户试验　user test　07.408

优先时差　priority phase offset　03.252

优先通行车道　priority lane　03.189

油包　oil pox　02.612

油石比　bitumen aggregate ratio　08.248

油压千斤顶张拉　hydraulic jack prestressing
　04.721

油毡防水层　water proof asphalt-felt　04.296

游离碳含量　free carbon content　08.240

游离碳含量试验　free carbon content test　09.076

有机结合料　organic binder　08.002

有机聚合物　organic polymer　08.034

有机物含量试验　organic matter content test
　09.034

有机质含量　organic matter content　08.218

有机[质]土　organic soil　01.286

有路面公路　paved highway　02.039

有限预应力混凝土　limited prestressed concrete
　04.614

有效绿灯时间　effective green time　03.247

有效碾压深度　effective rolling depth　02.303

右转弯导向线　right turn guide line　03.419

右转弯匝道　right turn ramp　02.243
诱导标志　induction sign　03.362
诱增交通量　induced traffic [volume]　03.124
迂回运输　roundabout transport　06.027
淤积　silting　04.120
淤泥　mud, silt　01.276
余振自谱分析　residual vibration auto-spectrum analysis　09.117
逾期交付　delayed delivery　06.190
逾期提货　delayed pick-up　06.191
* 雨水口　inlet　02.537
雨天事故　wet traffic accident　03.232
预防维护　preventive maintenance　05.257
预防性修理　preventive repair　04.796
预防性养护　preventive maintenance　02.566
预拱度　camber　04.719
预可行性研究　pre-feasibility study　01.039
预留沉落量　reserve settlement　02.310
预热　preheating　05.087
预弯法预应力　prestressing without tendon by pre-bending　04.619
预压法　preloading method　02.316
预应力度　degree of prestressing　04.611
预应力混凝土桥　prestressed concrete bridge　04.184
预应力混凝土桩　prestressed concrete pile　04.420
预应力筋　tendon　08.125
预应力损失　loss of prestress　04.629
预应力台座　prestressing bed　04.730
预制场　fabrication yard, casting yard　04.709
预制件挂车　prefab trailer　05.041
预制桩　precast pile　04.418
预钻插桩法　socketing pile in prebored hole　04.750
原件收付　original piece receipt and delivery　06.093
原状土　parent soil　01.293
原状土样　undisturbed sample　01.335
圆缓点　point of curve to spiral, CS　02.100
圆盘给料机　disk feeder　07.093
圆曲线　circular curve　02.082

圆石路面　cobble stone pavement　02.428
圆直点　point of tangent, PT　02.097
圆锥式碎石机　cone crusher　07.082
缘石标记　curb marking　03.429
约束运行　constrained operation　03.183
越岭线　ridge crossing line　02.066
月平均日交通量　mouthly average daily traffic volume, MADT volume　03.059
匀砂　sand brooming　02.661
允许间隙　permissible clearance　05.303
允许磨损　permissible wear　05.176
运达期限　time limit of shipment arrival　06.094
* 运管费　transport administration fee　06.386
运价　tariff, rate　06.442
运价比差　rate ratio　06.462
运价加成　rate addition　06.456
运价减成　rate reduction　06.457
运量　volume　06.341
运量波动系数　freight volume fluctuation coefficient　06.196
运量预测　transport volume forecast　06.337
运输包装　packing for transport　06.159
运输标志　transport mark　06.162
运输服务业　transport service industry　06.406
运输工程学　transportation engineering　01.005
运输管理费　transport administration fee　06.386
运输结构　transport structure　06.006
运输经济学　transportation economics　01.015
运输里程　carriage mileage　06.450
运输能力　transport capacity　06.008
运输市场　transport market　06.399
运输市场管理　transport market control　06.400
运输弹性系数　transport elasticity coefficient　06.382
运输体系　transport system　06.007
运输系数　transport coefficient　06.383
运输责任期　transport liability period　06.189
运输重量测定　shipping weight measurement　07.413
运行车速　operating speed　03.077

Z

匝道 ramp 02.237

匝道标线 ramp marking 03.426

匝道集成系统控制 ramp integrated system control 03.275

匝道交通调节 ramp metering 03.191

匝道连接处 ramp junction 03.180

匝道桥 ramp bridge 04.162

匝道通行能力 ramp capacity 03.170

载货汽车 [motor] truck 05.005

载客量利用率 bus capacity utilization rate 06.372

载重量利用率 truck loading capacity utilization rate 06.371

再生沥青混合料 reclaimed asphalt mixture 08.097

在用汽车 vehicle in use 05.054

暂定金额 provisional sums 01.139

暂停 standing 03.103

凿岩机 rock drill, jack hammer 07.018

凿岩钻车 drill wagon for jack hammer 07.028

早强混凝土 early strength concrete 08.103

早强剂 early strength agent 08.148

噪声等级 noise level 10.042

噪声等值线 noise isoline 10.043

噪声检定值 noise rating number, NRN 10.041

噪声容限 noise margin 10.039

噪声试验 noise level test 07.449

噪声污染 noise pollution 10.040

噪声系数 noise factor, NF 10.044

增长交通量 increment of traffic [volume] 03.123

增长率法 growth rate method 03.120

憎水性集料 hydrophobic aggregate 08.073

渣油 residual oil 08.033

渣油路面 residue oil pavement 02.422

闸门桥 water gate bridge 04.247

栅式进水口 grated inlet 02.538

窄桥标志 narrow bridge sign 03.383

展线 route development 02.068

栈道 trestle road along cliff 02.015

* 栈桥 viaduct, trestle 04.154

张拉程序 stressing sequence, tensioning procedure 04.725

张拉控制应力 control stress for prestressing 04.729

张拉千斤顶 tensioning jack, drawing jack 07.295

胀缝 expansion joint 02.460

招标 calling for tenders 01.099

招标文件 bidding documents 01.108

沼泽 swamp 01.277

照明 lighting 03.455

照明灯具 lamps and lanterns 03.471

照明灯柱 lighting standard 03.458

照明过渡 lighting transition 03.460

照明过渡段 lighting transition section 03.461

照明设备 lighting facilities 03.472

照明适应段 lighting adaptation section 03.462

罩面 overlay 02.351

褶皱 fold 01.327

真缝 true joint 02.465

真空脱水工艺 vacuum dewatering technique 02.478

真空吸尘清扫车 vacuum sweeper 07.222

真空吸水机械 vacuum water sucker 07.293

针入度 penetration 08.223

针入度试验 penetration test 09.061

针入度指数 penetration index 08.224

诊断参数 diagnostic parameters 05.237

诊断方法 diagnostic method 05.229

诊断工艺 diagnostic technology 05.228

诊断规范 diagnostic norms 05.238

诊断周期 interval between diagnosis 05.239

诊断专家系统 diagnostic expert system 05.233

震动沉桩法 pile vibrosinking method 04.761

震害 seismic hazard 02.654

振荡压路机 oscillatory roller 07.068

振动 vibration 10.064

振动沉拔桩机 vibration pile driver extractor 07.247

振动冲击夯 vibratory tamper 07.072

振动给料机　vibrating feeder　07.094

振动轮激振力测定　vibration drum exciting force measurement　07.461

振动轮频率测定　drum vibration frequency measurement　07.462

振动轮振幅测定　drum vibration amplitude measurement　07.463

振动平板夯　vibratory plate compactor　07.071

振动筛　vibrating screen　07.085

振动压路机　vibratory roller　07.061

振动压实　vibrating compaction　02.302

振抖　fluttering　05.208

阵发性客流　intermittent passenger traffic　06.040

蒸发池　evaporation pond　02.524

蒸发污染物　evaporative pollutants　10.014

蒸馏试验　distillation test　09.069

蒸气式沥青熔化装置　asphalt steam pipe melter　07.175

蒸汽养生　steam curing　02.486

整车货物运输　truck-load transport　06.165

整车运价　truck load rate　06.465

整平层　leveling course　02.361

整体式衬砌　integral lining　04.886

整箱货　full container load, FCL　06.268

整形　shaping　02.475

整形修理　form correction repair　05.284

正铲挖掘机　front shovel　07.039

正常磨损　normal wear　05.174

正点　on schedule　06.071

正己烷当量　hexane equivalent　10.026

正交叉　right-angled intersection　02.200

正交桥　right bridge　04.158

正交异性板　orthotropic slab　04.281

正时灯　timing light　05.341

正式试验　official test　07.402

正响应区　positive response zone　09.113

＊正向分析　progressive analysis for bridge erection　04.778

正循环钻机　circulation drill　07.253

正循环钻孔法　circulation boring method　04.756

支撑式桥台　supported type abutment　04.395

支撑应力　bearing stress　04.549

支承式伸缩装置　support type expansion installation　04.301

支承桩　bearing pile　04.436

支挡桩　soldier pile　04.431

支拱板条　lagging　04.698

支线公路　branch highway, feeder highway　02.036

支座　bearing　04.444

直达客运　through bus transport　06.058

直达运输　through transport　06.014

直道　straight road [Qin Dynasty]　02.016

直缓点　point of tangent to spiral, TS　02.098

直剪仪　direct shear apparatus　09.162

直接剪切试验　direct shear test　09.044

直接视野　direct field of vision　03.504

直接效益　direct benefit　01.053

[直升]导管浇筑混凝土法　tremie concreting　04.771

直线　straight line　02.075

直线形码头　straight line type wharf　04.955

直线行驶性能试验　straight-line running test　07.422

直圆点　point of curve, PC　02.096

植被　vegetation　01.171

植被固坡　planting protect slope　10.077

植草　grass planting　02.655

植物保护　plant protection　10.082

植物固沙　sand consolidation by planting　10.078

执行调度　operation dispatching　06.206

＊pH值　pH value　08.214

值班调度　routine dispatching　06.208

指路标志　directional sign, guide sign　03.357

指示标志　mandatory sign　03.356

止冲流速　non-scouring velocity　04.059

纸上定线　paper location　02.071

致命故障　critical failure　05.190

制动　braking　05.095

制动墩　braking pier, abutment pier　04.373

制动反应时间　brake reaction time　03.516

制动鼓车床　brake drum lathe　05.356

制动距离　brake distance　03.517

制动力　braking force　04.491

制动试验台　brake tester　05.327

[制动]踏板力计　foot brake pedal pressure gauge　05.348

制动蹄磨床　brake shoe grinder　05.357

制动性能试验　braking ability test　07.424

智能车路系统　intelligent vehicle highway system, IVHS　03.316

*智能汽车控制系统　advanced vehicle control system, AVCS, IVCS　03.318

质量事故　quality accident　06.217

质量指标　quality index　06.339

质心位置测定　mass center test　07.415

滞后响应区　hysteresis response zone　09.116

中标　award of contract　01.121

中标通知书　notification of award　01.122

中长隧道　medium tunnel　04.855

中承式桥　half-through bridge　04.156

中等交通　medium traffic　02.385

中断性交通流　interrupted traffic flow　03.048

中级路面　intermediate class pavement　02.336

中间车道　center lane　02.177

中间带　median　02.147

中间带排水　median drainage　02.513

中裂乳化沥青　medium breaking emulsified bitumen　08.010

中凝液体沥青　medium-curing liquid asphalt　08.018

中桥　medium bridge　04.170

中砂　medium sand　08.079

中湿类型　median dampness type　01.165

中水位桥　medium water level bridge　04.173

中线　center line　01.187

中线测量　center line survey　01.188

中线桩　center line stake　01.189

中心岛　center island　02.218

中心岛标线　center island marking　03.423

中央导洞法　central pilot tunneling method　04.874

中央分隔带　median separator　02.148

中央分隔带护栏　median barrier　03.445

中央分隔带开口　median opening　02.149

中转劳务包干费　transshipment service package charge　06.504

中桩填挖高度　height of cut and fill at center stake　01.212

终止合同　termination of contract　01.152

*重车里程　loaded mileage　06.358

重车行程　loaded mileage　06.358

重锤夯实法　heavy tamping method　02.300

重点货物　priority goods　06.152

重点物资运输计划　priority materials transport plan　06.335

重交通　heavy traffic　02.384

重力式挡土墙　gravity retaining wall　02.268

重力式斗式提升机　gravity bucket elevator　07.104

重力式锚碇　gravity anchor　04.337

重力式[桥]墩　gravity pier　04.364

重力式桥台　gravity abutment　04.390

重丘区　hilling terrain　01.223

重箱里程　loaded container mileage　06.452

重型击实试验　modified compaction test, modified Proctor test　09.014

重型平板挂车　heavy haul trailer　05.029

重油　heavy oil　08.032

仲裁　arbitration　01.157

舟桥　pontoon bridge, bateau bridge　04.248

周期长　cycle length　03.243

周期性闪光　cyclic flashing　03.468

周转量　turnover　06.342

周转总成　replacement unit　05.309

轴流式通风　ventilation by axial flow fan　04.915

轴流通风机　axial-flow blower　07.315

轴载　axle load　01.090

轴重检测器　axle weight detector　03.309

轴重仪　axle load meter　05.330

皱纹套管　spirally wound sheath　08.127

逐跨施工法　span by span construction method　04.666

主槽　main channel　04.086

主动安全性　active safety　05.142

主控制机　master controller　03.304

主跨　main span　04.010

主梁　main beam, main girder　04.263

主桥　main bridge　04.007

主线收费站　mainline toll station　03.339

主轴承镗床　main bearing boring machine　05.350

主桩　key pile　04.417

柱　column　04.277

柱板式挡土墙　pile-plank retaining wall　02.272

柱式举升机　post lift　05.359

柱式轮廓标　post delineator　03.452

柱式[桥]墩　column pier, shaft pier　04.367

筑岛沉井　sinking open caisson on built island　04.743

筑路机械　road machinery　01.016

筑路机械大修　major repair of road machine　07.327

筑路机械全过程管理　all-life period management of road machine　07.317

筑路机械小修　current repair of road machine　07.325

筑路机械中修　medium repair of road machine　07.326

驻地监理工程师　resident supervising engineer　01.107

抓铲挖掘机　clam shell excavator　07.041

爪形冲击锤　cross chipper hammer　07.261

专用车道　accommodation lane　02.182

专用公路　accommodation highway　02.034

专用汽车　special purpose vehicle　05.013

砖块路面　brick pavement　02.425

转点　turning point, TP　01.196

转斗式提升机　rotary bucket elevator　07.105

转垛　stacking area adjustment　06.304

转盘式钻机　rotary table rig　07.251

转体架桥法　bridge erection by swinging method　04.673

转弯车道　turning lane　02.168

转向　turning　05.092

转向性能试验　turning test　07.441

转移交通量　diverted traffic [volume]　03.125

转子式除雪机　rotary snow remover　07.217

转子式沥青乳化机　rotary bitumen emulsifying machine　07.179

转子式碎石机　rotator crusher　07.080

转子式稳定土搅拌机　rotary type soil stabilizer, rotary type stabilized soil road mixer　07.118

桩　pile　04.412

[桩]动测法　dynamic measurement of pile　09.142

* 桩负摩擦力　downdrag, negative friction of pile　04.497

桩贯入试验　pile penetration test　09.138

桩横向荷载试验　lateral loading test of pile　09.137

桩基c值法　subsoil reaction modulus c method for laterally loaded pile　04.654

桩基m值法　subsoil reaction modulus m method for laterally loaded pile　04.653

桩基础　pile foundation　04.409

桩架　pile frame　07.244

* 桩群　pile group　04.414

桩入土长度　embeded length of pile　04.034

桩完整性试验　pile integrity test　09.139

桩轴向荷载试验　axial loading test of pile　09.136

装拆式钢桥　fabricated and detachable steel bridge　04.259

装货落空损失费　compensation for failure of loading　06.490

装货网络　loading network　06.245

装配式桥　fabricated bridge　04.258

装箱　stuffing　06.262

装卸定额　handling quota　06.321

装卸方式　handling mode　06.274

装卸工班定额　handling shift quota　06.323

装卸工班效率　handling shift efficiency　06.324

装卸工人生产率　handling worker's productivity　06.328

装卸工时产量　handling hourly output　06.325

装卸工时定额　handling hourly quota　06.322

装卸工序　handling procedure　06.293

装卸工艺　handling technology　06.286

装卸工艺流程　handling technological process　06.290

装卸工艺设计　handling technology design　06.289

装卸工作成本　handling operation cost　06.330

装卸工作停歇时间　handling intermittence time　06.329

装卸机械　handling machinery　06.277

装卸机械化方案　handling mechanization scheme　06.288

装卸机械化系统　handling mechanization system　06.287

装卸机械计时包用费　chartered handling equipment time based charge　06.494

装卸机械生产率　handling machinery productivity　06.327

装卸机械效率指标　handling machinery efficiency

index 06.326

装卸机械走行费 travel expense of handling equipment 06.495

装卸距离 handling distance 06.314

装卸效率 handling efficiency 06.313

装卸直接费用 handling direct cost 06.437

装卸质量标准 handling quality standard 06.294

装卸质量合格率 loading and unloading quality conformity rate 06.235

装卸作业线 handling operation line 06.301

装卸作业指标 indexes of handling operation 06.315

装载机 loader 07.053

装载挖掘机 loader-excavator 07.056

撞人事故 pedestrian accidents 03.219

状况监测维护 status monitoring maintenance 05.253

锥形交通路标 traffic cone 03.483

锥形块石 Telford stone 08.071

锥形锚具 cone anchorage, Freyssinet cone anchorage 04.641

准正交各向异性板法 quasi-orthotropic slab method 04.582

自办站 self-owned terminal 06.128

自动操作 automatic operation 07.360

自动称重 automatic weighing 07.362

自动划线装置 autoset road marking 07.370

自动控制 automatic control 07.361

自动喷水灭火机 automatic water spraying fire extinguisher 04.946

自动弯沉仪 autodeflectometer 09.187

自动找平装置 automatic leveling device 07.366

自理行包 hand luggage 06.086

自落式混凝土搅拌设备 gravity type concrete mixing plant 07.189

自落式稳定土搅拌设备 gravity stabilized soil mixing plant 07.112

自然通风 natural ventilation 04.907

自升式模板 self-climbing formwork 04.717

自适应信号控制系统 adaptive traffic signal control system 03.268

自吸式水泵 self-prime pump 07.270

自卸挂车 dump trailer 05.036

自卸货车 dump truck 05.011

自行车车道 bike lane 02.195

自行车道 bike-way, cycle track 02.010

自行式铲运机 motor scraper 07.008

自行式沥青混凝土摊铺机 self-propelled asphalt paver 07.160

自行式平地机 motor grader 07.014

自行式桥梁检测架 self-propelled bridge inspection cradle 09.199

自应力水泥 self-stressing cement 08.044

自用车辆 private vehicle 05.057

* 自用运输 own account transport 06.020

自由交通流 free traffic flow 03.041

自由流车速 free flow speed 03.073

自由膨胀率 free swelling rate 08.215

综合法测图 panimatric photo 01.239

综合交通管理 comprehensive traffic management 03.205

综合稳定基层 comprehensive stabilized base 02.496

综合养护车 combined maintenance truck 07.214

综合运输系统 comprehensive transportation system 03.140

综合诊断 general inspection and diagnosis 05.234

总车吨位日 total vehicle-ton-days 06.356

总车客位日 total vehicle-seat-days 06.357

总车日 total vehicle-days 06.348

总成大修 unit major repair 05.280

总成互换修理法 unit exchange repair method 05.308

总成小修 unit current repair 05.281

总成修理 unit repair 05.279

总悬浮微粒 total suspended particles 10.023

纵断面 profile 02.111

纵断面测量 profile survey 01.203

纵断面分析仪 longitudinal profile analyzer 09.182

纵缝 longitudinal joint 02.458

纵列式停车 parallel parking 03.198

纵面线形 vertical alignment 02.110

纵坡 longitudinal gradient 02.112

纵坡折减 grade compensation 02.113

纵伸挂车 adjustable-wheelbase trailer 05.045

纵向连续带光源 longitudinal continuous band

illuminant 03.466

纵向裂缝 longitudinal crack 02.605

纵向排水 longitudinal drainage 02.514

纵向通风 longitudinal ventilation 04.909

纵向拖拉法 erection by longitudinal pulling method 04.672

走合 running-in 05.297

走合期驾驶 driving in running-in period 05.115

走合维护 running-in maintenance 05.256

走廊交通管理计划 corridor traffic management program 03.203

足尺试验 full-scale test 07.377

阻挡式护栏 block out type safety fence 03.443

阻塞密度 jam density 03.086

阻水面积 current-obstruction area 04.053

组分 constituent 08.244

组分试验 constituent test 09.074

组合衬砌 composite lining 04.888

组合梁 composite beam 04.267

组合梁桥 composite beam bridge 04.213

组合梁式斜拉桥 composite deck cable stayed bridge 04.238

组合式桥台 composite abutment 04.396

组合式压路机 combination roller 07.067

组合式纵向通风 composite longitudinal ventilation 04.912

组合应力 combined stress 04.550

组货服务费 freight sales charge 06.507

*钻爆法 mine tunneling method 04.868

钻斗 drilling bucket 07.257

钻孔垂[直]度检测 bored hole verticality measurement 09.141

钻孔扩端法 boring-and-underreaming method 04.752

钻孔泥浆试验 boring slurry test 09.143

钻孔潜望镜 borehole periscope 09.235

钻孔照相机 borehole camera 09.234

钻孔直径检测 bored hole diameter measurement 09.140

钻孔柱状图 boring log 01.311

钻孔桩 bored pile 04.425

钻探 boring test 01.253

最大安全速度 maximum safe speed 03.286

最大超高率 maximum superelevation rate 02.140

最大服务交通量 maximum service volume 03.176

最大服务流率 maximum service flow rate 03.177

最大干密度 maximum dry density 08.202

最大摊铺生产率测定 maximum paving capacity measurement 07.480

最低水位 lowest water level, LWL 04.093

最低稳定行驶速度测定 minimum steady travelling speed measurement 07.421

最低限速标志 minimum stated-speed sign 03.388

最高车速 maximum speed 03.078

最高水位 highest water level, HWL 04.092

最高行驶速度测定 maximum travelling speed measurement 07.420

最佳含水量 optimum water content 08.196

最佳级配 optimum gradation 08.192

最佳密度 optimum density 03.085

最佳速度 optimum speed 03.083

最小填土高度 minimum height of fill 02.282

左转弯导向线 left turn guide line 03.418

左转弯匝道 left turn ramp 02.242

作业性能试验 operating performance test 07.439

作业循环时间测定 determination of operating cycle 07.444

坐标法 coordinate method 01.200